# TRANSACTIONS

## OF THE

# AMERICAN PHILOSOPHICAL SOCIETY

### HELD AT PHILADELPHIA
### FOR PROMOTING USEFUL KNOWLEDGE

NEW SERIES—VOLUME 44
1954

THE AMERICAN PHILOSOPHICAL SOCIETY
INDEPENDENCE SQUARE
PHILADELPHIA 6

1954

# CONTENTS OF VOLUME 44

# TRANSACTIONS

OF THE

# AMERICAN PHILOSOPHICAL SOCIETY

HELD AT PHILADELPHIA
FOR PROMOTING USEFUL KNOWLEDGE

---

NEW SERIES—VOLUME 44, PART 6
1954

---

# THE ILLUMINATION AND POLARIZATION OF THE SUNLIT SKY ON RAYLEIGH SCATTERING

S. CHANDRASEKHAR AND DONNA D. ELBERT

*Yerkes Observatory, University of Chicago*

THE AMERICAN PHILOSOPHICAL SOCIETY
INDEPENDENCE SQUARE
PHILADELPHIA 6

DECEMBER, 1954

Library of Congress Catalog
Card No. 54–12909

# THE ILLUMINATION AND POLARIZATION OF THE SUNLIT SKY ON RAYLEIGH SCATTERING

S. Chandrasekhar and Donna D. Elbert

## CONTENTS

## 1. INTRODUCTION

Since 1871 when Lord Rayleigh first accounted for the principal features of the brightness and polarization of the sunlit sky in terms of the laws of scattering now associated with his name, it has been generally recognized that a problem of fundamental importance both for meteorological optics and for theories of planetary illumination is the following:

A parallel beam of radiation in a given state of polarization is incident on a plane-parallel atmosphere of optical thickness $\tau_1$ in some specified direction. Each element of the atmosphere scatters radiation in accordance with Rayleigh's laws. It is required to find the distribution of intensity and polarization of the light diffusely transmitted by the atmosphere below $\tau = \tau_1$ and of the light diffusely reflected by the atmosphere above $\tau = 0$.

In the theory of planetary illumination one is principally interested in the reflected light while in the theory of sky illumination one is similarly interested in the transmitted light. In this paper we shall be concerned only with the latter.

It is clear that an exact treatment of the foregoing problem in the theory of diffuse reflection and transmission will require the formulation and solution of the appropriate equations of radiative transfer. This was accomplished six years ago[1] and the theory is described and briefly illustrated in the book *Radiative Transfer* (Oxford, 1950) by one of us. A general account of the theory, together with a comparison of its predictions with observations particularly those relating to the polarization of the sunlit sky, was published in 1951 in a brief article.[2]

In this paper we shall present the calculations which we have made (at intervals) during the past five years with the object of giving the theory a concrete form. At one time it was our hope to present our calculations based on the exact mathematical solution of the problem with detailed comparisons not only with the available observational data but also with the calculations of the earlier investigators based on approximations of various kinds. But pressure of time and circumstance have forced us to abandon this plan: this paper will be restricted to giving the results of our calculations with only such comparisons with observations as seemed to us of particular interest.

## 2. THE SOLUTION OF THE FUNDAMENTAL PROBLEM

As we have stated, the problem in the theory of diffuse reflection and transmission formulated in § 1 has been exactly solved; the solution is given in *Radiative Transfer* (§§ 69–73). We shall not describe in any detail how the solution was obtained. But a few explanatory remarks on the parameters in terms of which the solution was obtained and on the structure of the solution itself may be useful in the present connection.

Since on Rayleigh's laws light gets partially plane-polarized whenever it is scattered, it is clear that in formulating the equations of radiative transfer we must allow for the partial plane-polarization of the radiation field. Now to describe a radiation field which is partially plane-polarized we need three parameters to specify the intensity, the degree of polarization, and the plane of polarization. It would scarcely be expected that one could include such diverse quantities as an intensity, a ratio, and an angle in any satisfactory way in formulating the basic equations of the problem. It appears that for these latter purposes the most convenient representation of polarized light is a set of parameters first introduced by Stokes[3] in 1852. The meaning of these parameters for a partially plane-polarized beam is simple: Let $l$ and $r$ refer to two arbitrarily chosen directions at right-angles to one another in the plane transverse to the direction of propagation of the beam. The intensity $I(\psi)$ in a direction making an angle $\psi$ (measured clock-wise) to the direction of $l$ can be expressed in the form

$$I(\psi) = I_l \cos^2 \psi + I_r \sin^2 \psi + \tfrac{1}{2} U \sin 2\psi. \quad (1)$$

The coefficients $I_l$, $I_r$ and $U$ in this representation are the Stokes parameters. In terms of these parameters the angle $\chi$ which the plane of polarization makes with the direction $l$ and the degree of polarization, $\delta$, are given by

$$\tan 2\chi = U/(I_l - I_r) \quad (2)$$

---

[1] S. Chandrasekhar, On the radiative equilibrium of a stellar atmosphere. XXII. (V. Rayleigh scattering), *Astrophys. Jour.* 107: 199, 1947.

[2] S. Chandrasekhar and Donna Elbert, Polarization of the sunlit sky, *Nature* 167: 51, 1951.

[3] G. G. Stokes, On the composition and resolution of streams of polarized light from different sources, *Trans. Camb. Phil. Soc.* 9: 399, 1852.

and

$$\delta = (I_l - I_r) \sec 2\chi / (I_l + I_r). \qquad (3)$$

The additive property of the Stokes parameters which makes them so convenient for treating problems of radiative transfer is evident from the representation (1): If two independent streams of polarized light are mixed, then the Stokes parameter characterizing the mixture is the sum of the Stokes parameters of the individual streams.

In terms of the Stokes parameters a law of scattering is specified by a matrix, since an elementary act of scattering results in a linear transformation of the parameters. Consequently, by considering the intensity as a vector $I$ with the components $I_l$, $I_r$ and $U$ (where $l$ and $r$ from now on refer to directions parallel and perpendicular, respectively, to the meridian through the point under consideration and in the plane containing the directions of the beam and of the normal to the plane of stratification of the atmosphere) and by replacing the "phase function" commonly introduced to describe the angular distribution of the scattered radiation by a *phase matrix*, $P$, we can

formulate the basic equation of transfer without any difficulty of principle. In this manner we find that the equation we have to solve is[4]

$$\mu \frac{dI(\tau, \mu, \varphi)}{d\tau} = I(\tau, \mu, \varphi)$$

$$- \frac{1}{4\pi} \int_{-1}^{+1} \int_0^{2\pi} P(\mu, \varphi; \mu', \varphi') I(\tau, \mu', \varphi') \, d\mu' \, d\varphi'$$

$$- \frac{1}{4} e^{-\tau/\mu_0} P(\mu, \varphi; -\mu_0, \varphi_0) F. \qquad (4)$$

where

$$F = (F_l, F_r, F_U) \qquad (5)$$

is the (Stokes) vector which represents the parallel beam of radiation incident on the atmosphere in the direction $(-\mu_0, \varphi_0)$: $\pi F_l$, $\pi F_r$ and $\pi F_U$ denote the net fluxes per unit area normal to the beam in the three Stokes parameters. Further, in equation (4) $\mu$ denotes the cosine of the angle to the outward normal and the azimuthal angle. And, finally, for the case of Rayleigh scattering the phase matrix $P(\mu, \varphi; \mu', \varphi')$ has the explicit form:

where

$$P(\mu, \varphi; \mu', \varphi') = Q[P^{(0)}(\mu, \mu') + (1 - \mu^2)^{\frac{1}{2}}(1 - \mu'^2)^{\frac{1}{2}} P^{(1)}(\mu, \varphi; \mu', \varphi') + P^{(2)}(\mu, \varphi; \mu', \varphi')], \qquad (6)$$

$$Q = \begin{bmatrix} 1 & 0 & 0 \\ 0 & 1 & 0 \\ 0 & 0 & 2 \end{bmatrix}, \qquad (7)$$

$$P^{(0)}(\mu, \mu') = \frac{3}{4} \begin{bmatrix} 2(1 - \mu^2)(1 - \mu'^2) + \mu^2\mu'^2 & \mu^2 & 0 \\ \mu'^2 & 1 & 0 \\ 0 & 0 & 0 \end{bmatrix}, \qquad (8)$$

$$P^{(1)}(\mu, \varphi; \mu', \varphi') = \frac{3}{4} \begin{bmatrix} 4\mu\mu' \cos(\varphi' - \varphi) & 0 & 2\mu \sin(\varphi' - \varphi) \\ 0 & 0 & 0 \\ -2\mu' \sin(\varphi' - \varphi) & 0 & \cos(\varphi' - \varphi) \end{bmatrix} \qquad (9)$$

and

$$P^{(2)}(\mu, \varphi; \mu', \varphi') = \frac{3}{4} \begin{bmatrix} \mu^2\mu'^2 \cos 2(\varphi' - \varphi) & -\mu^2 \cos 2(\varphi' - \varphi) & \mu^2\mu' \sin 2(\varphi' - \varphi) \\ -\mu'^2 \cos 2(\varphi' - \varphi) & \cos 2(\varphi' - \varphi) & -\mu' \sin 2(\varphi' - \varphi) \\ -\mu\mu'^2 \sin 2(\varphi' - \varphi) & \mu \sin 2(\varphi' - \varphi) & \mu\mu' \cos 2(\varphi' - \varphi) \end{bmatrix}. \qquad (10)$$

The solution of equation (4) appropriate to the problem on hand must satisfy the boundary conditions

$$\left. \begin{array}{l} I(0, -\mu, \varphi) \equiv 0 \quad (0 < \mu \leqslant 1, \ 0 \leqslant \varphi \leqslant 2\pi) \\ \text{and} \\ I(\tau_1, +\mu, \varphi) \equiv 0 \quad (0 < \mu \leqslant 1, \ 0 \leqslant \varphi \leqslant 2\pi), \end{array} \right\} \qquad (11)$$

since there is no diffuse radiation in any inward direction at $\tau = 0$ and in any outward direction at $\tau = \tau_1$. And the solution to the problem of diffuse reflection and transmission will be completed when we specify the angular distribution and the state of polarization of the diffuse light which emerges from $\tau = 0$ and $\tau = \tau_1$.

The laws of diffuse reflection and transmission by a plane-parallel atmosphere are generally expressed

(cf. *Radiative Transfer*, 44) in terms of a *scattering matrix*, $S(\mu, \varphi; \mu_0, \varphi_0)$ and a *transmission matrix*, $T(\mu, \varphi; \mu_0, \varphi_0)$ such that the reflected and the transmitted intensities are given by

$$I(0; \mu, \varphi; \mu_0, \varphi_0) = \frac{1}{4\mu} S(\mu, \varphi; \mu_0, \varphi_0) F$$

and $\qquad (12)$

$$I(\tau_1; -\mu, \varphi; \mu_0, \varphi_0) = \frac{1}{4\mu} T(\mu, \varphi; \mu_0, \varphi_0) F.$$

Our problem, then, is to specify $S$ and $T$ for an atmosphere scattering radiation in accordance with Rayleigh's laws.

[4] The derivation of the equations which follow will be found in *Radiative Transfer* § 16: 35–45.

The elements of $S$ and $T$ are clearly functions of the four variables $\mu$, $\varphi$, $\mu_0$ and $\varphi_0$ in addition, of course, to the optical thickness $\tau_1$ which may, however, be treated as a parameter. If the variables $\mu$, $\varphi$, $\mu_0$ and $\varphi_0$ were not separable the problem of tabulating $S$ and $T$ may indeed be considered as impracticable. But the essential feature of the solution for $S$ and $T$ (which we shall presently write down) which makes the problem a practicable one is that $S$ and $T$ involve only four pairs of functions $X_l(\mu)$, $Y_l(\mu)$; $X_r(\mu)$, $Y_r(\mu)$; $X^{(1)}(\mu)$, $Y^{(1)}(\mu)$; and $X^{(2)}(\mu)$, $Y^{(2)}(\mu)$ all of the single variable $\mu$. Further, these four pairs of functions belong to a general class (the $X$- and $Y$-functions) which satisfy a simultaneous pair of integral equations of the form[5]

$$X(\mu) = 1 + \mu \int_0^1 \frac{\Psi(\mu')}{\mu + \mu'}$$
$$\times [X(\mu)X(\mu') - Y(\mu)Y(\mu')]d\mu' \quad (13)$$

and

$$Y(\mu) = e^{-\tau_1/\mu} + \mu \int_0^1 \frac{\Psi(\mu')}{\mu - \mu'}$$
$$\times [Y(\mu)X(\mu') - X(\mu)Y(\mu')]d\mu', \quad (14)$$

where the *characteristic function* $\Psi(\mu)$ is in problems of radiative transfer, an even polynomial in $\mu$ satisfying the condition

$$\int_0^1 \Psi(\mu)\,d\mu \leqslant \tfrac{1}{2}. \quad (15)$$

The case when equality occurs in (15) (the so-called *conservative case*) is special: The solutions of equations (13) and (14) are, then, no longer unique; they form instead a one-parameter family. In conservative cases one therefore defines what are called *standard solutions* which have the property:

$$\int_0^1 X(\mu)\Psi(\mu)\,d\mu = 1$$

and

$$\int_0^1 Y(\mu)\Psi(\mu)\,d\mu = 0. \quad (16)$$

As we have already stated, the solutions for $S$ and $T$ (for an atmosphere scattering radiation in accordance with Rayleigh's laws) involve only four pairs of $X$- and $Y$-functions: $X_l$, $Y_l$; $X_r$, $Y_r$; $X^{(1)}$, $Y^{(1)}$; $X^{(2)}$, $Y^{(2)}$; and the characteristic functions in terms of which

these are defined are:

$$\Psi_l(\mu) = \tfrac{3}{4}(1 - \mu^2);$$
$$\Psi_r(\mu) = \tfrac{3}{8}(1 - \mu^2),$$
$$\Psi^{(1)}(\mu) = \tfrac{3}{8}(1 - \mu^2)(1 + 2\mu^2)$$

and

$$\Psi^{(2)}(\mu) = \tfrac{3}{16}(1 + \mu^2)^2, \quad (17)$$

respectively. The function $\Psi_l(\mu)$ belongs to the conservative class; accordingly, in this case we define $X_l(\mu)$, $Y_l(\mu)$ as the standard solutions having the property

$$\tfrac{3}{4} \int_0^1 X_l(\mu)(1 - \mu^2)\,d\mu = 1$$

and

$$\int_0^1 Y_l(\mu)(1 - \mu^2)\,d\mu = 0. \quad (18)$$

After these explanatory remarks we shall now write down the solutions for $S$ and $T$ given in *Radiative Transfer*:

The scattering and the transmission matrices allow a decomposition into azimuth independent and azimuth dependent terms in the same manner as the phase matrix [equation (6)] and have the forms:

$$S(\mu, \varphi; \mu_0, \varphi_0) = Q[\tfrac{3}{4}S^{(0)}(\mu; \mu_0) + (1 - \mu^2)^{\frac{1}{2}}(1 - \mu_0^2)^{\frac{1}{2}}S^{(1)}(\mu, \varphi; \mu_0, \varphi_0) + S^{(2)}(\mu, \varphi; \mu_0, \varphi_0)] \quad (19)$$

and

$$T(\mu, \varphi; \mu_0, \varphi_0) = Q[\tfrac{3}{4}T^{(0)}(\mu; \mu_0) + (1 - \mu^2)^{\frac{1}{2}}(1 - \mu_0^2)^{\frac{1}{2}}T^{(1)}(\mu, \varphi; \mu_0, \varphi_0) + T^{(2)}(\mu, \varphi; \mu_0, \varphi_0)]. \quad (20)$$

The dependence of the azimuth dependent terms $(S^{(1)}, T^{(1)})$ and $(S^{(2)}, T^{(2)})$ on $\varphi_0 - \varphi$ are essentially the same as $P^{(1)}$ and $P^{(2)}$; indeed, we have

$$\left(\frac{1}{\mu_0} + \frac{1}{\mu}\right) S^{(i)} = [X^{(i)}(\mu)X^{(i)}(\mu_0) - Y^{(i)}(\mu)Y^{(i)}(\mu_0)]P^{(i)}(\mu, \varphi; -\mu_0, \varphi_0)$$

and

$$\left(\frac{1}{\mu_0} - \frac{1}{\mu}\right) T^{(i)} = [Y^{(i)}(\mu)X^{(i)}(\mu_0) - X^{(i)}(\mu)Y^{(i)}(\mu_0)]P^{(i)}(-\mu, \varphi; -\mu_0, \varphi_0)$$
$$(i = 1, 2). \quad (21)$$

In contrast, the solutions for the azimuth independent terms $S^{(0)}$ and $T^{(0)}$ are very complicated. They are given by

$$\left(\frac{1}{\mu_0} + \frac{1}{\mu}\right) S^{(0)}(\mu; \mu_0) = \begin{pmatrix} \psi(\mu) & 2^{\frac{1}{2}}\phi(\mu) & 0 \\ \chi(\mu) & 2^{\frac{1}{2}}\zeta(\mu) & 0 \\ 0 & 0 & 0 \end{pmatrix} \begin{pmatrix} \psi(\mu_0) & \chi(\mu_0) & 0 \\ 2^{\frac{1}{2}}\phi(\mu_0) & 2^{\frac{1}{2}}\zeta(\mu_0) & 0 \\ 0 & 0 & 0 \end{pmatrix}$$

$$- \begin{pmatrix} \xi(\mu) & 2^{\frac{1}{2}}\eta(\mu) & 0 \\ \sigma(\mu) & 2^{\frac{1}{2}}\theta(\mu) & 0 \\ 0 & 0 & 0 \end{pmatrix} \begin{pmatrix} \xi(\mu_0) & \sigma(\mu_0) & 0 \\ 2^{\frac{1}{2}}\eta(\mu_0) & 2^{\frac{1}{2}}\theta(\mu_0) & 0 \\ 0 & 0 & 0 \end{pmatrix} \quad (22)$$

---

[5] For the theory of the $X$- and $Y$-functions see *Radiative Transfer*, chap. VIII.

and

$$\left(\frac{1}{\mu_0} - \frac{1}{\mu}\right) T^{(0)}(\mu; \mu_0) = \begin{pmatrix} \xi(\mu) & 2^{\frac{1}{2}}\eta(\mu) & 0 \\ \sigma(\mu) & 2^{\frac{1}{2}}\theta(\mu) & 0 \\ 0 & 0 & 0 \end{pmatrix} \begin{pmatrix} \psi(\mu_0) & \chi(\mu_0) & 0 \\ 2^{\frac{1}{2}}\phi(\mu_0) & 2^{\frac{1}{2}}\zeta(\mu_0) & 0 \\ 0 & 0 & 0 \end{pmatrix}$$

$$- \begin{pmatrix} \psi(\mu) & 2^{\frac{1}{2}}\phi(\mu) & 0 \\ \chi(\mu) & 2^{\frac{1}{2}}\zeta(\mu) & 0 \\ 0 & 0 & 0 \end{pmatrix} \begin{pmatrix} \xi(\mu_0) & \sigma(\mu_0) & 0 \\ 2^{\frac{1}{2}}\eta(\mu_0) & 2^{\frac{1}{2}}\theta(\mu_0) & 0 \\ 0 & 0 & 0 \end{pmatrix}, \quad (23)$$

where $\psi$, $\phi$, $\chi$, etc., are eight functions expressible in terms of the two pairs of $X$- and $Y$-functions, $X_l$, $Y_l$ and $X_r$, $Y_r$ in the forms

$$\psi(\mu) = \mu[\nu_1 Y_l(\mu) - \nu_2 X_l(\mu)],$$
$$\xi(\mu) = \mu[\nu_2 Y_l(\mu) - \nu_1 X_l(\mu)],$$
$$\phi(\mu) = (1 + \nu_4\mu)X_l(\mu) - \nu_3\mu Y_l(\mu),$$
$$\eta(\mu) = (1 - \nu_4\mu)Y_l(\mu) + \nu_3\mu X_l(\mu),$$
$$\chi(\mu) = (1 - u_4\mu)X_r(\mu) + u_3\mu Y_r(\mu) + Q(u_4 - u_3)\mu^2[X_r(\mu) - Y_r(\mu)],$$
$$\sigma(\mu) = (1 + u_4\mu)Y_r(\mu) - u_3\mu X_r(\mu) - Q(u_4 - u_3)\mu^2[X_r(\mu) - Y_r(\mu)],$$
$$\zeta(\mu) = \tfrac{1}{2}\mu[\nu_1 Y_r(\mu) - \nu_2 X_r(\mu)] + \tfrac{1}{2}Q(\nu_2 - \nu_1)\mu^2[X_r(\mu) - Y_r(\mu)],$$
$$\theta(\mu) = \tfrac{1}{2}\mu[\nu_2 Y_r(\mu) - \nu_1 X_r(\mu)] - \tfrac{1}{2}Q(\nu_2 - \nu_1)\mu^2[X_r(\mu) - Y_r(\mu)], \quad (24)$$

where the constants $\nu_1$, $\nu_2$, $\nu_3$, $\nu_4$, $u_3$, $u_4$ and $Q$ are to be determined by the following formulae:

$$\nu_2 + \nu_1 = 2\Delta_1(\kappa_1\delta_1 - \kappa_2\delta_2); \quad \nu_2 - \nu_1 = 2\Delta_2(\kappa_1\delta_1 - \kappa_2\delta_2),$$
$$\nu_4 + \nu_3 = \Delta_1(d_1\kappa_1 - d_0\kappa_2); \quad \nu_4 - \nu_3 = \Delta_2[c_1\delta_1 - c_0\delta_2 - 2Q(d_0\delta_1 - d_1\delta_2)],$$
$$u_4 + u_3 = \Delta_1(c_1\delta_1 - c_0\delta_2); \quad u_4 - u_3 = \Delta_2(d_1\kappa_1 - d_0\kappa_2),$$
$$\Delta_1 = (d_0\delta_1 - d_1\delta_2)^{-1}; \quad \Delta_2 = [c_0\kappa_1 - c_1\kappa_2 - 2Q(d_1\kappa_1 - d_0\kappa_2)]^{-1},$$
$$Q = (c_0 - c_2)[(d_0 - d_2)\tau_1 + 2(d_1 - d_3)]^{-1},$$
$$c_0 = A_0 + B_0 - \frac{8}{3}; \quad d_0 = A_0 - B_0 - \frac{8}{3},$$
$$c_n = A_n + B_n; \quad d_n = A_n - B_n; \quad \kappa_n = \alpha_n + \beta_n; \quad \delta_n = \alpha_n - \beta_n \quad (n = 1, 2, 3, \cdots), \quad (25)$$

$\alpha_n$, $\beta_n$, $A_n$ and $B_n$ are the moments of order $n$ of $X_l$, $Y_l$, $X_r$ and $Y_r$, respectively. (It may be recalled here that $X_l$ and $Y_l$ are the standard solutions for the case.)

In the theory of the illumination of the sky we are interested in the transmitted light in the case of incident natural light. In this latter case $F_l = F_r = \frac{1}{2}F$ (where $\pi F$ denotes the net flux of the incident natural light) and $F_U = 0$. The equations governing the intensity and polarization of the sky as witnessed by an observer at $\tau = \tau_1$ in these circumstances readily follow from the solutions already given; thus by setting $F = \frac{1}{2}(F_l, F_r, 0)$ in equation (12) and combining equations (20), (21), and (23) appropriately, we find:

$$I_l(\tau_1; -\mu, \varphi; \mu_0, \varphi_0) = \tfrac{3}{32}[\{\psi(\mu_0) + \chi(\mu_0)\}\xi(\mu) + 2\{\phi(\mu_0) + \zeta(\mu_0)\}\eta(\mu) - \{\xi(\mu_0) + \sigma(\mu_0)\}\psi(\mu)$$
$$- 2\{\theta(\mu_0) + \eta(\mu_0)\}\phi(\mu) + 4\mu\mu_0(1 - \mu^2)^{\frac{1}{2}}(1 - \mu_0^2)^{\frac{1}{2}}\{X^{(1)}(\mu_0)Y^{(1)}(\mu) - Y^{(1)}(\mu_0)X^{(1)}(\mu)\} \cos(\varphi_0 - \varphi)$$
$$- \mu^2(1 - \mu_0^2)\{X^{(2)}(\mu_0)Y^{(2)}(\mu) - Y^{(2)}(\mu_0)X^{(2)}(\mu)\} \cos 2(\varphi_0 - \varphi)]\frac{F\mu_0}{\mu - \mu_0},$$

$$I_r(\tau_1; -\mu, \varphi; \mu_0, \varphi_0) = \tfrac{3}{32}[\{\psi(\mu_0) + \chi(\mu_0)\}\sigma(\mu) + 2\{\phi(\mu_0) + \zeta(\mu_0)\}\theta(\mu) - \{\xi(\mu_0) + \sigma(\mu_0)\}\chi(\mu)$$
$$- 2\{\theta(\mu_0) + \eta(\mu_0)\}\zeta(\mu) + (1 - \mu_0^2)\{X^{(2)}(\mu_0)Y^{(2)}(\mu) - Y^{(2)}(\mu_0)X^{(2)}(\mu)\} \cos 2(\varphi_0 - \varphi)]\frac{F\mu_0}{\mu - \mu_0},$$

and

$$U(\tau_1; -\mu, \varphi; \mu_0, \varphi_0) = \tfrac{3}{16}[2(1 - \mu^2)^{\frac{1}{2}}(1 - \mu_0^2)^{\frac{1}{2}}\mu_0\{X^{(1)}(\mu_0)Y^{(1)}(\mu) - Y^{(1)}(\mu_0)X^{(1)}(\mu)\} \sin(\varphi_0 - \varphi)$$
$$- \mu(1 - \mu_0^2)\{X^{(2)}(\mu_0)Y^{(2)}(\mu) - Y^{(2)}(\mu_0)X^{(2)}(\mu)\} \sin 2(\varphi_0 - \varphi)]\frac{F\mu_0}{\mu - \mu_0}. \quad (26)$$

## 3. THE EFFECT OF REFLECTION BY THE GROUND

Before we can apply the solution for $S$ and $T$ given in § 2 to the problem of the illumination of the sky, we must consider the effect of the ground at $\tau = \tau_1$. The solution for $S$ and $T$ given in § 2 was derived on the assumption that at $\tau = \tau_1$ there is no diffuse radiation in the outward direction [cf. equation (11)]. The presence of the ground will alter this. However, if the law of reflection by the ground is specified then it is not a difficult matter to relate the solution of the problem when there is a ground to the solution of the problem when there is no ground. This reduction is particularly simple if the ground reflects according to Lambert's law with a certain albedo $\lambda_0$; that is, if the light reflected by the ground is unpolarized and uniform in the outward hemisphere independently of the state of polarization and the angular distribution of the incident light, and if, further, the outward flux of

the reflected light is always a certain fixed fraction, $\lambda_0$, of the inward flux of the radiation incident on the surface. Under these latter circumstances it can be shown (*Radiative Transfer*, § 73) that the effect of the ground is to increase the diffuse intensities emergent at $\tau = 0$ and directed inward at $\tau = \tau_1$ by amounts $I^*(0; \mu, \varphi)$ and $I^*(\tau_1; -\mu, \varphi)$ given by

$$I^*(0; \mu, \varphi) = \frac{\lambda_0 \mu_0}{2(1 - \lambda_0 \delta)} \Gamma(\mu; \mu_0)F \qquad (27)$$

and

$$I^*(\tau_1; -\mu, \varphi) = \frac{\lambda_0 \mu_0}{2(1 - \lambda_0 \delta)} \Lambda(\mu; \mu_0)F, \qquad (28)$$

where $\delta$ is a constant (to be defined presently) and

$$\Gamma(\mu; \mu_0) = \begin{bmatrix} \gamma_l(\mu)\gamma_l(\mu_0) & \gamma_l(\mu)\gamma_r(\mu_0) & 0 \\ \gamma_r(\mu)\gamma_l(\mu_0) & \gamma_r(\mu)\gamma_r(\mu_0) & 0 \\ 0 & 0 & 0 \end{bmatrix} \qquad (29)$$

and

$$\Lambda(\mu; \mu_0) = \begin{bmatrix} \{1 - \gamma_l(\mu)\}\gamma_l(\mu_0) & \{1 - \gamma_l(\mu)\}\gamma_r(\mu_0) & 0 \\ \{1 - \gamma_r(\mu)\}\gamma_l(\mu_0) & \{1 - \gamma_r(\mu)\}\gamma_r(\mu_0) & 0 \\ 0 & 0 & 0 \end{bmatrix}. \qquad (30)$$

In equations (29) and (30) $\gamma_l(\mu)$ and $\gamma_r(\mu)$ are two functions which are related to $X_l$, $Y_l$, $X_r$ and $Y_r$ by

$$\gamma_l(\mu) = \tfrac{3}{8}Q(\nu_2 - \nu_1)(d_0 - d_2)[X_l(\mu) + Y_l(\mu)], \qquad (31)$$

and

$$\gamma_r(\mu) = \tfrac{3}{8}Q(d_0 - d_2)$$
$$\times \big[ (u_4 - u_3)\{X_r(\mu) + Y_r(\mu)\} - u_5\mu\{X_r(\mu) + Y_r(\mu)\} \big], \qquad (32)$$

where

$$u_5 = \Delta_2(c_0\kappa_1 - c_1\kappa_2), \qquad (33)$$

and the remaining constants have the same meanings as in equations (25). Finally, the constant $\delta$ in equations (27) and (28) is given by

$$\delta = 1 - \tfrac{3}{8}Q(d_0 - d_2)$$
$$\times \big[ (\nu_2 - \nu_1)\kappa_1 + (u_4 - u_3)c_1 - u_5d_2 \big]. \qquad (34)$$

Again, when the incident light is natural the corrections which have to be made to the intensities given by equations (26) to allow for a ground surface at $\tau = \tau_1$ which reflects according to Lambert's law with an albedo $\lambda_0$ are given by

$$I_l^*(\tau_1; -\mu, \varphi; \mu_0, \varphi_0)$$
$$= \frac{\lambda_0}{4(1 - \lambda_0 \delta)} \{\gamma_l(\mu_0) + \gamma_r(\mu_0)\}$$
$$\times \{1 - \gamma_l(\mu)\}\mu_0 F,$$

$$I_r^*(\tau_1; -\mu, \varphi; \mu_0, \varphi_0)$$
$$= \frac{\lambda_0}{4(1 - \lambda_0 \delta)} \{\gamma_l(\mu_0) + \gamma_r(\mu_0)\}$$
$$\times \{1 - \gamma_r(\mu)\}\mu_0 F,$$

$$U^*(\tau_1; -\mu, \varphi; \mu_0, \varphi_0) = 0. \qquad (35)$$

## 4. DESCRIPTION OF THE TABLES

The solution of the fundamental problem in the theory of the illumination of the sky given in the two preceding sections was obtained some six years ago. A detailed examination of its predictions had to await the tabulation of the basic eight functions $X_l$, $Y_l$, $X_r$, $Y_r$, $X^{(1)}$, $Y^{(1)}$, $X^{(2)}$, and $Y^{(2)}$ of the variable $\mu$ for various values of $\tau_1$. This tabulation has now been completed for $\tau_1 = 0.05, 0.10, 0.15, 0.20, 0.25, 0.50,$ and $1.00$ with the cooperation of the Watson Scientific Computing Laboratory (New York).[6] The solutions were obtained by a direct process of iteration applied to the governing integral equations. The iterations were started with the solutions in the corrected second approximation described in *Radiative Transfer*, Chapter VIII [§ 60, see particularly equations (117), (118), and (120)]. The corrected second

[6] While these calculations were in progress, Dr. Z. Sekera initiated a similar program at the Department of Meteorology of the University of California at Los Angeles in cooperation with the Institute for Numerical Analysis of the National Bureau of Standards at Los Angeles. Their Report No. 3 (prepared by the Air Material Command, Air Force Cambridge Research Center) provides some calculations for $\tau_1 = 0.15, 0.25,$ and $1.0$. Their calculations, while they are much less extensive than ours, do provide a valuable check.

approximations were computed by one of us (D. D. E.) at the Yerkes Observatory. The iterations were carried out at the Watson Scientific Laboratory with IBM pluggable sequence relay calculators by Miss Ann Franklin to whom and to Dr. Wallace Eckert we are very greatly indebted. The functions obtained after the iterations showed, however, a certain "raggedness" between $\mu = 0.9$ and 1.0. The solutions have therefore been "smoothed" by plotting the deviations of the iterated solutions from the corrected second approximations. Table 1 presents these smoothed solutions. While the solutions have been tabulated to five decimals the last place is definitely not reliable. But it is expected that if the tabulated solutions are rounded to one less place the solution may be trusted to two or three units in the surviving place. The solutions for $\tau_1 \leqslant 0.20$ are very probably more accurate than this while for $\tau_1 = 0.5$ and 1.0 they may be less accurate. Nevertheless, the solutions are given to five places since the functions as tabulated do have smooth differences and as such they can be used for further iterations to improve their accuracy if the need for it should arise.

The moments $\alpha_n$, $\beta_n$, $A_n$ and $B_n$ of order $n$ of $X_l$, $Y_l$, $X_r$ and $Y_r$, respectively, which are needed in the evaluation of the various terms of $S$ and $T$ are given in table 2. The theory of the $X$- and $Y$-functions leads to a number of identical relations which must exist between their moments; these relations among the moments listed in table 2 have been verified within the accuracy of the tabulated values. Table 2 also includes the values of the constants $\nu_1$, $\nu_2$, $\nu_3$, $\nu_4$, $u_3$, $u_4$ and $Q$ which occur in the definitions of the functions $\psi$, $\phi$, etc. [equations (24) and (25)]; the value of the constant $\delta$ [equation (34)] which occurs in the expressions for the ground corrections [equations (27) and (28)] is also listed in this table.

According to equations (22) and (23) the calculation of $S$ and $T$ in a given case can be most easily carried out in terms of the auxiliary functions $\psi$, $\phi$, $\chi$, $\zeta$, $\xi$, $\eta$, $\sigma$ and $\theta$. These functions computed with the aid of tables 1 and 2 are given in table 3. Similarly, table 4 gives the functions $\gamma_l(\mu)$ and $\gamma_r(\mu)$ which are needed to allow for reflection by the ground [cf. equations (31) and (32)].

With the basic functions tabulated we can calculate the theoretical illumination and polarization of the sky on Rayleigh scattering for plane-parallel atmospheres. To illustrate the use of tables 1–4 we have made some model calculations to which the remaining tables are devoted.

The most extensive calculations were made for $\tau_1 = 0.15$; this is approximately the value of the optical thickness at $\lambda 4500A$. For $\tau_1 = 0.15$, angles of incidence (or, zenith distances) $\theta_0 = 90°$, 85.4°, 76.1°, 58.7°, 50.1°, 43.9°, 36.9°, 19.95°, and 0° (corresponding to $\mu_0 = 0$, 0.08, 0.24, 0.52, 0.64, 0.72, 0.80, 0.94, and 1.00) and azimuthal differences $\varphi_0 - \varphi$ (denoted,

simply, by $\varphi$ in the tables) $= 0°(10°)90°$ were considered. The results of the calculations are summarized in table 5; it gives the total intensity, $I_l + I_r$, in units of $F$, the inclination, $\chi$, of the plane of polarization with the meridian through the direction of observation and the degree of polarization, $\delta$. Less extensive calculations were made for $\tau_1 = 0.10$ and 0.20. For these two values of the optical thicknesses the total intensity $(I_l + I_r)$ and the degree of polarization, $\delta$, were found only in the principal meridian $(\varphi_0 - \varphi = 0°)$ containing the sun and in the plane at right-angles $(\varphi_0 - \varphi = 90°)$ for zenith distances $\theta_0 = 90°$, 80.8°, 60°, 30.7°, and 0° (corresponding to $\mu_0 = 0$, 0.16, 0.50, 0.86, and 1.00). The results of the calculations are given in tables 6 and 7.

In the calculations presented in tables 5–7 no allowance has been made for the reflection by the ground surface. This is taken into account in tables 8–10 in accordance with equations (35). Again, the most extensive calculations were made for $\tau_1 = 0.15$. For angles of incidence corresponding to $\mu_0 = 0$, 0.24, 0.52, 0.64, 0.72, 0.80, and 1.00 the effect of a ground reflecting according to Lambert's law for two values of the albedo, $\lambda_0$, on the intensity and the degree of polarization in the principal meridian $(\varphi_0 - \varphi = 0°)$ was determined. For $\mu_0 = 0.64$ the effect on the intensity for $\varphi_0 - \varphi = 0°(10°)90°$ and $\lambda_0 = 0.10$ was determined; and for the same angle of incidence the effect on the intensity for $\varphi_0 - \varphi = 90°$ was also found for $\lambda_0 = 0.25$. The results of all these calculations are given in table 8. For $\tau_1 = 0.10$ and 0.20 the effect of ground reflection on the intensity and polarization in the principal meridian is illustrated for $\lambda_0 = 0.10$ and 0.20 and for angles of incidence corresponding to $\mu_0 = 0$, 0.16, 0.50, 0.86, and 1.00. The results of these calculations are given in tables 9 and 10.

Table 11 gives the positions of the neutral points. The results given in this table will be described and discussed in the following section.

Finally the supplementary table 12 gives the functions $\psi$, $\phi$, $\chi$, $\zeta$, $\xi$, $\eta$, $\sigma$, $\theta$, $X^{(1)}$, $Y^{(1)}$, $X^{(2)}$, $Y^{(2)}$, $\gamma_l$, and $\gamma_r$ for $\tau_1 = 0.01(0.01)0.20(0.05)0.50(0.10)1.0$ and for representative values of $\mu$. At the head of the table for each value $\tau_1$, the value of $\delta$ (needed for the evaluation of the ground correction) is also given. The values of the functions listed in this table are *not* based on the solutions of the basic $X$- and $Y$-functions derived from the integral equations they satisfy; they are based, instead, on the corrected second approximations (*cf. Radiative Transfer*, § 60) for the relevant $X$- and $Y$-functions. However, for $\tau_1 \leqslant 0.25$ the table should suffice to calculate $S$ and $T$ to well within a fraction of a per cent; for the larger values of $\tau_1$ accuracy within a few per cent may be expected. But one could, if one wished, obtain from these tables values of considerably higher precision by differencing the values of the functions for $\tau_1 = 0.05$, 0.10, 0.15,

0.20, 0.25, 0.50, and 1.0 given in tables 3 and 12 and interpolating among these differences to estimate the corrections (to the values given by the corrected second approximation) for any other intermediate value of $\tau_1$. With these supplementary tables, then, the theory developed and described in *Radiative Transfer* has been, finally, brought to a point where it is capable of giving numerical values for any of the desired quantities under most conditions in which they are likely to be of interest.

## 5. THE POLARIZATION OF THE SUNLIT SKY: THE THEORY OF THE NEUTRAL POINTS AND LINES

As we have already stated in the introductory section we shall not attempt in this paper any detailed comparison between the calculations presented here

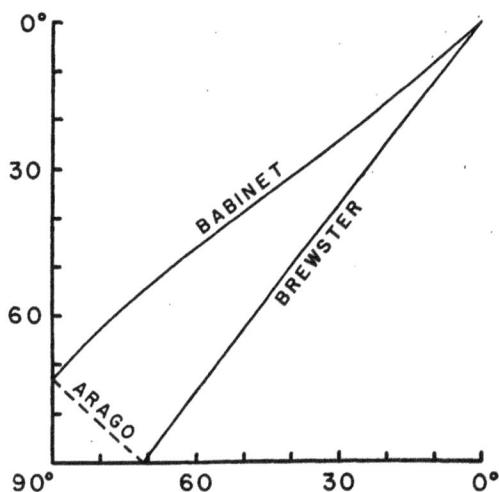

Fig. 2. Calculated positions of the neutral points for various angles of incidence for an atmosphere of optical thickness $\tau_1 = 0.15$. The abscissa and the ordinate have otherwise the same meanings as in fig. 1.

Fig. 1. Calculated positions of the neutral points for various angles of incidence for an atmosphere of optical thickness $\tau_1 = 0.10$. The abscissa gives the zenith distance of the sun and the ordinate gives the corresponding positions of the neutral points. The Arago point occurs on the side of the horizon opposite the sun; to emphasize this its position on the sky is indicated by the dashed curve.

and those of earlier investigators based on approximations of various kinds. But an exception might be made with regard to the quantitative explanation which the present theory affords for the phenomena associated with the *neutral points* of Arago, Babinet, and Brewster. The phenomena in question are these: The neutral points are the points of zero polarization; from symmetry we should, of course, expect them to occur in the principal meridian. And for a long time it has been known that there are in general two

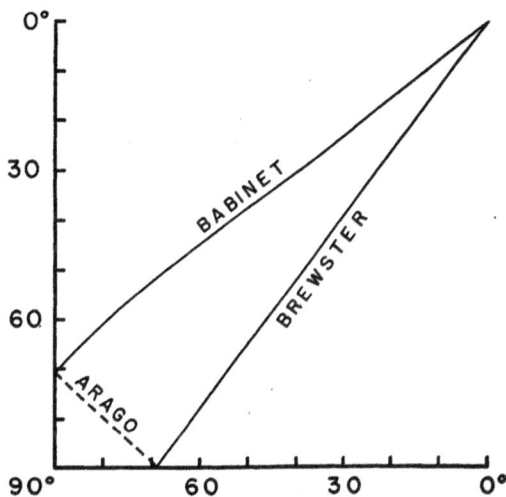

such neutral points. For angles of incidence not exceeding 70° these neutral points occur between 0° and 20° above and below the sun; these are the neutral points of Babinet and Brewster, respectively. But when the sun is low, the neutral point occurs about 20° above the anti-solar point in the opposite sky: this is the Arago point. These facts concerning the

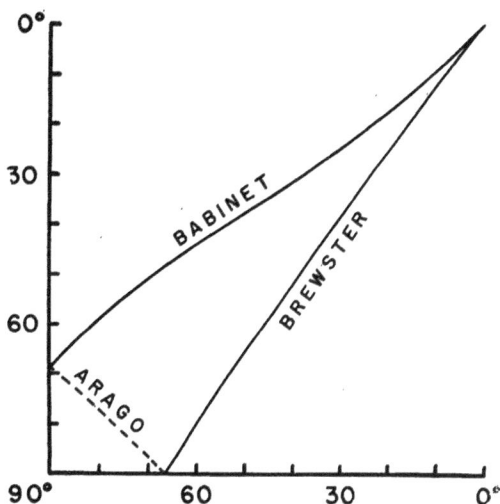

Fig. 3. Calculated positions of the neutral points for various angles of incidence for an atmosphere of optical thickness $\tau_1 = 0.20$. The abscissa and the ordinate have otherwise the same meanings as in fig. 1.

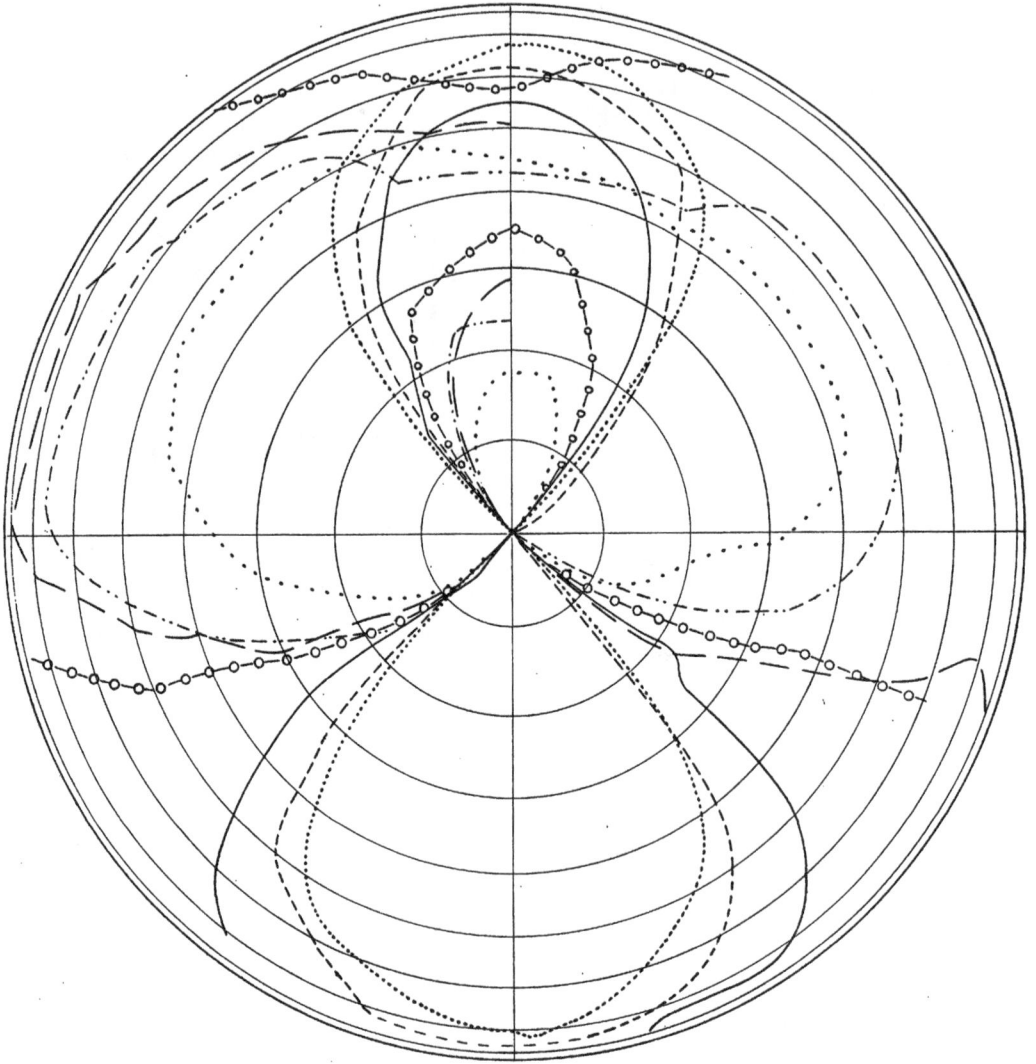

FIG. 4. Dorno's observations of the neutral lines on May 17, 1917, at Davos. The various curves were determined during the following times when the zenith distance of the sun varied by the amounts given:

| | | |
|---|---|---|
| ·············· | 7$^p$ 19– 7$^p$ 28 | $\theta_0 = 87° 10'- 0°$ |
| - - - - - - - - | 6$^p$ 13– 6$^p$ 33 | $\theta_0 = 78° 14'-81° 30'$ |
| —————— | 5$^p$ 14– 5$^p$ 34 | $\theta_0 = 68° 20'-71° 44'$ |
| –o–o–o–o– | 3$^p$ 15– 3$^p$ 42 | $\theta_0 = 48° 12'-52° 40'$ |
| — — — — | 9$^a$ 9– 9$^a$ 34 | $\theta_0 = 44° 21'-40° 33'$ |
| —·—··—·· | 9$^a$ 44–10$^a$ 11 | $\theta_0 = 39° 5'-35° 26'$ |
| ········· | 12$^p$ 56– 1$^p$ 14 | $\theta_0 = 29° 50'-31° 26'$ |

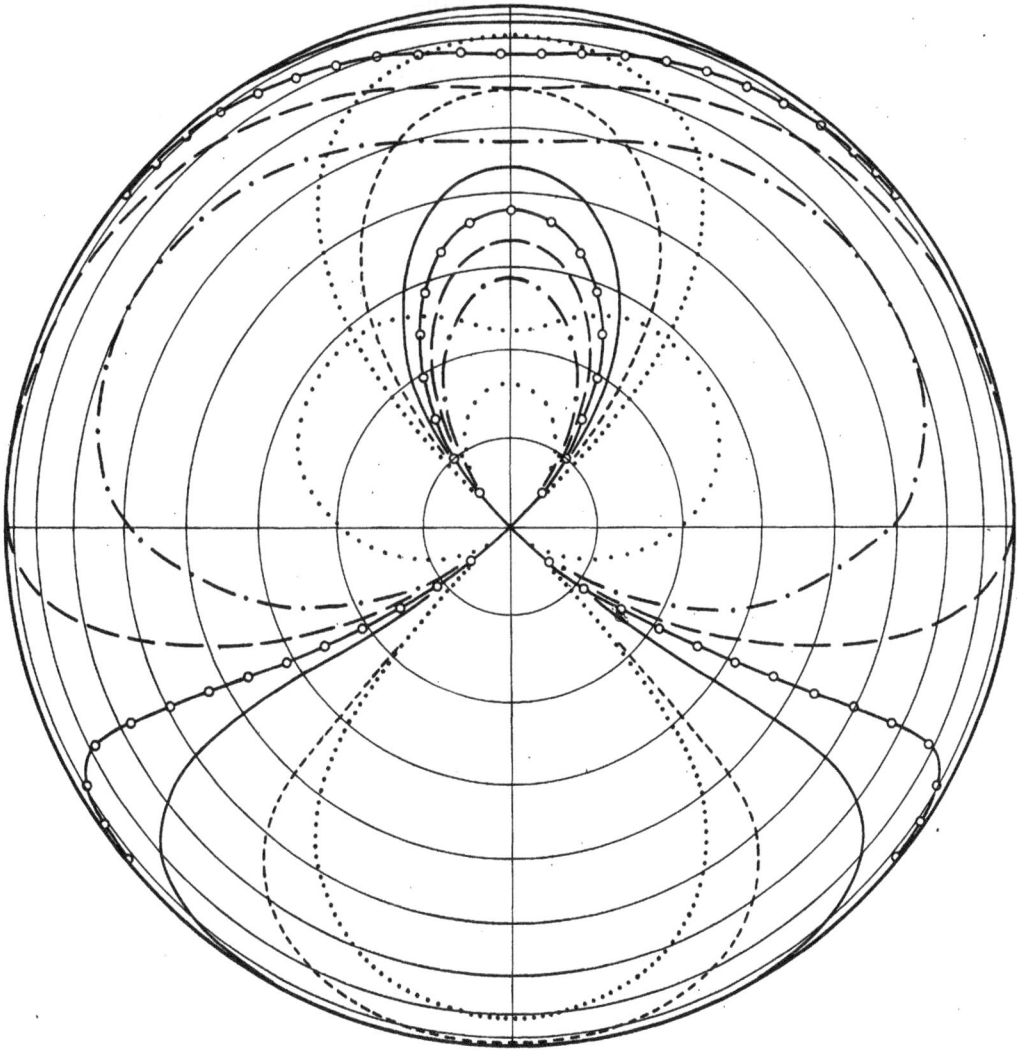

FIG. 5.  The neutral lines as predicted by the theory.  The various curves refer to the following zenith distances of the sun:

| | | |
|---|---|---|
| ·········· = 90°; | ———— = 58.7°; | ———— = 43.9°; |
| -------- = 76.1°; | -o-o-o-o- = 50.2°; | —·—·— = 36.9°; |
| | ·········· = 19.9°. | |

The curves in this figure which roughly correspond
to Dorno's observations are marked similarly.

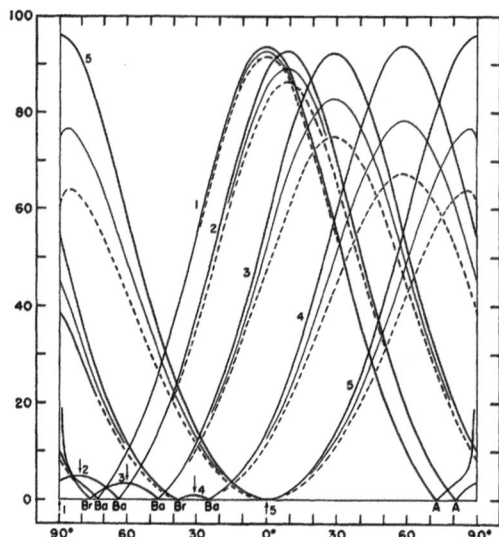

FIG. 6. Variation of the degree of polarization in the principal meridian for various angles of incidence for an atmosphere of optical thickness $\tau_1 = 0.10$. The abscissa gives the zenith distance and the ordinate gives the degree of polarization in per cent. The curves marked 1, 2, 3, 4, and 5 represent the variation for the angles of incidence $\theta_0 = 90°$, $80.8°$, $60.0°$, $30.7°$ and $0°$, respectively. The thick solid curves are obtained before any ground corrections have been applied. The dashed curves are obtained if we allow for a ground reflecting according to Lambert's law with an albedo $\lambda_0 = 0.20$. The thin intermediate curves are obtained if $\lambda_0 = 0.10$. The positions of the neutral points are also indicated.

points of zero polarization should be contrasted with what should be expected on the laws of single scattering, namely that the polarization should tend to zero as we approach the direction towards the sun.

During the nineteenth century the existence of these neutral points and their behavior with the direction of the sun were regarded as among the most remarkable phenomena in meteorological optics. As such they were studied with great care and attention and by none more than Carl Dorno whose monumental work on the subject[7] contains a wealth of information painstakingly gathered. Dorno not only observed the neutral points on the principal meridian, but he also investigated in detail the continuation of these neutral points over the entire hemisphere along what he called the neutral lines. These lines separate the regions of positive from the regions of negative polarization;[8] they show a remarkable dependence on the

[7] C. Dorno, Himmelshelligkeit, Himmelspolarisation und Sonnenintensität in Davos 1911 bis 1918, *Veröffentl. Preuss. Met. Inst.*, No. 303, Berlin, 1919.

[8] The polarization is assumed positive if $I_l$ is less than $I_r$ and negative if the reverse is true. On this convention the polarization is negative on the principal meridian between the neutral points of Babinet and Brewster; these points, therefore, separate

direction of the sun. A diagram representing Dorno's principal results is reproduced in figure 4. It will be seen from this figure that when the sun is nearly on the horizon the neutral line connects the Babinet and the Arago points by a closed symmetrical curve of the shape of a lemniscate: this is the so-called lemniscate of Busch. As the sun rises the lemniscate becomes more and more asymmetrical and when the angle of incidence exceeds about 70° the lemniscate opens out and a part of the locus appears on the horizon below the sun and passes through the Brewster point which has now risen. The neutral line consists of two such separated curves until the angle of incidence becomes about 45° when the opposite ends join together to form a closed re-entrant curve. For still smaller angles of incidence the neutral line collapses towards the center and finally reduces to a point when the sun is at the zenith.

We shall now see how this entire range of phenomena associated with the neutral points and lines are faithfully reproduced by our calculations. In

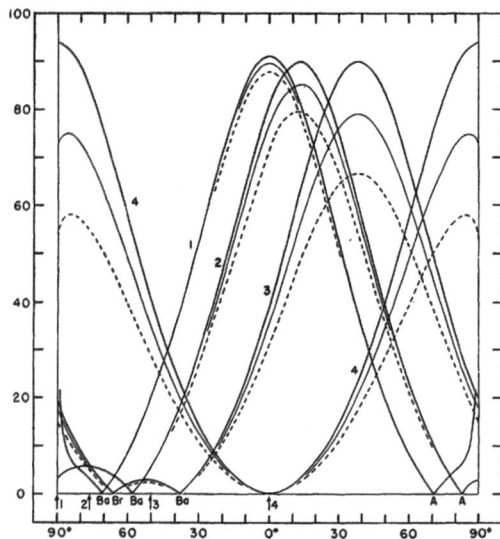

FIG. 7. Variation of the degree of polarization in the principal meridian for various angles of incidence for an atmosphere of optical thickness $\tau_1 = 0.15$. The abscissa gives the zenith distance and the ordinate gives the degree of polarization in per cent. The curves marked 1, 2, 3 and 4 represent the variation for the angles of incidence $\theta_0 = 90°$, $76.1°$, $50.2°$ and $0°$, respectively. The thick solid curves are obtained before any ground corrections have been applied. The dashed curves are obtained if we allow for a ground reflecting according to Lambert's law with an albedo $\lambda_0 = 20.5$. The thin intermediate curves are obtained if $\lambda_0 = 0.10$. The positions of the neutral points are also indicated.

the regions of positive from the regions of negative polarization on this meridian. The neutral lines of Dorno do the same for the entire hemisphere.

table 11 we have collected all the information contained in tables 5–10 regarding the points at which the polarization changes sign for various zenith distances of the sun; they are further illustrated in figures 1, 2, 3, and 5. Considering first figures 1–3 (which refer to the principal meridian), we observe that the calculations predict the occurrence of the neutral points as observed. In particular it will be noticed that the Brewster point sets when the angle of incidence is about 70°: its dependence on the values of the optical thickness in the range of interest is not pronounced. Also as the Brewster point sets the Arago point rises in the opposite sky. And as the sun sinks lower, the Arago point continues to rise until, when the sun sets, the Babinet and the Arago points are both at an equal

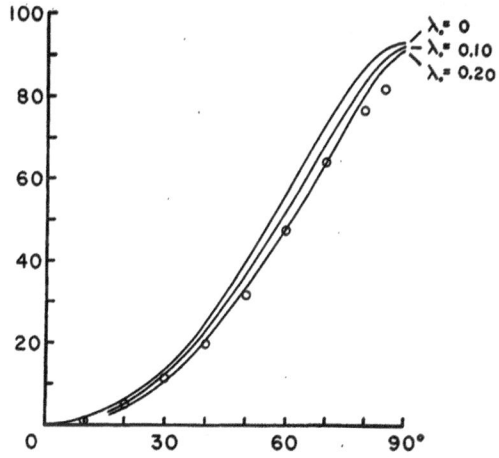

FIG. 8. Variation of the degree of polarization in the principal meridian for various angles of incidence for an atmosphere of optical thickness $\tau_1 = 0.20$. The abscissa gives the zenith distance and the ordinate gives the degree of polarization in per cent. The curves marked 1, 2, 3, 4, and 5 represent the variation for the angles of incidence $\theta_0 = 90°$, $80.8°$, $60.0°$, $30.7°$, and $0°$, respectively. The thick solid curves are obtained before any ground corrections have been applied. The dashed curves are obtained if we allow for a ground reflecting according to Lambert's law with an albedo $\lambda_0 = 0.20$. The thin intermediate curves are obtained if $\lambda_0 = 0.10$. The positions of the neutral points are also indicated.

elevation of about 20° (it varies from 17° to 21° for $\tau_1$ in the range $0.10 \leqslant \tau_1 \leqslant 0.20$) from the horizon.

With the calculations for the different values of $\varphi_0 - \varphi$ for $\tau_1 = 0.15$ given in table 5, we can draw an entire system of calculated neutral lines. This has been done in figure 5. Comparing it with the results of Dorno's observations (fig. 4) we observe how well the two sets of curves match.

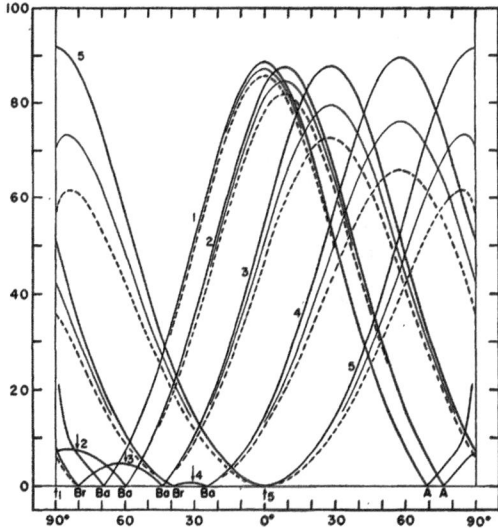

FIG. 9. Variation of the degree of polarization at the zenith for various angles of incidence and for an atmosphere of optical thickness $\tau_1 = 0.10$. The curve $\lambda_0 = 0$ allows for no ground reflection. The other two curves allow for reflection by a ground with the albedos indicated. The circles represent the observations of Tousey and Hulburt for values of $\tau_1$ and $\lambda_0$ estimated at 0.10 and 0.20, respectively.

Turning next to the calculated degrees of polarization, we have illustrated its variation on the principal meridian for various angles of incidence and for the three values of the optical thickness for which calculations have been made in figures 6, 7, and 8. It will be noticed from these figures that the effect of a ground surface with an albedo even as high as 0.25 does not

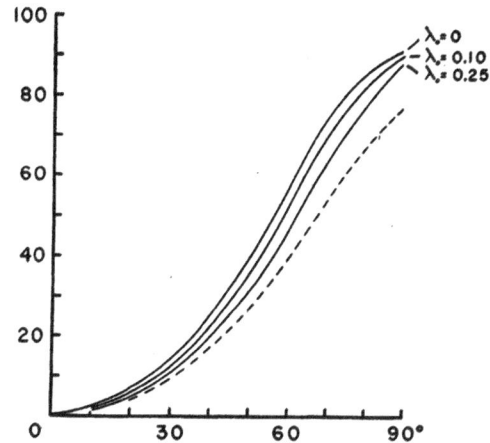

FIG. 10. Variation of the degree of polarization at the zenith for various angles of incidence and for an atmosphere of optical thickness $\tau_1 = 0.15$. The curve $\lambda_0 = 0$ allows for no ground reflection. The other two curves allow for reflection by a ground with the albedos indicated. The dashed curve represents the observations of Tichanowsky.

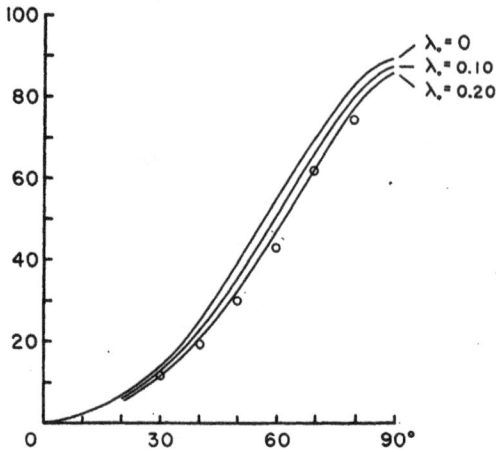

FIG. 11. Variation of the degree of polarization at the zenith for various angles of incidence and for an atmosphere of optical thickness $\tau_1 = 0.20$. The curve $\lambda_0 = 0$ allows for no ground reflection. The other two curves allow for reflection by a ground with the albedos indicated. The circles represent the observations of Richardson and Hulburt for a value of $\tau_1$ estimated at 0.3.

make any essential difference to the predicted positions of the neutral points. This independence of the neutral points (and lines) on ground reflection is not difficult to understand. As is well known, the laws of Rayleigh scattering give the maximum polarization for the scattered light; all other laws give much less polarization. A ground reflecting according to Lambert's law can, therefore, hardly compete with Rayleigh scattering for producing polarization. The effect of reflection by the ground is therefore essentially one of adding a component of natural light to the polarized light already present. This last statement is, of course, not strictly true. The difference between $\gamma_l$ and $\gamma_r$ is precisely a measure of the polarization of

the ground contribution to the sky brightness; but as is apparent from the tables the difference between $\gamma_l$ and $\gamma_r$ is generally very small. For these same reasons we should not expect that the direction of maximum polarization will be influenced by ground reflection. This is, indeed, the case: the direction of maximum polarization always occurs in a direction which is very nearly at right-angles to the direction of the sun. In contrast, ground reflection has a very pronounced effect on the *degree* of maximum polarization: In the absence of ground corrections, the maximum polarization varies between 94 and 89 per cent depending on $\tau_1$ but is very nearly independent of the altitude of the sun (though there is a slight increase for higher altitudes). However, when the effect of ground reflection is taken into account the maximum polarization shows a very marked decrease with altitude. Observations do show such a behavior and we conclude that this is probably due to the effect of ground reflection.

Finally, in figures 9, 10, and 11 we have compared the degree of polarization at the zenith for various angles of incidence with the observations of Tichanowsky,[9] Tousey and Hulburt,[10] and Richardson and Hulburt.[11] It will be seen that the agreement between the theory and the observations is as good as one might expect.

In concluding this paper we should again like to record our thanks to Dr. Wallace Eckert and Miss Ann Franklin of the Watson Scientific Computing Laboratory for their generous co-operation in obtaining the solutions for the basic $X$- and $Y$-functions.

[9] J. J. Tichanowsky, Resultate der Messungen der Himmelspolarisation in verschiedenen Spektrumabschnitten, *Meteor. Ztschr.* **43**: 288, 1926.

[10] R. Tousey and E. O. Hulburt, Brightness and polarization of the daylight sky at various altitudes above sea level, *Jour. Opt. Soc. Amer.* **37**: 78, 1947.

[11] R. A. Richardson and E. O. Hulburt, Sky-brightness measurements near Bocaiuva, Brazil, *Jour. Geophys. Research* **54**: 215, 1949.

6. THE TABLES

## ERRATUM

Page 686.   For $\mu = 0.80$ and ($\mu_0 = 0.24$; $\varphi = 40°$) for $I_l + I_r$

read 0.0323 instead of 0.0325

TABLE 1

THE FUNCTIONS $X_1(\mu)$, $Y_1(\mu)$; $X_r(\mu)$, $Y_r(\mu)$; $X^{(1)}(\mu)$, $Y^{(1)}(\mu)$; AND $X^{(2)}(\mu)$, $Y^{(2)}(\mu)$

FOR VARIOUS VALUES OF $\tau_i$ OBTAINED BY ITERATING THE INTEGRAL EQUATIONS THEY SATISFY

$$\tau_i = 0.05$$

| $\mu$ | $X_1$ | $Y_1$ | $X_r$ | $Y_r$ | $X^{(1)}$ | $Y^{(1)}$ | $X^{(2)}$ | $Y^{(2)}$ |
|---|---|---|---|---|---|---|---|---|
| 0 | 1.00000 | 0 | 1.00000 | 0 | 1.00000 | 0 | 1.00000 | 0 |
| 0.01 | 1.04591 | +0.01497 | 1.01618 | 0.01589 | 1.01825 | 0.01785 | 1.01135 | 0.01440 |
| 0.02 | 1.08187 | +0.09608 | 1.02621 | 0.10059 | 1.02999 | 0.10421 | 1.01906 | 0.09710 |
| 0.03 | 1.11580 | +0.20067 | 1.03282 | 0.21459 | 1.03779 | 0.21943 | 1.02427 | 0.20941 |
| 0.04 | 1.14850 | +0.28965 | 1.03733 | 0.31749 | 1.04315 | 0.32318 | 1.02786 | 0.31103 |
| 0.05 | 1.17187 | +0.35806 | 1.04057 | 0.40279 | 1.04700 | 0.40911 | 1.03045 | 0.39585 |
| 0.06 | 1.19940 | +0.40893 | 1.04300 | 0.47251 | 1.04989 | 0.47930 | 1.03240 | 0.46432 |
| 0.07 | 1.22638 | +0.44601 | 1.04488 | 0.52982 | 1.05213 | 0.53698 | 1.03391 | 0.52103 |
| 0.08 | 1.25297 | +0.47246 | 1.04638 | 0.57744 | 1.05392 | 0.58490 | 1.03511 | 0.56817 |
| 0.09 | 1.27929 | +0.49061 | 1.04760 | 0.61750 | 1.05538 | 0.62520 | 1.03610 | 0.60782 |
| 0.10 | 1.30541 | +0.50226 | 1.04862 | 0.65159 | 1.05659 | 0.65948 | 1.03692 | 0.64157 |
| 0.11 | 1.33135 | +0.50874 | 1.04947 | 0.68090 | 1.05761 | 0.68897 | 1.03760 | 0.67060 |
| 0.12 | 1.35719 | +0.51102 | 1.05020 | 0.70636 | 1.05848 | 0.71457 | 1.03819 | 0.69581 |
| 0.13 | 1.38293 | +0.50988 | 1.05083 | 0.72865 | 1.05923 | 0.73699 | 1.03870 | 0.71789 |
| 0.14 | 1.40858 | +0.50592 | 1.05138 | 0.74833 | 1.05989 | 0.75678 | 1.03915 | 0.73739 |
| 0.15 | 1.43417 | +0.49960 | 1.05187 | 0.76583 | 1.06047 | 0.77438 | 1.03954 | 0.75472 |
| 0.16 | 1.45969 | +0.49130 | 1.05230 | 0.78148 | 1.06098 | 0.79011 | 1.03989 | 0.77022 |
| 0.17 | 1.48519 | +0.48129 | 1.05268 | 0.79556 | 1.06144 | 0.80427 | 1.04020 | 0.78417 |
| 0.18 | 1.51064 | +0.46984 | 1.05303 | 0.80829 | 1.06185 | 0.81707 | 1.04048 | 0.79679 |
| 0.19 | 1.53604 | +0.45714 | 1.05334 | 0.81986 | 1.06222 | 0.82870 | 1.04073 | 0.80825 |
| 0.20 | 1.56145 | +0.44335 | 1.05362 | 0.83041 | 1.06256 | 0.83932 | 1.04096 | 0.81870 |
| 0.21 | 1.58681 | +0.42859 | 1.05388 | 0.84008 | 1.06287 | 0.84904 | 1.04117 | 0.82828 |
| 0.22 | 1.61215 | +0.41300 | 1.05411 | 0.84897 | 1.06315 | 0.85797 | 1.04136 | 0.83709 |
| 0.23 | 1.63747 | +0.39668 | 1.05433 | 0.85717 | 1.06341 | 0.86622 | 1.04153 | 0.84522 |
| 0.24 | 1.66277 | +0.37970 | 1.05453 | 0.86476 | 1.06365 | 0.87384 | 1.04170 | 0.85274 |
| 0.25 | 1.68808 | +0.36212 | 1.05472 | 0.87180 | 1.06387 | 0.88092 | 1.04185 | 0.85971 |
| 0.26 | 1.71336 | +0.34402 | 1.05489 | 0.87835 | 1.06407 | 0.88751 | 1.04198 | 0.86620 |
| 0.27 | 1.73862 | +0.32545 | 1.05505 | 0.88446 | 1.06427 | 0.89365 | 1.04211 | 0.87226 |
| 0.28 | 1.76388 | +0.30645 | 1.05520 | 0.89018 | 1.06444 | 0.89939 | 1.04223 | 0.87792 |
| 0.29 | 1.78913 | +0.28707 | 1.05533 | 0.89553 | 1.06461 | 0.90477 | 1.04235 | 0.88322 |
| 0.30 | 1.81436 | +0.26734 | 1.05546 | 0.90055 | 1.06477 | 0.90983 | 1.04245 | 0.88820 |
| 0.31 | 1.83959 | +0.24729 | 1.05559 | 0.90528 | 1.06491 | 0.91458 | 1.04255 | 0.89289 |
| 0.32 | 1.86482 | +0.22694 | 1.05570 | 0.90973 | 1.06505 | 0.91906 | 1.04265 | 0.89730 |
| 0.33 | 1.89003 | +0.20633 | 1.05581 | 0.91394 | 1.06518 | 0.92328 | 1.04273 | 0.90147 |
| 0.34 | 1.91525 | +0.18547 | 1.05591 | 0.91791 | 1.06530 | 0.92728 | 1.04282 | 0.90541 |
| 0.35 | 1.94046 | +0.16438 | 1.05601 | 0.92168 | 1.06542 | 0.93106 | 1.04289 | 0.90914 |
| 0.36 | 1.96567 | +0.14308 | 1.05610 | 0.92525 | 1.06553 | 0.93465 | 1.04297 | 0.91267 |
| 0.37 | 1.99085 | +0.12160 | 1.05619 | 0.92864 | 1.06563 | 0.93806 | 1.04304 | 0.91603 |
| 0.38 | 2.01604 | +0.09993 | 1.05627 | 0.93186 | 1.06573 | 0.94130 | 1.04311 | 0.91923 |
| 0.39 | 2.04123 | +0.07810 | 1.05635 | 0.93493 | 1.06582 | 0.94438 | 1.04317 | 0.92227 |
| 0.40 | 2.06643 | +0.05611 | 1.05642 | 0.93785 | 1.06591 | 0.94732 | 1.04323 | 0.92517 |
| 0.41 | 2.09160 | +0.03399 | 1.05649 | 0.94064 | 1.06600 | 0.95012 | 1.04329 | 0.92793 |
| 0.42 | 2.11678 | +0.01173 | 1.05656 | 0.94331 | 1.06608 | 0.95280 | 1.04334 | 0.93057 |
| 0.43 | 2.14195 | −0.01066 | 1.05663 | 0.94586 | 1.06615 | 0.95537 | 1.04340 | 0.93310 |
| 0.44 | 2.16712 | −0.03316 | 1.05669 | 0.94830 | 1.06623 | 0.95782 | 1.04345 | 0.93551 |
| 0.45 | 2.19229 | −0.05578 | 1.05675 | 0.95063 | 1.06630 | 0.96017 | 1.04349 | 0.93783 |
| 0.46 | 2.21748 | −0.07851 | 1.05681 | 0.95287 | 1.06637 | 0.96242 | 1.04354 | 0.94005 |
| 0.47 | 2.24268 | −0.10132 | 1.05686 | 0.95502 | 1.06643 | 0.96458 | 1.04358 | 0.94218 |
| 0.48 | 2.26780 | −0.12422 | 1.05691 | 0.95709 | 1.06650 | 0.96665 | 1.04363 | 0.94423 |
| 0.49 | 2.29296 | −0.14721 | 1.05696 | 0.95907 | 1.06656 | 0.96865 | 1.04367 | 0.94620 |
| 0.50 | 2.31812 | −0.17029 | 1.05701 | 0.96098 | 1.06661 | 0.97057 | 1.04371 | 0.94809 |

TABLE 1 (continued)

$$\tau_1 = 0.05$$

| $\mu$ | $X_1$ | $Y_1$ | $X_r$ | $Y_r$ | $X^{(1)}$ | $Y^{(1)}$ | $X^{(2)}$ | $Y^{(2)}$ |
|---|---|---|---|---|---|---|---|---|
| 0.51 | 2.34327 | -0.19343 | 1.05706 | 0.96282 | 1.06667 | 0.97242 | 1.04374 | 0.94991 |
| 0.52 | 2.36843 | -0.21665 | 1.05710 | 0.96459 | 1.06672 | 0.97420 | 1.04378 | 0.95167 |
| 0.53 | 2.39358 | -0.23994 | 1.05714 | 0.96630 | 1.06677 | 0.97591 | 1.04381 | 0.95336 |
| 0.54 | 2.41873 | -0.26329 | 1.05719 | 0.96795 | 1.06682 | 0.97757 | 1.04385 | 0.95499 |
| 0.55 | 2.44388 | -0.28670 | 1.05723 | 0.96954 | 1.06687 | 0.97917 | 1.04388 | 0.95657 |
| 0.56 | 2.46903 | -0.31017 | 1.05726 | 0.97107 | 1.06692 | 0.98071 | 1.04391 | 0.95809 |
| 0.57 | 2.49419 | -0.33370 | 1.05730 | 0.97256 | 1.06696 | 0.98220 | 1.04394 | 0.95956 |
| 0.58 | 2.51933 | -0.35728 | 1.05734 | 0.97399 | 1.06700 | 0.98364 | 1.04397 | 0.96098 |
| 0.59 | 2.54448 | -0.38091 | 1.05737 | 0.97538 | 1.06705 | 0.98504 | 1.04400 | 0.96236 |
| 0.60 | 2.56961 | -0.40457 | 1.05741 | 0.97673 | 1.06709 | 0.98639 | 1.04403 | 0.96369 |
| 0.61 | 2.59476 | -0.42830 | 1.05744 | 0.97803 | 1.06713 | 0.98770 | 1.04405 | 0.96498 |
| 0.62 | 2.61990 | -0.45206 | 1.05747 | 0.97929 | 1.06716 | 0.98897 | 1.04408 | 0.96623 |
| 0.63 | 2.64505 | -0.47587 | 1.05750 | 0.98051 | 1.06720 | 0.99020 | 1.04410 | 0.96744 |
| 0.64 | 2.67018 | -0.49971 | 1.05753 | 0.98170 | 1.06724 | 0.99139 | 1.04413 | 0.96862 |
| 0.65 | 2.69533 | -0.52360 | 1.05756 | 0.98285 | 1.06727 | 0.99255 | 1.04415 | 0.96976 |
| 0.66 | 2.72047 | -0.54752 | 1.05759 | 0.98397 | 1.06730 | 0.99367 | 1.04417 | 0.97087 |
| 0.67 | 2.74561 | -0.57147 | 1.05762 | 0.98505 | 1.06734 | 0.99476 | 1.04420 | 0.97194 |
| 0.68 | 2.77075 | -0.59546 | 1.05764 | 0.98611 | 1.06737 | 0.99582 | 1.04422 | 0.97299 |
| 0.69 | 2.79588 | -0.61947 | 1.05767 | 0.98713 | 1.06740 | 0.99685 | 1.04424 | 0.97400 |
| 0.70 | 2.82102 | -0.64352 | 1.05769 | 0.98813 | 1.06743 | 0.99785 | 1.04426 | 0.97499 |
| 0.71 | 2.84616 | -0.66760 | 1.05772 | 0.98910 | 1.06746 | 0.99883 | 1.04428 | 0.97595 |
| 0.72 | 2.87129 | -0.69170 | 1.05774 | 0.99004 | 1.06749 | 0.99978 | 1.04430 | 0.97689 |
| 0.73 | 2.89643 | -0.71583 | 1.05776 | 0.99096 | 1.06751 | 1.00070 | 1.04432 | 0.97780 |
| 0.74 | 2.92156 | -0.73999 | 1.05779 | 0.99186 | 1.06754 | 1.00160 | 1.04433 | 0.97869 |
| 0.75 | 2.94669 | -0.76417 | 1.05781 | 0.99273 | 1.06757 | 1.00248 | 1.04435 | 0.97955 |
| 0.76 | 2.97182 | -0.78837 | 1.05783 | 0.99358 | 1.06759 | 1.00333 | 1.04437 | 0.98039 |
| 0.77 | 2.99695 | -0.81260 | 1.05785 | 0.99441 | 1.06762 | 1.00416 | 1.04439 | 0.98122 |
| 0.78 | 3.02209 | -0.83685 | 1.05787 | 0.99521 | 1.06764 | 1.00497 | 1.04440 | 0.98202 |
| 0.79 | 3.04722 | -0.86112 | 1.05789 | 0.99600 | 1.06766 | 1.00577 | 1.04442 | 0.98279 |
| 0.80 | 3.07235 | -0.88541 | 1.05791 | 0.99677 | 1.06769 | 1.00654 | 1.04443 | 0.98356 |
| 0.81 | 3.09747 | -0.90972 | 1.05793 | 0.99752 | 1.06771 | 1.00729 | 1.04445 | 0.98430 |
| 0.82 | 3.12260 | -0.93405 | 1.05794 | 0.99825 | 1.06773 | 1.00803 | 1.04446 | 0.98503 |
| 0.83 | 3.14774 | -0.95840 | 1.05796 | 0.99897 | 1.06775 | 1.00875 | 1.04448 | 0.98574 |
| 0.84 | 3.17286 | -0.98276 | 1.05798 | 0.99967 | 1.06777 | 1.00945 | 1.04449 | 0.98643 |
| 0.85 | 3.19800 | -1.00715 | 1.05800 | 1.00035 | 1.06779 | 1.01014 | 1.04451 | 0.98711 |
| 0.86 | 3.22313 | -1.03155 | 1.05801 | 1.00102 | 1.06781 | 1.01081 | 1.04452 | 0.98777 |
| 0.87 | 3.24825 | -1.05596 | 1.05803 | 1.00167 | 1.06783 | 1.01146 | 1.04453 | 0.98842 |
| 0.88 | 3.27338 | -1.08039 | 1.05805 | 1.00231 | 1.06785 | 1.01211 | 1.04455 | 0.98905 |
| 0.89 | 3.29852 | -1.10485 | 1.05806 | 1.00293 | 1.06787 | 1.01273 | 1.04456 | 0.98967 |
| 0.90 | 3.32364 | -1.12930 | 1.05808 | 1.00355 | 1.06789 | 1.01335 | 1.04457 | 0.99027 |
| 0.91 | 3.34876 | -1.15377 | 1.05809 | 1.00414 | 1.06791 | 1.01395 | 1.04458 | 0.99086 |
| 0.92 | 3.37388 | -1.17826 | 1.05811 | 1.00473 | 1.06792 | 1.01454 | 1.04460 | 0.99145 |
| 0.93 | 3.39900 | -1.20275 | 1.05812 | 1.00530 | 1.06794 | 1.01511 | 1.04461 | 0.99201 |
| 0.94 | 3.42413 | -1.22727 | 1.05813 | 1.00586 | 1.06796 | 1.01568 | 1.04462 | 0.99257 |
| 0.95 | 3.44926 | -1.25180 | 1.05815 | 1.00641 | 1.06797 | 1.01623 | 1.04463 | 0.99311 |
| 0.96 | 3.47438 | -1.27633 | 1.05816 | 1.00695 | 1.06799 | 1.01677 | 1.04464 | 0.99365 |
| 0.97 | 3.49951 | -1.30088 | 1.05817 | 1.00748 | 1.06800 | 1.01730 | 1.04465 | 0.99417 |
| 0.98 | 3.52465 | -1.32545 | 1.05819 | 1.00799 | 1.06802 | 1.01782 | 1.04466 | 0.99468 |
| 0.99 | 3.54977 | -1.35002 | 1.05820 | 1.00850 | 1.06804 | 1.01833 | 1.04467 | 0.99519 |
| 1.00 | 3.57488 | -1.37459 | 1.05821 | 1.00900 | 1.06805 | 1.01883 | 1.04468 | 0.99568 |

TABLE 1 (continued)

$$\tau_1 = 0.10$$

| μ | $X_1$ | $Y_1$ | $X_r$ | $Y_r$ | $X^{(1)}$ | $Y^{(1)}$ | $X^{(2)}$ | $Y^{(2)}$ |
|---|---|---|---|---|---|---|---|---|
| 0 | 1.00000 | 0 | 1.00000 | 0 | 1.00000 | 0 | 1.00000 | 0 |
| 0.01 | 1.04663 | +0.00480 | 1.01651 | 0.00654 | 1.01882 | 0.00844 | 1.01162 | 0.00631 |
| 0.02 | 1.08343 | +0.01715 | 1.02787 | 0.02068 | 1.03238 | 0.02455 | 1.02052 | 0.01976 |
| 0.03 | 1.11694 | +0.03079 | 1.03697 | 0.03740 | 1.04338 | 0.06314 | 1.02796 | 0.03540 |
| 0.04 | 1.14836 | +0.09873 | 1.04426 | 0.11104 | 1.05227 | 0.11841 | 1.03407 | 0.10788 |
| 0.05 | 1.17829 | +0.14976 | 1.05013 | 0.17063 | 1.05946 | 0.17937 | 1.03905 | 0.16637 |
| 0.06 | 1.20713 | +0.19771 | 1.05490 | 0.22964 | 1.06534 | 0.23951 | 1.04313 | 0.22437 |
| 0.07 | 1.23516 | +0.24005 | 1.05883 | 0.28509 | 1.07019 | 0.29593 | 1.04652 | 0.27893 |
| 0.08 | 1.26259 | +0.27607 | 1.06212 | 0.33597 | 1.07426 | 0.34762 | 1.04936 | 0.32903 |
| 0.09 | 1.28956 | +0.30599 | 1.06489 | 0.38214 | 1.07770 | 0.39450 | 1.05178 | 0.37451 |
| 0.10 | 1.31615 | +0.33034 | 1.06727 | 0.42385 | 1.08065 | 0.43682 | 1.05385 | 0.41562 |
| 0.11 | 1.34243 | +0.34973 | 1.06932 | 0.46152 | 1.08321 | 0.47500 | 1.05565 | 0.45276 |
| 0.12 | 1.36849 | +0.36476 | 1.07112 | 0.49557 | 1.08545 | 0.50951 | 1.05722 | 0.48633 |
| 0.13 | 1.39435 | +0.37596 | 1.07269 | 0.52641 | 1.08741 | 0.54077 | 1.05861 | 0.51676 |
| 0.14 | 1.42005 | +0.38381 | 1.07409 | 0.55445 | 1.08915 | 0.56917 | 1.05983 | 0.54441 |
| 0.15 | 1.44562 | +0.38872 | 1.07534 | 0.57999 | 1.09071 | 0.59504 | 1.06093 | 0.56960 |
| 0.16 | 1.47107 | +0.39107 | 1.07645 | 0.60334 | 1.09210 | 0.61868 | 1.06192 | 0.59265 |
| 0.17 | 1.49644 | +0.39113 | 1.07747 | 0.62476 | 1.09337 | 0.64036 | 1.06281 | 0.61378 |
| 0.18 | 1.52173 | +0.38919 | 1.07838 | 0.64444 | 1.09452 | 0.66030 | 1.06362 | 0.63321 |
| 0.19 | 1.54691 | +0.38549 | 1.07921 | 0.66261 | 1.09556 | 0.67868 | 1.06436 | 0.65113 |
| 0.20 | 1.57206 | +0.38020 | 1.07998 | 0.67941 | 1.09652 | 0.69568 | 1.06503 | 0.66771 |
| 0.21 | 1.59714 | +0.37350 | 1.08067 | 0.69499 | 1.09740 | 0.71144 | 1.06565 | 0.68310 |
| 0.22 | 1.62218 | +0.36552 | 1.08132 | 0.70947 | 1.09820 | 0.72610 | 1.06622 | 0.69739 |
| 0.23 | 1.64718 | +0.35641 | 1.08192 | 0.72297 | 1.09895 | 0.73975 | 1.06675 | 0.71071 |
| 0.24 | 1.67214 | +0.34628 | 1.08247 | 0.73557 | 1.09964 | 0.75249 | 1.06724 | 0.72315 |
| 0.25 | 1.69706 | +0.33522 | 1.08298 | 0.74736 | 1.10029 | 0.76443 | 1.06769 | 0.73479 |
| 0.26 | 1.72195 | +0.32331 | 1.08346 | 0.75842 | 1.10088 | 0.77561 | 1.06812 | 0.74571 |
| 0.27 | 1.74682 | +0.31064 | 1.08390 | 0.76881 | 1.10144 | 0.78612 | 1.06852 | 0.75597 |
| 0.28 | 1.77165 | +0.29726 | 1.08432 | 0.77859 | 1.10197 | 0.79601 | 1.06889 | 0.76563 |
| 0.29 | 1.79648 | +0.28323 | 1.08472 | 0.78780 | 1.10246 | 0.80533 | 1.06924 | 0.77473 |
| 0.30 | 1.82126 | +0.26865 | 1.08509 | 0.79652 | 1.10293 | 0.81414 | 1.06956 | 0.78332 |
| 0.31 | 1.84604 | +0.25351 | 1.08544 | 0.80475 | 1.10336 | 0.82246 | 1.06987 | 0.79145 |
| 0.32 | 1.87082 | +0.23786 | 1.08576 | 0.81254 | 1.10377 | 0.83036 | 1.07016 | 0.79915 |
| 0.33 | 1.89556 | +0.22176 | 1.08607 | 0.81994 | 1.10416 | 0.83783 | 1.07044 | 0.80647 |
| 0.34 | 1.92029 | +0.20525 | 1.08637 | 0.82696 | 1.10453 | 0.84494 | 1.07070 | 0.81340 |
| 0.35 | 1.94501 | +0.18833 | 1.08665 | 0.83365 | 1.10488 | 0.85169 | 1.07095 | 0.82000 |
| 0.36 | 1.96972 | +0.17105 | 1.08691 | 0.84001 | 1.10521 | 0.85812 | 1.07118 | 0.82628 |
| 0.37 | 1.99440 | +0.15343 | 1.08717 | 0.84607 | 1.10552 | 0.86425 | 1.07140 | 0.83226 |
| 0.38 | 2.01910 | +0.13550 | 1.08741 | 0.85185 | 1.10583 | 0.87010 | 1.07161 | 0.83798 |
| 0.39 | 2.04377 | +0.11727 | 1.08763 | 0.85738 | 1.10610 | 0.87569 | 1.07182 | 0.84343 |
| 0.40 | 2.06843 | +0.09877 | 1.08785 | 0.86266 | 1.10637 | 0.88103 | 1.07201 | 0.84865 |
| 0.41 | 2.09309 | +0.08000 | 1.08805 | 0.86772 | 1.10663 | 0.88614 | 1.07219 | 0.85364 |
| 0.42 | 2.11774 | +0.06100 | 1.08825 | 0.87257 | 1.10688 | 0.89104 | 1.07237 | 0.85843 |
| 0.43 | 2.14239 | +0.04177 | 1.08844 | 0.87721 | 1.10712 | 0.89574 | 1.07254 | 0.86301 |
| 0.44 | 2.16701 | +0.02233 | 1.08862 | 0.88167 | 1.10735 | 0.90024 | 1.07270 | 0.86741 |
| 0.45 | 2.19165 | +0.00269 | 1.08880 | 0.88595 | 1.10757 | 0.90457 | 1.07285 | 0.87164 |
| 0.46 | 2.21627 | -0.01714 | 1.08896 | 0.89006 | 1.10778 | 0.90873 | 1.07300 | 0.87570 |
| 0.47 | 2.24088 | -0.03715 | 1.08913 | 0.89402 | 1.10798 | 0.91273 | 1.07314 | 0.87961 |
| 0.48 | 2.26549 | -0.05733 | 1.08928 | 0.89783 | 1.10817 | 0.91659 | 1.07328 | 0.88338 |
| 0.49 | 2.29011 | -0.07767 | 1.08943 | 0.90149 | 1.10836 | 0.92030 | 1.07341 | 0.88700 |
| 0.50 | 2.31471 | -0.09815 | 1.08957 | 0.90503 | 1.10854 | 0.92387 | 1.07354 | 0.89049 |

TABLE 1 (continued)

$$\tau_1 = 0.10$$

| μ | $X_1$ | $Y_1$ | $X_r$ | $Y_r$ | $X^{(1)}$ | $Y^{(1)}$ | $X^{(2)}$ | $Y^{(2)}$ |
|---|---|---|---|---|---|---|---|---|
| 0.51 | 2.33931 | -0.11879 | 1.08971 | 0.90844 | 1.10870 | 0.92732 | 1.07366 | 0.89386 |
| 0.52 | 2.36391 | -0.13955 | 1.08984 | 0.91174 | 1.10887 | 0.93065 | 1.07378 | 0.89712 |
| 0.53 | 2.38849 | -0.16044 | 1.08997 | 0.91492 | 1.10903 | 0.93386 | 1.07389 | 0.90026 |
| 0.54 | 2.41308 | -0.18145 | 1.09010 | 0.91799 | 1.10919 | 0.93697 | 1.07400 | 0.90329 |
| 0.55 | 2.43766 | -0.20257 | 1.09021 | 0.92096 | 1.10933 | 0.93997 | 1.07411 | 0.90623 |
| 0.56 | 2.46224 | -0.22381 | 1.09033 | 0.92383 | 1.10948 | 0.94288 | 1.07421 | 0.90907 |
| 0.57 | 2.48681 | -0.24515 | 1.09044 | 0.92662 | 1.10962 | 0.94569 | 1.07431 | 0.91182 |
| 0.58 | 2.51138 | -0.26658 | 1.09055 | 0.92931 | 1.10975 | 0.94841 | 1.07440 | 0.91448 |
| 0.59 | 2.53595 | -0.28812 | 1.09065 | 0.93192 | 1.10989 | 0.95105 | 1.07450 | 0.91706 |
| 0.60 | 2.56051 | -0.30975 | 1.09076 | 0.93445 | 1.11001 | 0.95361 | 1.07458 | 0.91956 |
| 0.61 | 2.58508 | -0.33146 | 1.09085 | 0.93691 | 1.11012 | 0.95609 | 1.07467 | 0.92199 |
| 0.62 | 2.60965 | -0.35325 | 1.09095 | 0.93929 | 1.11024 | 0.95850 | 1.07476 | 0.92434 |
| 0.63 | 2.63420 | -0.37512 | 1.09104 | 0.94160 | 1.11036 | 0.96083 | 1.07484 | 0.92662 |
| 0.64 | 2.65876 | -0.39707 | 1.09113 | 0.94385 | 1.11047 | 0.96310 | 1.07492 | 0.92884 |
| 0.65 | 2.68330 | -0.41909 | 1.09122 | 0.94603 | 1.11058 | 0.96531 | 1.07499 | 0.93100 |
| 0.66 | 2.70785 | -0.44118 | 1.09130 | 0.94815 | 1.11069 | 0.96745 | 1.07507 | 0.93309 |
| 0.67 | 2.73241 | -0.46334 | 1.09138 | 0.95021 | 1.11079 | 0.96954 | 1.07514 | 0.93513 |
| 0.68 | 2.75696 | -0.48556 | 1.09146 | 0.95222 | 1.11089 | 0.97156 | 1.07521 | 0.93711 |
| 0.69 | 2.78151 | -0.50784 | 1.09154 | 0.95417 | 1.11098 | 0.97354 | 1.07528 | 0.93904 |
| 0.70 | 2.80605 | -0.53018 | 1.09162 | 0.95607 | 1.11108 | 0.97545 | 1.07535 | 0.94091 |
| 0.71 | 2.83047 | -0.55252 | 1.09169 | 0.95792 | 1.11117 | 0.97732 | 1.07541 | 0.94274 |
| 0.72 | 2.85513 | -0.57502 | 1.09176 | 0.95972 | 1.11126 | 0.97915 | 1.07547 | 0.94452 |
| 0.73 | 2.87967 | -0.59752 | 1.09183 | 0.96148 | 1.11135 | 0.98092 | 1.07554 | 0.94626 |
| 0.74 | 2.90420 | -0.62007 | 1.09190 | 0.96319 | 1.11143 | 0.98265 | 1.07560 | 0.94795 |
| 0.75 | 2.92874 | -0.64267 | 1.09196 | 0.96486 | 1.11151 | 0.98434 | 1.07565 | 0.94960 |
| 0.76 | 2.95328 | -0.66532 | 1.09203 | 0.96649 | 1.11159 | 0.98598 | 1.07571 | 0.95121 |
| 0.77 | 2.97781 | -0.68801 | 1.09209 | 0.96807 | 1.11167 | 0.98759 | 1.07577 | 0.95278 |
| 0.78 | 3.00236 | -0.71075 | 1.09215 | 0.96962 | 1.11175 | 0.98915 | 1.07582 | 0.95431 |
| 0.79 | 3.02687 | -0.73352 | 1.09221 | 0.97114 | 1.11182 | 0.99068 | 1.07587 | 0.95580 |
| 0.80 | 3.05141 | -0.75684 | 1.09227 | 0.97262 | 1.11190 | 0.99218 | 1.07592 | 0.95726 |
| 0.81 | 3.07593 | -0.77919 | 1.09233 | 0.97406 | 1.11197 | 0.99364 | 1.07597 | 0.95869 |
| 0.82 | 3.10044 | -0.80207 | 1.09238 | 0.97547 | 1.11204 | 0.99506 | 1.07602 | 0.96008 |
| 0.83 | 3.12498 | -0.82501 | 1.09244 | 0.97685 | 1.11211 | 0.99645 | 1.07607 | 0.96144 |
| 0.84 | 3.14950 | -0.84796 | 1.09249 | 0.97820 | 1.11217 | 0.99782 | 1.07612 | 0.96278 |
| 0.85 | 3.17403 | -0.87096 | 1.09254 | 0.97951 | 1.11224 | 0.99915 | 1.07616 | 0.96408 |
| 0.86 | 3.19856 | -0.89399 | 1.09259 | 0.98080 | 1.11230 | 1.00045 | 1.07621 | 0.96535 |
| 0.87 | 3.22308 | -0.91705 | 1.09264 | 0.98206 | 1.11236 | 1.00173 | 1.07625 | 0.96660 |
| 0.88 | 3.24760 | -0.94014 | 1.09269 | 0.98330 | 1.11242 | 1.00297 | 1.07630 | 0.96782 |
| 0.89 | 3.27214 | -0.96327 | 1.09274 | 0.98451 | 1.11248 | 1.00419 | 1.07634 | 0.96901 |
| 0.90 | 3.29663 | -0.98641 | 1.09278 | 0.98569 | 1.11254 | 1.00539 | 1.07638 | 0.97018 |
| 0.91 | 3.32115 | -1.00959 | 1.09283 | 0.98685 | 1.11260 | 1.00656 | 1.07642 | 0.97132 |
| 0.92 | 3.34567 | -1.03279 | 1.09287 | 0.98798 | 1.11265 | 1.00770 | 1.07646 | 0.97244 |
| 0.93 | 3.37019 | -1.05602 | 1.09291 | 0.98909 | 1.11271 | 1.00883 | 1.07650 | 0.97354 |
| 0.94 | 3.39471 | -1.07927 | 1.09296 | 0.99018 | 1.11276 | 1.00993 | 1.07653 | 0.97462 |
| 0.95 | 3.41923 | -1.10256 | 1.09300 | 0.99125 | 1.11281 | 1.01100 | 1.07657 | 0.97567 |
| 0.96 | 3.44373 | -1.12585 | 1.09304 | 0.99229 | 1.11286 | 1.01206 | 1.07661 | 0.97670 |
| 0.97 | 3.46825 | -1.14918 | 1.09308 | 0.99332 | 1.11291 | 1.01310 | 1.07664 | 0.97772 |
| 0.98 | 3.49278 | -1.17254 | 1.09312 | 0.99432 | 1.11296 | 1.01411 | 1.07668 | 0.97871 |
| 0.99 | 3.51728 | -1.19591 | 1.09316 | 0.99531 | 1.11301 | 1.01511 | 1.07671 | 0.97968 |
| 1.00 | 3.54180 | -1.21930 | 1.09320 | 0.99628 | 1.11306 | 1.01609 | 1.07674 | 0.98064 |

TABLE 1 (continued)

$$\tau_1 = 0.15$$

| $\mu$ | $X_1$ | $Y_1$ | $X_r$ | $Y_r$ | $X^{(1)}$ | $Y^{(1)}$ | $X^{(2)}$ | $Y^{(2)}$ |
|---|---|---|---|---|---|---|---|---|
| 0 | 1.00000 | 0 | 1.00000 | 0 | 1.00000 | 0 | 1.00000 | 0 |
| 0.01 | 1.04716 | +0.00302 | 1.01681 | 0.00515 | 1.01920 | 0.00697 | 1.01180 | 0.00544 |
| 0.02 | 1.08394 | +0.00757 | 1.02843 | 0.01150 | 1.03311 | 0.01528 | 1.02095 | 0.01183 |
| 0.03 | 1.11766 | +0.01821 | 1.03810 | 0.02412 | 1.04496 | 0.02989 | 1.02899 | 0.02420 |
| 0.04 | 1.14954 | +0.03881 | 1.04636 | 0.04760 | 1.05524 | 0.05531 | 1.03609 | 0.04718 |
| 0.05 | 1.18007 | +0.06714 | 1.05343 | 0.08045 | 1.06412 | 0.08995 | 1.04232 | 0.07936 |
| 0.06 | 1.20955 | +0.09923 | 1.05949 | 0.11893 | 1.07180 | 0.13003 | 1.04774 | 0.11713 |
| 0.07 | 1.23820 | +0.13189 | 1.06473 | 0.15985 | 1.07846 | 0.17238 | 1.05245 | 0.15732 |
| 0.08 | 1.26618 | +0.16310 | 1.06926 | 0.20105 | 1.08426 | 0.21487 | 1.05657 | 0.19783 |
| 0.09 | 1.29365 | +0.19174 | 1.07322 | 0.24123 | 1.08931 | 0.25620 | 1.06019 | 0.23735 |
| 0.10 | 1.32066 | +0.21734 | 1.07669 | 0.27971 | 1.09376 | 0.29570 | 1.06338 | 0.27522 |
| 0.11 | 1.34731 | +0.23971 | 1.07976 | 0.31612 | 1.09771 | 0.33302 | 1.06622 | 0.31106 |
| 0.12 | 1.37366 | +0.25887 | 1.08248 | 0.35035 | 1.10122 | 0.36808 | 1.06874 | 0.34477 |
| 0.13 | 1.39975 | +0.27498 | 1.08493 | 0.38241 | 1.10436 | 0.40088 | 1.07099 | 0.37633 |
| 0.14 | 1.42565 | +0.28822 | 1.08712 | 0.41237 | 1.10718 | 0.43150 | 1.07303 | 0.40583 |
| 0.15 | 1.45134 | +0.29881 | 1.08909 | 0.44034 | 1.10974 | 0.46007 | 1.07487 | 0.43339 |
| 0.16 | 1.47690 | +0.30696 | 1.09089 | 0.46646 | 1.11206 | 0.48675 | 1.07655 | 0.45913 |
| 0.17 | 1.50229 | +0.31288 | 1.09253 | 0.49086 | 1.11417 | 0.51166 | 1.07808 | 0.48318 |
| 0.18 | 1.52759 | +0.31676 | 1.09403 | 0.51368 | 1.11610 | 0.53495 | 1.07948 | 0.50566 |
| 0.19 | 1.55279 | +0.31877 | 1.09540 | 0.53505 | 1.11788 | 0.55674 | 1.08076 | 0.52671 |
| 0.20 | 1.57788 | +0.31910 | 1.09667 | 0.55506 | 1.11952 | 0.57715 | 1.08194 | 0.54645 |
| 0.21 | 1.60288 | +0.31788 | 1.09783 | 0.57385 | 1.12103 | 0.59631 | 1.08304 | 0.56497 |
| 0.22 | 1.62783 | +0.31523 | 1.09892 | 0.59151 | 1.12243 | 0.61431 | 1.08405 | 0.58237 |
| 0.23 | 1.65272 | +0.31129 | 1.09993 | 0.60813 | 1.12374 | 0.63125 | 1.08500 | 0.59876 |
| 0.24 | 1.67755 | +0.30614 | 1.10087 | 0.62379 | 1.12495 | 0.64720 | 1.08588 | 0.61420 |
| 0.25 | 1.70233 | +0.29992 | 1.10174 | 0.63857 | 1.12608 | 0.66226 | 1.08670 | 0.62877 |
| 0.26 | 1.72706 | +0.29270 | 1.10256 | 0.65254 | 1.12714 | 0.67649 | 1.08747 | 0.64254 |
| 0.27 | 1.75176 | +0.28454 | 1.10333 | 0.66576 | 1.12815 | 0.68995 | 1.08819 | 0.65557 |
| 0.28 | 1.77639 | +0.27553 | 1.10405 | 0.67828 | 1.12908 | 0.70270 | 1.08887 | 0.66793 |
| 0.29 | 1.80100 | +0.26574 | 1.10473 | 0.69016 | 1.12995 | 0.71479 | 1.08951 | 0.67964 |
| 0.30 | 1.82559 | +0.25522 | 1.10537 | 0.70144 | 1.13079 | 0.72628 | 1.09012 | 0.69076 |
| 0.31 | 1.85017 | +0.24400 | 1.10598 | 0.71217 | 1.13157 | 0.73721 | 1.09068 | 0.70134 |
| 0.32 | 1.87470 | +0.23216 | 1.10655 | 0.72237 | 1.13231 | 0.74760 | 1.09122 | 0.71141 |
| 0.33 | 1.89919 | +0.21976 | 1.10709 | 0.73211 | 1.13302 | 0.75750 | 1.09173 | 0.72100 |
| 0.34 | 1.92368 | +0.20679 | 1.10760 | 0.74139 | 1.13368 | 0.76696 | 1.09222 | 0.73016 |
| 0.35 | 1.94814 | +0.19332 | 1.10810 | 0.75026 | 1.13432 | 0.77597 | 1.09267 | 0.73891 |
| 0.36 | 1.97259 | +0.17938 | 1.10856 | 0.75873 | 1.13492 | 0.78460 | 1.09311 | 0.74726 |
| 0.37 | 1.99702 | +0.16498 | 1.10900 | 0.76683 | 1.13550 | 0.79285 | 1.09353 | 0.75525 |
| 0.38 | 2.02142 | +0.15017 | 1.10942 | 0.77460 | 1.13604 | 0.80075 | 1.09393 | 0.76291 |
| 0.39 | 2.04580 | +0.13498 | 1.10983 | 0.78204 | 1.13657 | 0.80832 | 1.09431 | 0.77025 |
| 0.40 | 2.07020 | +0.11942 | 1.11021 | 0.78917 | 1.13707 | 0.81557 | 1.09468 | 0.77728 |
| 0.41 | 2.09456 | +0.10351 | 1.11058 | 0.79603 | 1.13754 | 0.82255 | 1.09503 | 0.78404 |
| 0.42 | 2.11891 | +0.08728 | 1.11092 | 0.80261 | 1.13800 | 0.82924 | 1.09536 | 0.79053 |
| 0.43 | 2.14324 | +0.07074 | 1.11126 | 0.80893 | 1.13844 | 0.83568 | 1.09567 | 0.79678 |
| 0.44 | 2.16757 | +0.05392 | 1.11159 | 0.81502 | 1.13887 | 0.84188 | 1.09598 | 0.80279 |
| 0.45 | 2.19189 | +0.03682 | 1.11190 | 0.82089 | 1.13927 | 0.84784 | 1.09628 | 0.80857 |
| 0.46 | 2.21622 | +0.01946 | 1.11220 | 0.82654 | 1.13966 | 0.85358 | 1.09656 | 0.81414 |
| 0.47 | 2.24052 | +0.00186 | 1.11249 | 0.83198 | 1.14003 | 0.85912 | 1.09683 | 0.81951 |
| 0.48 | 2.26482 | -0.01597 | 1.11276 | 0.83723 | 1.14038 | 0.86447 | 1.09709 | 0.82469 |
| 0.49 | 2.28909 | -0.03401 | 1.11303 | 0.84231 | 1.14073 | 0.86962 | 1.09735 | 0.82969 |
| 0.50 | 2.31337 | -0.05227 | 1.11328 | 0.84721 | 1.14107 | 0.87460 | 1.09759 | 0.83452 |

TABLE 1 (continued)

$$\mathcal{T}_1 = 0.15$$

| $\mu$ | $X_1$ | $Y_1$ | $X_r$ | $Y_r$ | $X^{(1)}$ | $Y^{(1)}$ | $X^{(2)}$ | $Y^{(2)}$ |
|---|---|---|---|---|---|---|---|---|
| 0.51 | 2.33763 | -0.07072 | 1.11353 | 0.85194 | 1.14139 | 0.87941 | 1.09782 | 0.83919 |
| 0.52 | 2.36189 | -0.08936 | 1.11377 | 0.85651 | 1.14170 | 0.88408 | 1.09805 | 0.84371 |
| 0.53 | 2.38614 | -0.10817 | 1.11400 | 0.86094 | 1.14200 | 0.88858 | 1.09827 | 0.84808 |
| 0.54 | 2.41040 | -0.12716 | 1.11422 | 0.86523 | 1.14229 | 0.89294 | 1.09848 | 0.85231 |
| 0.55 | 2.43464 | -0.14631 | 1.11444 | 0.86939 | 1.14257 | 0.89716 | 1.09868 | 0.85640 |
| 0.56 | 2.45889 | -0.16562 | 1.11464 | 0.87341 | 1.14283 | 0.90125 | 1.09888 | 0.86037 |
| 0.57 | 2.48312 | -0.18506 | 1.11484 | 0.87731 | 1.14310 | 0.90522 | 1.09907 | 0.86422 |
| 0.58 | 2.50735 | -0.20466 | 1.11504 | 0.88109 | 1.14335 | 0.90906 | 1.09926 | 0.86795 |
| 0.59 | 2.53158 | -0.22438 | 1.11523 | 0.88476 | 1.14360 | 0.91280 | 1.09944 | 0.87157 |
| 0.60 | 2.55580 | -0.24424 | 1.11541 | 0.88832 | 1.14384 | 0.91642 | 1.09961 | 0.87509 |
| 0.61 | 2.58000 | -0.26420 | 1.11558 | 0.89178 | 1.14406 | 0.91993 | 1.09977 | 0.87850 |
| 0.62 | 2.60420 | -0.28429 | 1.11576 | 0.89515 | 1.14429 | 0.92335 | 1.09993 | 0.88182 |
| 0.63 | 2.62843 | -0.30451 | 1.11593 | 0.89842 | 1.14451 | 0.92668 | 1.10009 | 0.88504 |
| 0.64 | 2.65263 | -0.32482 | 1.11609 | 0.90159 | 1.14472 | 0.92991 | 1.10025 | 0.88818 |
| 0.65 | 2.67684 | -0.34524 | 1.11625 | 0.90468 | 1.14492 | 0.93305 | 1.10040 | 0.89122 |
| 0.66 | 2.70100 | -0.36574 | 1.11640 | 0.90769 | 1.14511 | 0.93611 | 1.10054 | 0.89419 |
| 0.67 | 2.72519 | -0.38636 | 1.11655 | 0.91062 | 1.14531 | 0.93908 | 1.10068 | 0.89708 |
| 0.68 | 2.74938 | -0.40706 | 1.11670 | 0.91347 | 1.14550 | 0.94198 | 1.10082 | 0.89989 |
| 0.69 | 2.77358 | -0.42786 | 1.11684 | 0.91624 | 1.14568 | 0.94480 | 1.10095 | 0.90263 |
| 0.70 | 2.79776 | -0.44873 | 1.11698 | 0.91895 | 1.14586 | 0.94756 | 1.10109 | 0.90530 |
| 0.71 | 2.82191 | -0.46967 | 1.11712 | 0.92159 | 1.14604 | 0.95024 | 1.10120 | 0.90791 |
| 0.72 | 2.84609 | -0.49071 | 1.11725 | 0.92416 | 1.14621 | 0.95286 | 1.10133 | 0.91045 |
| 0.73 | 2.87028 | -0.51183 | 1.11737 | 0.92667 | 1.14637 | 0.95540 | 1.10145 | 0.91293 |
| 0.74 | 2.89445 | -0.53301 | 1.11750 | 0.92911 | 1.14654 | 0.95789 | 1.10156 | 0.91534 |
| 0.75 | 2.91862 | -0.55426 | 1.11762 | 0.93150 | 1.14669 | 0.96032 | 1.10168 | 0.91770 |
| 0.76 | 2.94275 | -0.57556 | 1.11774 | 0.93383 | 1.14685 | 0.96269 | 1.10179 | 0.92000 |
| 0.77 | 2.96691 | -0.59695 | 1.11785 | 0.93611 | 1.14700 | 0.96500 | 1.10190 | 0.92224 |
| 0.78 | 2.99108 | -0.61840 | 1.11797 | 0.93833 | 1.14714 | 0.96726 | 1.10200 | 0.92444 |
| 0.79 | 3.01525 | -0.63991 | 1.11808 | 0.94050 | 1.14729 | 0.96947 | 1.10211 | 0.92658 |
| 0.80 | 3.03940 | -0.66147 | 1.11818 | 0.94263 | 1.14743 | 0.97163 | 1.10221 | 0.92868 |
| 0.81 | 3.06352 | -0.68308 | 1.11829 | 0.94470 | 1.14754 | 0.97374 | 1.10231 | 0.93073 |
| 0.82 | 3.08767 | -0.70476 | 1.11839 | 0.94673 | 1.14768 | 0.97580 | 1.10240 | 0.93273 |
| 0.83 | 3.11183 | -0.72649 | 1.11849 | 0.94872 | 1.14781 | 0.97782 | 1.10250 | 0.93469 |
| 0.84 | 3.13599 | -0.74828 | 1.11859 | 0.95066 | 1.14793 | 0.97979 | 1.10259 | 0.93660 |
| 0.85 | 3.16012 | -0.77010 | 1.11869 | 0.95256 | 1.14806 | 0.98172 | 1.10268 | 0.93848 |
| 0.86 | 3.18427 | -0.79198 | 1.11878 | 0.95442 | 1.14818 | 0.98361 | 1.10277 | 0.94032 |
| 0.87 | 3.20843 | -0.81391 | 1.11887 | 0.95624 | 1.14830 | 0.98546 | 1.10286 | 0.94211 |
| 0.88 | 3.23258 | -0.83588 | 1.11896 | 0.95802 | 1.14842 | 0.98727 | 1.10294 | 0.94387 |
| 0.89 | 3.25671 | -0.85789 | 1.11905 | 0.95977 | 1.14853 | 0.98905 | 1.10302 | 0.94559 |
| 0.90 | 3.28086 | -0.87995 | 1.11913 | 0.96148 | 1.14864 | 0.99079 | 1.10310 | 0.94728 |
| 0.91 | 3.30499 | -0.90204 | 1.11922 | 0.96315 | 1.14875 | 0.99249 | 1.10318 | 0.94894 |
| 0.92 | 3.32914 | -0.92418 | 1.11930 | 0.96479 | 1.14886 | 0.99416 | 1.10326 | 0.95056 |
| 0.93 | 3.35328 | -0.94636 | 1.11938 | 0.96640 | 1.14896 | 0.99579 | 1.10334 | 0.95214 |
| 0.94 | 3.37741 | -0.96857 | 1.11946 | 0.96798 | 1.14906 | 0.99740 | 1.10341 | 0.95370 |
| 0.95 | 3.40155 | -0.99082 | 1.11954 | 0.96953 | 1.14916 | 0.99897 | 1.10349 | 0.95523 |
| 0.96 | 3.42569 | -1.01311 | 1.11961 | 0.97105 | 1.14926 | 1.00051 | 1.10356 | 0.95673 |
| 0.97 | 3.44982 | -1.03543 | 1.11969 | 0.97254 | 1.14936 | 1.00202 | 1.10363 | 0.95820 |
| 0.98 | 3.47396 | -1.05778 | 1.11976 | 0.97400 | 1.14945 | 1.00351 | 1.10370 | 0.95964 |
| 0.99 | 3.49809 | -1.08016 | 1.11983 | 0.97543 | 1.14955 | 1.00496 | 1.10376 | 0.96105 |
| 1.00 | 3.52222 | -1.10258 | 1.11990 | 0.97684 | 1.14964 | 1.00639 | 1.10383 | 0.96244 |

TABLE 1 (continued)

$$\tau_1 = 0.20$$

| $\mu$ | $X_1$ | $Y_1$ | $X_r$ | $Y_r$ | $X^{(1)}$ | $Y^{(1)}$ | $X^{(2)}$ | $Y^{(2)}$ |
|---|---|---|---|---|---|---|---|---|
| 0 | 1.00000 | 0 | 1.00000 | 0 | 1.00000 | 0 | 1.00000 | 0 |
| 0.01 | 1.04671 | +0.00227 | 1.01694 | 0.00429 | 1.01934 | 0.00595 | 1.01195 | 0.00488 |
| 0.02 | 1.08379 | +0.00516 | 1.02870 | 0.00909 | 1.03351 | 0.01265 | 1.02125 | 0.01015 |
| 0.03 | 1.11772 | +0.00977 | 1.03857 | 0.01558 | 1.04572 | 0.02116 | 1.02952 | 0.01688 |
| 0.04 | 1.14980 | +0.01881 | 1.04718 | 0.02675 | 1.05657 | 0.03435 | 1.03702 | 0.02812 |
| 0.05 | 1.16060 | +0.03346 | 1.05479 | 0.04425 | 1.06628 | 0.05384 | 1.04383 | 0.04549 |
| 0.06 | 1.21041 | +0.05282 | 1.06152 | 0.06756 | 1.07497 | 0.07900 | 1.04999 | 0.06852 |
| 0.07 | 1.23941 | +0.07504 | 1.06751 | 0.09508 | 1.08274 | 0.10828 | 1.05553 | 0.09568 |
| 0.08 | 1.26775 | +0.09846 | 1.07284 | 0.12523 | 1.08967 | 0.14003 | 1.06053 | 0.12540 |
| 0.09 | 1.29552 | +0.12181 | 1.07760 | 0.15668 | 1.09590 | 0.17296 | 1.06502 | 0.15641 |
| 0.10 | 1.32283 | +0.14419 | 1.08186 | 0.18847 | 1.10149 | 0.20612 | 1.06908 | 0.18775 |
| 0.11 | 1.34974 | +0.16505 | 1.08569 | 0.21994 | 1.10653 | 0.23884 | 1.07274 | 0.21878 |
| 0.12 | 1.37632 | +0.18406 | 1.08915 | 0.25066 | 1.11109 | 0.27071 | 1.07607 | 0.24907 |
| 0.13 | 1.40259 | +0.20106 | 1.09228 | 0.28037 | 1.11524 | 0.30146 | 1.07909 | 0.27834 |
| 0.14 | 1.42865 | +0.21599 | 1.09513 | 0.30889 | 1.11900 | 0.33096 | 1.08184 | 0.30647 |
| 0.15 | 1.45449 | +0.22886 | 1.09773 | 0.33616 | 1.12244 | 0.35913 | 1.08437 | 0.33335 |
| 0.16 | 1.48014 | +0.23976 | 1.10011 | 0.36217 | 1.12560 | 0.38596 | 1.08668 | 0.35899 |
| 0.17 | 1.50564 | +0.24873 | 1.10230 | 0.38691 | 1.12850 | 0.41146 | 1.08881 | 0.38339 |
| 0.18 | 1.53099 | +0.25591 | 1.10432 | 0.41043 | 1.13118 | 0.43569 | 1.09079 | 0.40658 |
| 0.19 | 1.55621 | +0.26138 | 1.10619 | 0.43267 | 1.13367 | 0.45868 | 1.09261 | 0.42860 |
| 0.20 | 1.58136 | +0.26523 | 1.10791 | 0.45397 | 1.13596 | 0.48049 | 1.09431 | 0.44951 |
| 0.21 | 1.60637 | +0.26762 | 1.10952 | 0.47411 | 1.13810 | 0.50121 | 1.09588 | 0.46938 |
| 0.22 | 1.63131 | +0.26860 | 1.11102 | 0.49324 | 1.14009 | 0.52087 | 1.09735 | 0.48824 |
| 0.23 | 1.65615 | +0.26828 | 1.11241 | 0.51142 | 1.14195 | 0.53955 | 1.09873 | 0.50618 |
| 0.24 | 1.68096 | +0.26673 | 1.11372 | 0.52871 | 1.14369 | 0.55732 | 1.10002 | 0.52323 |
| 0.25 | 1.70568 | +0.26405 | 1.11494 | 0.54516 | 1.14532 | 0.57422 | 1.10124 | 0.53945 |
| 0.26 | 1.73033 | +0.26034 | 1.11609 | 0.56083 | 1.14686 | 0.59030 | 1.10236 | 0.55491 |
| 0.27 | 1.75496 | +0.25562 | 1.11718 | 0.57576 | 1.14834 | 0.60562 | 1.10343 | 0.56963 |
| 0.28 | 1.77950 | +0.24999 | 1.11820 | 0.59000 | 1.14967 | 0.62023 | 1.10444 | 0.58368 |
| 0.29 | 1.80402 | +0.24351 | 1.11916 | 0.60359 | 1.15096 | 0.63417 | 1.10540 | 0.59708 |
| 0.30 | 1.82850 | +0.23621 | 1.12008 | 0.61656 | 1.15218 | 0.64748 | 1.10631 | 0.60988 |
| 0.31 | 1.85293 | +0.22818 | 1.12094 | 0.62897 | 1.15333 | 0.66020 | 1.10715 | 0.62213 |
| 0.32 | 1.87735 | +0.21942 | 1.12176 | 0.64084 | 1.15443 | 0.67237 | 1.10797 | 0.63383 |
| 0.33 | 1.90171 | +0.21002 | 1.12254 | 0.65221 | 1.15547 | 0.68402 | 1.10874 | 0.64504 |
| 0.34 | 1.92604 | +0.20000 | 1.12328 | 0.66309 | 1.15646 | 0.69519 | 1.10948 | 0.65578 |
| 0.35 | 1.95037 | +0.18939 | 1.12398 | 0.67353 | 1.15740 | 0.70589 | 1.11018 | 0.66608 |
| 0.36 | 1.97467 | +0.17824 | 1.12465 | 0.68356 | 1.15830 | 0.71616 | 1.11082 | 0.67596 |
| 0.37 | 1.99894 | +0.16657 | 1.12529 | 0.69317 | 1.15915 | 0.72602 | 1.11146 | 0.68545 |
| 0.38 | 2.02316 | +0.15443 | 1.12590 | 0.70242 | 1.15997 | 0.73548 | 1.11207 | 0.69458 |
| 0.39 | 2.04739 | +0.14181 | 1.12649 | 0.71130 | 1.16075 | 0.74459 | 1.11266 | 0.70335 |
| 0.40 | 2.07160 | +0.12878 | 1.12705 | 0.71985 | 1.16150 | 0.75336 | 1.11322 | 0.71178 |
| 0.41 | 2.09579 | +0.11532 | 1.12759 | 0.72809 | 1.16222 | 0.76179 | 1.11375 | 0.71990 |
| 0.42 | 2.11996 | +0.10149 | 1.12810 | 0.73603 | 1.16291 | 0.76992 | 1.11427 | 0.72773 |
| 0.43 | 2.14412 | +0.08729 | 1.12859 | 0.74368 | 1.16357 | 0.77775 | 1.11476 | 0.73529 |
| 0.44 | 2.16825 | +0.07276 | 1.12906 | 0.75105 | 1.16421 | 0.78531 | 1.11523 | 0.74257 |
| 0.45 | 2.19240 | +0.05788 | 1.12952 | 0.75818 | 1.16482 | 0.79260 | 1.11569 | 0.74959 |
| 0.46 | 2.21649 | +0.04270 | 1.12996 | 0.76506 | 1.16540 | 0.79965 | 1.11611 | 0.75638 |
| 0.47 | 2.24060 | +0.02723 | 1.13038 | 0.77171 | 1.16598 | 0.80645 | 1.11653 | 0.76294 |
| 0.48 | 2.26469 | +0.01147 | 1.13079 | 0.77813 | 1.16652 | 0.81304 | 1.11694 | 0.76928 |
| 0.49 | 2.28878 | -0.00456 | 1.13118 | 0.78435 | 1.16704 | 0.81940 | 1.11733 | 0.77542 |
| 0.50 | 2.31283 | -0.02082 | 1.13155 | 0.79036 | 1.16755 | 0.82556 | 1.11771 | 0.78135 |

663

TABLE 1 (continued)

$$\mathcal{T}_1 = 0.20$$

| μ | $x_1$ | $y_1$ | $x_r$ | $y_r$ | $x^{(1)}$ | $y^{(1)}$ | $x^{(2)}$ | $y^{(2)}$ |
|---|---|---|---|---|---|---|---|---|
| 0.51 | 2.33689 | -0.03734 | 1.13192 | 0.79619 | 1.16804 | 0.83152 | 1.11808 | 0.78710 |
| 0.52 | 2.36094 | -0.05410 | 1.13227 | 0.80184 | 1.16851 | 0.83730 | 1.11843 | 0.79267 |
| 0.53 | 2.38499 | -0.07107 | 1.13262 | 0.80731 | 1.16896 | 0.84290 | 1.11877 | 0.79806 |
| 0.54 | 2.40902 | -0.08824 | 1.13294 | 0.81261 | 1.16941 | 0.84833 | 1.11910 | 0.80330 |
| 0.55 | 2.43303 | -0.10561 | 1.13326 | 0.81776 | 1.16983 | 0.85359 | 1.11942 | 0.80837 |
| 0.56 | 2.45704 | -0.12319 | 1.13357 | 0.82275 | 1.17025 | 0.85871 | 1.11972 | 0.81331 |
| 0.57 | 2.48104 | -0.14093 | 1.13387 | 0.82761 | 1.17065 | 0.86367 | 1.12002 | 0.81809 |
| 0.58 | 2.50503 | -0.15886 | 1.13416 | 0.83231 | 1.17104 | 0.86849 | 1.12031 | 0.82274 |
| 0.59 | 2.52903 | -0.17696 | 1.13444 | 0.83689 | 1.17142 | 0.87317 | 1.12059 | 0.82726 |
| 0.60 | 2.55302 | -0.19522 | 1.13472 | 0.84134 | 1.17178 | 0.87772 | 1.12086 | 0.83164 |
| 0.61 | 2.57698 | -0.21362 | 1.13497 | 0.84567 | 1.17214 | 0.88215 | 1.12113 | 0.83592 |
| 0.62 | 2.60095 | -0.23218 | 1.13523 | 0.84988 | 1.17248 | 0.88646 | 1.12139 | 0.84008 |
| 0.63 | 2.62492 | -0.25088 | 1.13548 | 0.85397 | 1.17281 | 0.89065 | 1.12164 | 0.84412 |
| 0.64 | 2.64888 | -0.26971 | 1.13572 | 0.85797 | 1.17314 | 0.89473 | 1.12188 | 0.84805 |
| 0.65 | 2.67285 | -0.28868 | 1.13596 | 0.86185 | 1.17345 | 0.89870 | 1.12212 | 0.85188 |
| 0.66 | 2.69680 | -0.30777 | 1.13618 | 0.86563 | 1.17376 | 0.90258 | 1.12232 | 0.85561 |
| 0.67 | 2.72075 | -0.32698 | 1.13641 | 0.86931 | 1.17405 | 0.90635 | 1.12255 | 0.85925 |
| 0.68 | 2.74468 | -0.34630 | 1.13663 | 0.87291 | 1.17435 | 0.91002 | 1.12277 | 0.86280 |
| 0.69 | 2.76860 | -0.36574 | 1.13683 | 0.87642 | 1.17463 | 0.91360 | 1.12298 | 0.86626 |
| 0.70 | 2.79255 | -0.38530 | 1.13704 | 0.87983 | 1.17490 | 0.91711 | 1.12318 | 0.86963 |
| 0.71 | 2.81648 | -0.40494 | 1.13724 | 0.88317 | 1.17517 | 0.92051 | 1.12338 | 0.87292 |
| 0.72 | 2.84043 | -0.42471 | 1.13743 | 0.88642 | 1.17543 | 0.92384 | 1.12358 | 0.87613 |
| 0.73 | 2.86435 | -0.44455 | 1.13762 | 0.88961 | 1.17569 | 0.92710 | 1.12377 | 0.87927 |
| 0.74 | 2.88827 | -0.46450 | 1.13781 | 0.89271 | 1.17594 | 0.93027 | 1.12396 | 0.88233 |
| 0.75 | 2.91220 | -0.48453 | 1.13799 | 0.89574 | 1.17618 | 0.93337 | 1.12414 | 0.88532 |
| 0.76 | 2.93611 | -0.50465 | 1.13816 | 0.89870 | 1.17641 | 0.93640 | 1.12432 | 0.88824 |
| 0.77 | 2.96003 | -0.52485 | 1.13834 | 0.90160 | 1.17665 | 0.93936 | 1.12449 | 0.89110 |
| 0.78 | 2.98394 | -0.54514 | 1.13851 | 0.90443 | 1.17687 | 0.94226 | 1.12466 | 0.89389 |
| 0.79 | 3.00784 | -0.56550 | 1.13867 | 0.90720 | 1.17709 | 0.94508 | 1.12482 | 0.89662 |
| 0.80 | 3.03174 | -0.58594 | 1.13883 | 0.90990 | 1.17731 | 0.94785 | 1.12498 | 0.89929 |
| 0.81 | 3.05563 | -0.60644 | 1.13899 | 0.91255 | 1.17752 | 0.95056 | 1.12513 | 0.90190 |
| 0.82 | 3.07952 | -0.62703 | 1.13913 | 0.91514 | 1.17772 | 0.95321 | 1.12528 | 0.90446 |
| 0.83 | 3.10342 | -0.64768 | 1.13929 | 0.91767 | 1.17792 | 0.95580 | 1.12544 | 0.90696 |
| 0.84 | 3.12730 | -0.66840 | 1.13943 | 0.92016 | 1.17812 | 0.95834 | 1.12558 | 0.90941 |
| 0.85 | 3.15120 | -0.68918 | 1.13958 | 0.92259 | 1.17830 | 0.96082 | 1.12573 | 0.91181 |
| 0.86 | 3.17506 | -0.71002 | 1.13972 | 0.92497 | 1.17849 | 0.96326 | 1.12587 | 0.91415 |
| 0.87 | 3.19896 | -0.73092 | 1.13985 | 0.92730 | 1.17868 | 0.96565 | 1.12600 | 0.91646 |
| 0.88 | 3.22285 | -0.75190 | 1.13999 | 0.92959 | 1.17886 | 0.96798 | 1.12614 | 0.91871 |
| 0.89 | 3.24672 | -0.77291 | 1.14012 | 0.93183 | 1.17903 | 0.97027 | 1.12627 | 0.92092 |
| 0.90 | 3.27063 | -0.79400 | 1.14025 | 0.93402 | 1.17920 | 0.97251 | 1.12640 | 0.92308 |
| 0.91 | 3.29449 | -0.81512 | 1.14037 | 0.93618 | 1.17938 | 0.97472 | 1.12653 | 0.92521 |
| 0.92 | 3.31836 | -0.83629 | 1.14049 | 0.93829 | 1.17954 | 0.97687 | 1.12665 | 0.92729 |
| 0.93 | 3.34225 | -0.85754 | 1.14062 | 0.94036 | 1.17970 | 0.97899 | 1.12677 | 0.92933 |
| 0.94 | 3.36612 | -0.87882 | 1.14074 | 0.94239 | 1.17986 | 0.98106 | 1.12689 | 0.93134 |
| 0.95 | 3.39002 | -0.90016 | 1.14085 | 0.94438 | 1.18002 | 0.98310 | 1.12701 | 0.93330 |
| 0.96 | 3.41390 | -0.92154 | 1.14097 | 0.94633 | 1.18017 | 0.98509 | 1.12712 | 0.93523 |
| 0.97 | 3.43776 | -0.94296 | 1.14107 | 0.94825 | 1.18031 | 0.98705 | 1.12724 | 0.93712 |
| 0.98 | 3.46163 | -0.96442 | 1.14118 | 0.95013 | 1.18046 | 0.98898 | 1.12735 | 0.93898 |
| 0.99 | 3.48548 | -0.98592 | 1.14129 | 0.95198 | 1.18061 | 0.99087 | 1.12745 | 0.94080 |
| 1.00 | 3.50934 | -1.00747 | 1.14140 | 0.95379 | 1.18075 | 0.99273 | 1.12756 | 0.94259 |

TABLE 1 (continued)

$$\tau_1 = 0.25$$

| μ | $X_1$ | $Y_1$ | $X_r$ | $Y_r$ | $X^{(1)}$ | $Y^{(1)}$ | $X^{(2)}$ | $Y^{(2)}$ |
|---|---|---|---|---|---|---|---|---|
| 0 | 1.00000 | 0 | 1.00000 | 0 | 1.00000 | 0 | 1.00000 | 0 |
| 0.01 | 1.04660 | +0.00208 | 1.01692 | 0.00369 | 1.01956 | 0.00542 | 1.01208 | 0.00447 |
| 0.02 | 1.08369 | +0.00410 | 1.02877 | 0.00773 | 1.03392 | 0.01124 | 1.02150 | 0.00921 |
| 0.03 | 1.11769 | +0.00668 | 1.03875 | 0.01236 | 1.04636 | 0.01782 | 1.02990 | 0.01445 |
| 0.04 | 1.14994 | +0.01127 | 1.04754 | 0.01884 | 1.05748 | 0.02630 | 1.03760 | 0.02140 |
| 0.05 | 1.18089 | +0.01905 | 1.05544 | 0.02877 | 1.06759 | 0.03823 | 1.04472 | 0.03164 |
| 0.06 | 1.21080 | +0.03043 | 1.06254 | 0.04288 | 1.07681 | 0.05430 | 1.05129 | 0.04589 |
| 0.07 | 1.23994 | +0.04478 | 1.06897 | 0.06090 | 1.08525 | 0.07421 | 1.05784 | 0.06392 |
| 0.08 | 1.26842 | +0.06129 | 1.07478 | 0.08205 | 1.09293 | 0.09719 | 1.06292 | 0.08499 |
| 0.09 | 1.29636 | +0.07896 | 1.08006 | 0.10548 | 1.09995 | 0.12234 | 1.06803 | 0.10827 |
| 0.10 | 1.32386 | +0.09697 | 1.08486 | 0.13037 | 1.10636 | 0.14885 | 1.07272 | 0.13298 |
| 0.11 | 1.35097 | +0.11468 | 1.08923 | 0.15608 | 1.11223 | 0.17607 | 1.07702 | 0.15845 |
| 0.12 | 1.37774 | +0.13163 | 1.09323 | 0.18209 | 1.11760 | 0.20349 | 1.08098 | 0.18420 |
| 0.13 | 1.40421 | +0.14751 | 1.09688 | 0.20801 | 1.12254 | 0.23072 | 1.08463 | 0.20985 |
| 0.14 | 1.43041 | +0.16212 | 1.10025 | 0.23356 | 1.12709 | 0.25751 | 1.08800 | 0.23513 |
| 0.15 | 1.45638 | +0.17531 | 1.10335 | 0.25856 | 1.13129 | 0.28366 | 1.09112 | 0.25986 |
| 0.16 | 1.48215 | +0.18704 | 1.10620 | 0.28287 | 1.13517 | 0.30905 | 1.09400 | 0.28391 |
| 0.17 | 1.50772 | +0.19730 | 1.10886 | 0.30643 | 1.13877 | 0.33362 | 1.09667 | 0.30720 |
| 0.18 | 1.53313 | +0.20611 | 1.11131 | 0.32917 | 1.14211 | 0.35731 | 1.09917 | 0.32968 |
| 0.19 | 1.55843 | +0.21346 | 1.11360 | 0.35108 | 1.14522 | 0.38011 | 1.10149 | 0.35134 |
| 0.20 | 1.58359 | +0.21940 | 1.11572 | 0.37215 | 1.14812 | 0.40201 | 1.10366 | 0.37218 |
| 0.21 | 1.60866 | +0.22404 | 1.11771 | 0.39239 | 1.15084 | 0.42305 | 1.10569 | 0.39219 |
| 0.22 | 1.63362 | +0.22740 | 1.11958 | 0.41184 | 1.15339 | 0.44322 | 1.10760 | 0.41140 |
| 0.23 | 1.65847 | +0.22954 | 1.12132 | 0.43048 | 1.15577 | 0.46257 | 1.10989 | 0.42983 |
| 0.24 | 1.68326 | +0.23052 | 1.12296 | 0.44838 | 1.15802 | 0.48112 | 1.11108 | 0.44751 |
| 0.25 | 1.70798 | +0.23040 | 1.12451 | 0.46554 | 1.16014 | 0.49890 | 1.11266 | 0.46448 |
| 0.26 | 1.73261 | +0.22926 | 1.12596 | 0.48201 | 1.16213 | 0.51596 | 1.11416 | 0.48075 |
| 0.27 | 1.75719 | +0.22713 | 1.12734 | 0.49781 | 1.16401 | 0.53232 | 1.11558 | 0.49636 |
| 0.28 | 1.78172 | +0.22408 | 1.12864 | 0.51297 | 1.16580 | 0.54801 | 1.11692 | 0.51135 |
| 0.29 | 1.80618 | +0.22017 | 1.12987 | 0.52754 | 1.16748 | 0.56307 | 1.11819 | 0.52574 |
| 0.30 | 1.83061 | +0.21543 | 1.13104 | 0.54153 | 1.16908 | 0.57754 | 1.11940 | 0.53956 |
| 0.31 | 1.85500 | +0.20991 | 1.13216 | 0.55497 | 1.17061 | 0.59143 | 1.12054 | 0.55284 |
| 0.32 | 1.87935 | +0.20365 | 1.13321 | 0.56790 | 1.17206 | 0.60479 | 1.12164 | 0.56561 |
| 0.33 | 1.90364 | +0.19670 | 1.13421 | 0.58033 | 1.17343 | 0.61763 | 1.12267 | 0.57790 |
| 0.34 | 1.92790 | +0.18911 | 1.13518 | 0.59229 | 1.17475 | 0.62998 | 1.12367 | 0.58971 |
| 0.35 | 1.95213 | +0.18088 | 1.13609 | 0.60381 | 1.17600 | 0.64187 | 1.12462 | 0.60108 |
| 0.36 | 1.97632 | +0.17207 | 1.13696 | 0.61490 | 1.17721 | 0.65333 | 1.12552 | 0.61205 |
| 0.37 | 2.00049 | +0.16270 | 1.13779 | 0.62559 | 1.17835 | 0.66437 | 1.12638 | 0.62261 |
| 0.38 | 2.02464 | +0.15280 | 1.13860 | 0.63591 | 1.17945 | 0.67501 | 1.12721 | 0.63280 |
| 0.39 | 2.04875 | +0.14240 | 1.13935 | 0.64585 | 1.18050 | 0.68527 | 1.12800 | 0.64262 |
| 0.40 | 2.07286 | +0.13152 | 1.14009 | 0.65546 | 1.18151 | 0.69517 | 1.12877 | 0.65212 |
| 0.41 | 2.09691 | +0.12021 | 1.14080 | 0.66473 | 1.18248 | 0.70474 | 1.12949 | 0.66127 |
| 0.42 | 2.12098 | +0.10847 | 1.14147 | 0.67368 | 1.18341 | 0.71398 | 1.13020 | 0.67012 |
| 0.43 | 2.14499 | +0.09632 | 1.14211 | 0.68235 | 1.18429 | 0.72291 | 1.13088 | 0.67868 |
| 0.44 | 2.16901 | +0.08378 | 1.14274 | 0.69073 | 1.18515 | 0.73155 | 1.13152 | 0.68695 |
| 0.45 | 2.19302 | +0.07087 | 1.14334 | 0.69882 | 1.18598 | 0.73989 | 1.13215 | 0.69496 |
| 0.46 | 2.21698 | +0.05761 | 1.14392 | 0.70667 | 1.18678 | 0.74798 | 1.13275 | 0.70271 |
| 0.47 | 2.24095 | +0.04402 | 1.14448 | 0.71427 | 1.18755 | 0.75582 | 1.13333 | 0.71022 |
| 0.48 | 2.26490 | +0.03011 | 1.14501 | 0.72163 | 1.18829 | 0.76341 | 1.13389 | 0.71750 |
| 0.49 | 2.28884 | +0.01590 | 1.14553 | 0.72876 | 1.18900 | 0.77076 | 1.13444 | 0.72455 |
| 0.50 | 2.31278 | +0.00138 | 1.14603 | 0.73569 | 1.18969 | 0.77790 | 1.13495 | 0.73138 |

TABLE 1 (continued)

$$\tau_1 = 0.25$$

| $\mu$ | $X_1$ | $Y_1$ | $X_r$ | $Y_r$ | $X^{(1)}$ | $Y^{(1)}$ | $X^{(2)}$ | $Y^{(2)}$ |
|---|---|---|---|---|---|---|---|---|
| 0.51 | 2.33669 | -0.01341 | 1.14651 | 0.74240 | 1.19036 | 0.78481 | 1.13546 | 0.73801 |
| 0.52 | 2.36057 | -0.02845 | 1.14698 | 0.74892 | 1.19100 | 0.79152 | 1.13595 | 0.74444 |
| 0.53 | 2.38446 | -0.04374 | 1.14743 | 0.75525 | 1.19163 | 0.79804 | 1.13642 | 0.75070 |
| 0.54 | 2.40835 | -0.05929 | 1.14787 | 0.76139 | 1.19223 | 0.80437 | 1.13687 | 0.75677 |
| 0.55 | 2.43221 | -0.07507 | 1.14829 | 0.76736 | 1.19281 | 0.81052 | 1.13732 | 0.76267 |
| 0.56 | 2.45608 | -0.09107 | 1.14871 | 0.77316 | 1.19338 | 0.81650 | 1.13775 | 0.76840 |
| 0.57 | 2.47990 | -0.10727 | 1.14911 | 0.77880 | 1.19393 | 0.82232 | 1.13817 | 0.77398 |
| 0.58 | 2.50374 | -0.12368 | 1.14949 | 0.78430 | 1.19446 | 0.82797 | 1.13857 | 0.77940 |
| 0.59 | 2.52758 | -0.14030 | 1.14986 | 0.78964 | 1.19498 | 0.83347 | 1.13896 | 0.78468 |
| 0.60 | 2.55140 | -0.15710 | 1.15022 | 0.79485 | 1.19548 | 0.83883 | 1.13934 | 0.78982 |
| 0.61 | 2.57521 | -0.17408 | 1.15058 | 0.79991 | 1.19596 | 0.84404 | 1.13971 | 0.79482 |
| 0.62 | 2.59902 | -0.19123 | 1.15092 | 0.80484 | 1.19644 | 0.84913 | 1.14007 | 0.79969 |
| 0.63 | 2.62284 | -0.20856 | 1.15125 | 0.80964 | 1.19690 | 0.85408 | 1.14042 | 0.80445 |
| 0.64 | 2.64663 | -0.22604 | 1.15157 | 0.81433 | 1.19734 | 0.85890 | 1.14076 | 0.80907 |
| 0.65 | 2.67040 | -0.24367 | 1.15190 | 0.81889 | 1.19779 | 0.86360 | 1.14109 | 0.81358 |
| 0.66 | 2.69419 | -0.26146 | 1.15220 | 0.82336 | 1.19821 | 0.86819 | 1.14140 | 0.81798 |
| 0.67 | 2.71797 | -0.27939 | 1.15249 | 0.82770 | 1.19862 | 0.87266 | 1.14172 | 0.82228 |
| 0.68 | 2.74172 | -0.29745 | 1.15279 | 0.83194 | 1.19902 | 0.87704 | 1.14202 | 0.82646 |
| 0.69 | 2.76549 | -0.31565 | 1.15307 | 0.83607 | 1.19942 | 0.88129 | 1.14232 | 0.83056 |
| 0.70 | 2.78928 | -0.33399 | 1.15335 | 0.84011 | 1.19980 | 0.88546 | 1.14261 | 0.83456 |
| 0.71 | 2.81303 | -0.35245 | 1.15362 | 0.84407 | 1.20016 | 0.88952 | 1.14290 | 0.83845 |
| 0.72 | 2.83681 | -0.37102 | 1.15387 | 0.84792 | 1.20053 | 0.89349 | 1.14316 | 0.84226 |
| 0.73 | 2.86055 | -0.38971 | 1.15414 | 0.85170 | 1.20088 | 0.89737 | 1.14343 | 0.84599 |
| 0.74 | 2.88427 | -0.40850 | 1.15438 | 0.85538 | 1.20123 | 0.90116 | 1.14369 | 0.84962 |
| 0.75 | 2.90800 | -0.42741 | 1.15463 | 0.85898 | 1.20156 | 0.90487 | 1.14395 | 0.85318 |
| 0.76 | 2.93174 | -0.44643 | 1.15486 | 0.86250 | 1.20189 | 0.90849 | 1.14420 | 0.85666 |
| 0.77 | 2.95549 | -0.46555 | 1.15510 | 0.86594 | 1.20221 | 0.91205 | 1.14445 | 0.86006 |
| 0.78 | 2.97922 | -0.48477 | 1.15532 | 0.86932 | 1.20253 | 0.91551 | 1.14468 | 0.86339 |
| 0.79 | 3.00294 | -0.50408 | 1.15555 | 0.87261 | 1.20283 | 0.91891 | 1.14491 | 0.86665 |
| 0.80 | 3.02668 | -0.52348 | 1.15576 | 0.87584 | 1.20313 | 0.92223 | 1.14514 | 0.86983 |
| 0.81 | 3.05040 | -0.54297 | 1.15597 | 0.87900 | 1.20342 | 0.92548 | 1.14536 | 0.87297 |
| 0.82 | 3.07411 | -0.56255 | 1.15618 | 0.88210 | 1.20371 | 0.92867 | 1.14558 | 0.87602 |
| 0.83 | 3.09781 | -0.58219 | 1.15638 | 0.88513 | 1.20399 | 0.93179 | 1.14579 | 0.87901 |
| 0.84 | 3.12153 | -0.60194 | 1.15657 | 0.88810 | 1.20426 | 0.93484 | 1.14600 | 0.88195 |
| 0.85 | 3.14525 | -0.62176 | 1.15677 | 0.89100 | 1.20453 | 0.93783 | 1.14621 | 0.88482 |
| 0.86 | 3.16894 | -0.64164 | 1.15697 | 0.89385 | 1.20480 | 0.94077 | 1.14640 | 0.88764 |
| 0.87 | 3.19264 | -0.66162 | 1.15714 | 0.89665 | 1.20505 | 0.94364 | 1.14659 | 0.89040 |
| 0.88 | 3.21634 | -0.68165 | 1.15733 | 0.89938 | 1.20530 | 0.94647 | 1.14678 | 0.89310 |
| 0.89 | 3.24005 | -0.70176 | 1.15750 | 0.90207 | 1.20555 | 0.94923 | 1.14697 | 0.89576 |
| 0.90 | 3.26374 | -0.72193 | 1.15768 | 0.90470 | 1.20579 | 0.95195 | 1.14716 | 0.89836 |
| 0.91 | 3.28742 | -0.74217 | 1.15784 | 0.90729 | 1.20603 | 0.95460 | 1.14734 | 0.90091 |
| 0.92 | 3.31113 | -0.76248 | 1.15802 | 0.90982 | 1.20626 | 0.95722 | 1.14751 | 0.90342 |
| 0.93 | 3.33480 | -0.78284 | 1.15818 | 0.91232 | 1.20649 | 0.95978 | 1.14769 | 0.90588 |
| 0.94 | 3.35851 | -0.80328 | 1.15834 | 0.91476 | 1.20671 | 0.96229 | 1.14785 | 0.90829 |
| 0.95 | 3.38218 | -0.82375 | 1.15850 | 0.91716 | 1.20693 | 0.96476 | 1.14801 | 0.91066 |
| 0.96 | 3.40587 | -0.84430 | 1.15865 | 0.91951 | 1.20715 | 0.96718 | 1.14818 | 0.91299 |
| 0.97 | 3.42955 | -0.86489 | 1.15880 | 0.92182 | 1.20736 | 0.96957 | 1.14834 | 0.91527 |
| 0.98 | 3.45324 | -0.88554 | 1.15895 | 0.92409 | 1.20756 | 0.97190 | 1.14849 | 0.91751 |
| 0.99 | 3.47692 | -0.90625 | 1.15909 | 0.92632 | 1.20776 | 0.97419 | 1.14865 | 0.91970 |
| 1.00 | 3.50058 | -0.92700 | 1.15924 | 0.92851 | 1.20797 | 0.97645 | 1.14880 | 0.92188 |

TABLE 1 (continued)

$$\tau_1 = 0.50$$

| $\mu$ | $X_1$ | $Y_1$ | $X_r$ | $Y_r$ | $X^{(1)}$ | $Y^{(1)}$ | $X^{(2)}$ | $Y^{(2)}$ |
|---|---|---|---|---|---|---|---|---|
| 0 | 1.00000 | 0 | 1.00000 | 0 | 1.00000 | 0 | 1.00000 | 0 |
| 0.01 | 1.04654 | +0.00023 | 1.01702 | 0.00192 | 1.01990 | 0.00352 | 1.01244 | 0.00319 |
| 0.02 | 1.08371 | +0.00112 | 1.02904 | 0.00408 | 1.03470 | 0.00720 | 1.02222 | 0.00650 |
| 0.03 | 1.11778 | +0.00214 | 1.03927 | 0.00647 | 1.04767 | 0.01115 | 1.03101 | 0.01002 |
| 0.04 | 1.14989 | +0.00327 | 1.04833 | 0.00893 | 1.05945 | 0.01526 | 1.03913 | 0.01367 |
| 0.05 | 1.18109 | +0.00444 | 1.05659 | 0.01161 | 1.07032 | 0.01963 | 1.04676 | 0.01749 |
| 0.06 | 1.21126 | +0.00595 | 1.06417 | 0.01462 | 1.08047 | 0.02436 | 1.05399 | 0.02157 |
| 0.07 | 1.24048 | +0.00804 | 1.07117 | 0.01812 | 1.09002 | 0.02972 | 1.06088 | 0.02620 |
| 0.08 | 1.26923 | +0.01069 | 1.07768 | 0.02241 | 1.09890 | 0.03589 | 1.06746 | 0.03154 |
| 0.09 | 1.29724 | +0.01413 | 1.08378 | 0.02768 | 1.10750 | 0.04310 | 1.07373 | 0.03779 |
| 0.10 | 1.32500 | +0.01838 | 1.08952 | 0.03405 | 1.11556 | 0.05138 | 1.07976 | 0.04505 |
| 0.11 | 1.35247 | +0.02340 | 1.09495 | 0.04151 | 1.12324 | 0.06077 | 1.08552 | 0.05338 |
| 0.12 | 1.37952 | +0.02914 | 1.10003 | 0.05008 | 1.13054 | 0.07127 | 1.09102 | 0.06274 |
| 0.13 | 1.40632 | +0.03546 | 1.10484 | 0.05967 | 1.13746 | 0.08276 | 1.09628 | 0.07307 |
| 0.14 | 1.43280 | +0.04216 | 1.10940 | 0.07018 | 1.14406 | 0.09512 | 1.10129 | 0.08424 |
| 0.15 | 1.45898 | +0.04919 | 1.11372 | 0.08151 | 1.15033 | 0.10827 | 1.10609 | 0.09619 |
| 0.16 | 1.48498 | +0.05635 | 1.11780 | 0.09353 | 1.15628 | 0.12210 | 1.11067 | 0.10877 |
| 0.17 | 1.51079 | +0.06353 | 1.12169 | 0.10610 | 1.16196 | 0.13643 | 1.11505 | 0.12187 |
| 0.18 | 1.53645 | +0.07060 | 1.12534 | 0.11914 | 1.16736 | 0.15118 | 1.11924 | 0.13540 |
| 0.19 | 1.56196 | +0.07745 | 1.12882 | 0.13256 | 1.17251 | 0.16625 | 1.12323 | 0.14925 |
| 0.20 | 1.58729 | +0.08403 | 1.13214 | 0.14624 | 1.17742 | 0.18154 | 1.12705 | 0.16333 |
| 0.21 | 1.61253 | +0.09025 | 1.13529 | 0.16009 | 1.18209 | 0.19695 | 1.13072 | 0.17757 |
| 0.22 | 1.63763 | +0.09605 | 1.13829 | 0.17407 | 1.18656 | 0.21244 | 1.13422 | 0.19190 |
| 0.23 | 1.66261 | +0.10137 | 1.14115 | 0.18811 | 1.19083 | 0.22793 | 1.13758 | 0.20625 |
| 0.24 | 1.68745 | +0.10620 | 1.14388 | 0.20214 | 1.19491 | 0.24336 | 1.14079 | 0.22058 |
| 0.25 | 1.71222 | +0.11047 | 1.14651 | 0.21612 | 1.19881 | 0.25872 | 1.14387 | 0.23484 |
| 0.26 | 1.73690 | +0.11424 | 1.14899 | 0.23004 | 1.20256 | 0.27395 | 1.14681 | 0.24900 |
| 0.27 | 1.76150 | +0.11745 | 1.15137 | 0.24383 | 1.20613 | 0.28904 | 1.14965 | 0.26303 |
| 0.28 | 1.78602 | +0.12007 | 1.15365 | 0.25747 | 1.20957 | 0.30393 | 1.15237 | 0.27689 |
| 0.29 | 1.81045 | +0.12214 | 1.15584 | 0.27096 | 1.21288 | 0.31862 | 1.15497 | 0.29058 |
| 0.30 | 1.83481 | +0.12362 | 1.15794 | 0.28426 | 1.21604 | 0.33309 | 1.15748 | 0.30408 |
| 0.31 | 1.85911 | +0.12454 | 1.15996 | 0.29738 | 1.21908 | 0.34734 | 1.15990 | 0.31738 |
| 0.32 | 1.88333 | +0.12489 | 1.16191 | 0.31029 | 1.22201 | 0.36132 | 1.16221 | 0.33045 |
| 0.33 | 1.90750 | +0.12467 | 1.16376 | 0.32299 | 1.22484 | 0.37506 | 1.16446 | 0.34329 |
| 0.34 | 1.93165 | +0.12391 | 1.16555 | 0.33547 | 1.22755 | 0.38857 | 1.16660 | 0.35592 |
| 0.35 | 1.95575 | +0.12259 | 1.16728 | 0.34773 | 1.23016 | 0.40182 | 1.16869 | 0.36832 |
| 0.36 | 1.97979 | +0.12075 | 1.16893 | 0.35975 | 1.23268 | 0.41481 | 1.17068 | 0.38046 |
| 0.37 | 2.00378 | +0.11839 | 1.17054 | 0.37155 | 1.23510 | 0.42752 | 1.17262 | 0.39239 |
| 0.38 | 2.02778 | +0.11551 | 1.17208 | 0.38213 | 1.23746 | 0.43999 | 1.17449 | 0.40407 |
| 0.39 | 2.05167 | +0.11216 | 1.17355 | 0.39447 | 1.23971 | 0.45221 | 1.17629 | 0.41552 |
| 0.40 | 2.07558 | +0.10829 | 1.17502 | 0.40559 | 1.24189 | 0.46417 | 1.17803 | 0.42673 |
| 0.41 | 2.09942 | +0.10399 | 1.17639 | 0.41649 | 1.24401 | 0.47590 | 1.17972 | 0.43772 |
| 0.42 | 2.12326 | +0.09921 | 1.17772 | 0.42716 | 1.24604 | 0.48736 | 1.18135 | 0.44847 |
| 0.43 | 2.14705 | +0.09399 | 1.17902 | 0.43761 | 1.24802 | 0.49859 | 1.18293 | 0.45900 |
| 0.44 | 2.17084 | +0.08834 | 1.18026 | 0.44785 | 1.24992 | 0.50958 | 1.18445 | 0.46932 |
| 0.45 | 2.19456 | +0.08225 | 1.18148 | 0.45788 | 1.25177 | 0.52032 | 1.18594 | 0.47942 |
| 0.46 | 2.21826 | +0.07577 | 1.18264 | 0.46770 | 1.25355 | 0.53086 | 1.18737 | 0.48930 |
| 0.47 | 2.24191 | +0.06889 | 1.18378 | 0.47731 | 1.25529 | 0.54115 | 1.18876 | 0.49898 |
| 0.48 | 2.26557 | +0.06164 | 1.18488 | 0.48672 | 1.25696 | 0.55122 | 1.19011 | 0.50846 |
| 0.49 | 2.28917 | +0.05398 | 1.18595 | 0.49593 | 1.25859 | 0.56109 | 1.19142 | 0.51773 |
| 0.50 | 2.31283 | +0.04597 | 1.18699 | 0.50496 | 1.26018 | 0.57073 | 1.19269 | 0.52681 |

TABLE 1 (continued)

$$\tau_1 = 0.50$$

| μ | $X_1$ | $Y_1$ | $X_r$ | $Y_r$ | $x^{(1)}$ | $Y^{(1)}$ | $X^{(2)}$ | $Y^{(2)}$ |
|---|---|---|---|---|---|---|---|---|
| 0.51 | 2.33640 | +0.03765 | 1.18799 | 0.51379 | 1.26172 | 0.58018 | 1.19393 | 0.53570 |
| 0.52 | 2.35998 | +0.02896 | 1.18896 | 0.52244 | 1.26323 | 0.58944 | 1.19512 | 0.54441 |
| 0.53 | 2.38353 | +0.01996 | 1.18991 | 0.53092 | 1.26468 | 0.59850 | 1.19629 | 0.55293 |
| 0.54 | 2.40703 | +0.01064 | 1.19082 | 0.53923 | 1.26609 | 0.60737 | 1.19743 | 0.56127 |
| 0.55 | 2.43053 | +0.00100 | 1.19171 | 0.54736 | 1.26747 | 0.61606 | 1.19852 | 0.56945 |
| 0.56 | 2.45401 | -0.00894 | 1.19258 | 0.55534 | 1.26880 | 0.62456 | 1.19960 | 0.57746 |
| 0.57 | 2.47750 | -0.01919 | 1.19343 | 0.56314 | 1.27010 | 0.63290 | 1.20065 | 0.58531 |
| 0.58 | 2.50094 | -0.02966 | 1.19425 | 0.57079 | 1.27137 | 0.64106 | 1.20166 | 0.59299 |
| 0.59 | 2.52440 | -0.04047 | 1.19505 | 0.57829 | 1.27260 | 0.64905 | 1.20265 | 0.60052 |
| 0.60 | 2.54786 | -0.05152 | 1.19583 | 0.58563 | 1.27380 | 0.65688 | 1.20362 | 0.60790 |
| 0.61 | 2.57128 | -0.06283 | 1.19658 | 0.59283 | 1.27496 | 0.66455 | 1.20456 | 0.61513 |
| 0.62 | 2.59468 | -0.07440 | 1.19733 | 0.59988 | 1.27610 | 0.67207 | 1.20548 | 0.62221 |
| 0.63 | 2.61808 | -0.08619 | 1.19804 | 0.60679 | 1.27721 | 0.67943 | 1.20638 | 0.62916 |
| 0.64 | 2.64146 | -0.09823 | 1.19876 | 0.61358 | 1.27828 | 0.68666 | 1.20726 | 0.63597 |
| 0.65 | 2.66482 | -0.11049 | 1.19944 | 0.62023 | 1.27935 | 0.69374 | 1.20810 | 0.64265 |
| 0.66 | 2.68818 | -0.12296 | 1.20012 | 0.62675 | 1.28038 | 0.70068 | 1.20893 | 0.64920 |
| 0.67 | 2.71157 | -0.13568 | 1.20076 | 0.63316 | 1.28138 | 0.70750 | 1.20975 | 0.65561 |
| 0.68 | 2.73486 | -0.14858 | 1.20141 | 0.63943 | 1.28236 | 0.71418 | 1.21054 | 0.66191 |
| 0.69 | 2.75821 | -0.16169 | 1.20203 | 0.64559 | 1.28331 | 0.72073 | 1.21131 | 0.66809 |
| 0.70 | 2.78153 | -0.17499 | 1.20263 | 0.65164 | 1.28425 | 0.72716 | 1.21208 | 0.67416 |
| 0.71 | 2.80484 | -0.18848 | 1.20323 | 0.65757 | 1.28517 | 0.73346 | 1.21281 | 0.68010 |
| 0.72 | 2.82815 | -0.20215 | 1.20380 | 0.66339 | 1.28606 | 0.73965 | 1.21354 | 0.68594 |
| 0.73 | 2.85145 | -0.21601 | 1.20438 | 0.66911 | 1.28693 | 0.74573 | 1.21425 | 0.69168 |
| 0.74 | 2.87480 | -0.23006 | 1.20495 | 0.67471 | 1.28779 | 0.75169 | 1.21495 | 0.69730 |
| 0.75 | 2.89807 | -0.24427 | 1.20547 | 0.68022 | 1.28863 | 0.75753 | 1.21562 | 0.70282 |
| 0.76 | 2.92138 | -0.25866 | 1.20600 | 0.68562 | 1.28944 | 0.76328 | 1.21628 | 0.70825 |
| 0.77 | 2.94465 | -0.27318 | 1.20652 | 0.69094 | 1.29024 | 0.76890 | 1.21693 | 0.71357 |
| 0.78 | 2.96790 | -0.28787 | 1.20703 | 0.69616 | 1.29103 | 0.77444 | 1.21757 | 0.71881 |
| 0.79 | 2.99117 | -0.30273 | 1.20752 | 0.70129 | 1.29179 | 0.77988 | 1.21820 | 0.72394 |
| 0.80 | 3.01445 | -0.31772 | 1.20801 | 0.70633 | 1.29256 | 0.78523 | 1.21881 | 0.72899 |
| 0.81 | 3.03771 | -0.33287 | 1.20849 | 0.71128 | 1.29328 | 0.79049 | 1.21940 | 0.73395 |
| 0.82 | 3.06095 | -0.34816 | 1.20895 | 0.71614 | 1.29400 | 0.79565 | 1.21998 | 0.73883 |
| 0.83 | 3.08420 | -0.36358 | 1.20941 | 0.72092 | 1.29471 | 0.80073 | 1.22057 | 0.74361 |
| 0.84 | 3.10747 | -0.37914 | 1.20985 | 0.72561 | 1.29539 | 0.80571 | 1.22113 | 0.74833 |
| 0.85 | 3.13071 | -0.39484 | 1.21030 | 0.73023 | 1.29607 | 0.81061 | 1.22167 | 0.75295 |
| 0.86 | 3.15398 | -0.41067 | 1.21072 | 0.73476 | 1.29674 | 0.81541 | 1.22221 | 0.75751 |
| 0.87 | 3.17720 | -0.42660 | 1.21115 | 0.73923 | 1.29741 | 0.82015 | 1.22275 | 0.76198 |
| 0.88 | 3.20045 | -0.44267 | 1.21157 | 0.74362 | 1.29804 | 0.82481 | 1.22327 | 0.76638 |
| 0.89 | 3.22370 | -0.45886 | 1.21197 | 0.74793 | 1.29865 | 0.82938 | 1.22377 | 0.77072 |
| 0.90 | 3.24690 | -0.47515 | 1.21238 | 0.75219 | 1.29927 | 0.83389 | 1.22428 | 0.77497 |
| 0.91 | 3.27010 | -0.49154 | 1.21278 | 0.75637 | 1.29987 | 0.83832 | 1.22477 | 0.77917 |
| 0.92 | 3.29333 | -0.50806 | 1.21317 | 0.76047 | 1.30046 | 0.84267 | 1.22525 | 0.78328 |
| 0.93 | 3.31655 | -0.52470 | 1.21355 | 0.76452 | 1.30104 | 0.84697 | 1.22573 | 0.78734 |
| 0.94 | 3.33976 | -0.54143 | 1.21392 | 0.76851 | 1.30160 | 0.85119 | 1.22619 | 0.79133 |
| 0.95 | 3.36297 | -0.55825 | 1.21429 | 0.77242 | 1.30217 | 0.85534 | 1.22666 | 0.79526 |
| 0.96 | 3.38618 | -0.57520 | 1.21465 | 0.77628 | 1.30271 | 0.85943 | 1.22710 | 0.79913 |
| 0.97 | 3.40938 | -0.59223 | 1.21501 | 0.78009 | 1.30326 | 0.86346 | 1.22755 | 0.80294 |
| 0.98 | 3.43257 | -0.60934 | 1.21536 | 0.78382 | 1.30379 | 0.86742 | 1.22798 | 0.80668 |
| 0.99 | 3.45578 | -0.62656 | 1.21570 | 0.78751 | 1.30430 | 0.87132 | 1.22840 | 0.81037 |
| 1.00 | 3.47895 | -0.64388 | 1.21605 | 0.79114 | 1.30481 | 0.87516 | 1.22882 | 0.81400 |

TABLE 1 (continued)

$$\tau_i = 1.00$$

| $\mu$ | $x_1$ | $Y_1$ | $x_r$ | $Y_r$ | $x^{(1)}$ | $Y^{(1)}$ | $x^{(2)}$ | $Y^{(2)}$ |
|---|---|---|---|---|---|---|---|---|
| 0 | 1.0000 | 0 | 1.00000 | 0 | 1.00000 | 0 | 1.00000 | 0 |
| 0.01 | 1.0463 | +0.0002 | 1.01715 | 0.00073 | 1.02025 | 0.00177 | 1.01045 | 0.00198 |
| 0.02 | 1.0837 | +0.0006 | 1.02928 | 0.00167 | 1.03537 | 0.00367 | 1.02288 | 0.00402 |
| 0.03 | 1.1178 | +0.0011 | 1.03956 | 0.00258 | 1.04864 | 0.00564 | 1.03198 | 0.00615 |
| 0.04 | 1.1499 | +0.0012 | 1.04872 | 0.00361 | 1.06071 | 0.00779 | 1.04046 | 0.00831 |
| 0.05 | 1.1805 | +0.0015 | 1.05707 | 0.00465 | 1.07196 | 0.00996 | 1.04845 | 0.01056 |
| 0.06 | 1.2104 | +0.0020 | 1.06472 | 0.00572 | 1.08257 | 0.01232 | 1.05605 | 0.01287 |
| 0.07 | 1.2398 | +0.0021 | 1.07186 | 0.00683 | 1.09258 | 0.01475 | 1.06333 | 0.01522 |
| 0.08 | 1.2685 | +0.0023 | 1.07856 | 0.00799 | 1.10209 | 0.01725 | 1.07036 | 0.01762 |
| 0.09 | 1.2966 | +0.0027 | 1.08483 | 0.00929 | 1.11108 | 0.01983 | 1.07711 | 0.02016 |
| 0.10 | 1.3243 | +0.0032 | 1.09076 | 0.01062 | 1.11973 | 0.02252 | 1.08365 | 0.02279 |
| 0.11 | 1.3517 | +0.0036 | 1.09640 | 0.01203 | 1.12804 | 0.02541 | 1.08996 | 0.02553 |
| 0.12 | 1.3789 | +0.0041 | 1.10178 | 0.01360 | 1.13606 | 0.02842 | 1.09608 | 0.02842 |
| 0.13 | 1.4057 | +0.0048 | 1.10689 | 0.01528 | 1.14384 | 0.03164 | 1.10203 | 0.03148 |
| 0.14 | 1.4320 | +0.0058 | 1.11178 | 0.01723 | 1.15126 | 0.03513 | 1.10781 | 0.03474 |
| 0.15 | 1.4584 | +0.0069 | 1.11648 | 0.01942 | 1.15848 | 0.03887 | 1.11341 | 0.03821 |
| 0.16 | 1.4846 | +0.0081 | 1.12098 | 0.02184 | 1.16543 | 0.04286 | 1.11888 | 0.04193 |
| 0.17 | 1.5105 | +0.0095 | 1.12531 | 0.02454 | 1.17215 | 0.04713 | 1.12420 | 0.04593 |
| 0.18 | 1.5361 | +0.0112 | 1.12945 | 0.02752 | 1.17866 | 0.05172 | 1.12939 | 0.05020 |
| 0.19 | 1.5614 | +0.0128 | 1.13345 | 0.03079 | 1.18496 | 0.05667 | 1.13442 | 0.05478 |
| 0.20 | 1.5868 | +0.0146 | 1.13731 | 0.03438 | 1.19108 | 0.06194 | 1.13934 | 0.05965 |
| 0.21 | 1.6118 | +0.0165 | 1.14100 | 0.03823 | 1.19698 | 0.06751 | 1.14412 | 0.06479 |
| 0.22 | 1.6369 | +0.0186 | 1.14457 | 0.04242 | 1.20274 | 0.07341 | 1.14878 | 0.07021 |
| 0.23 | 1.6619 | +0.0208 | 1.14802 | 0.04692 | 1.20834 | 0.07963 | 1.15333 | 0.07592 |
| 0.24 | 1.6867 | +0.0229 | 1.15136 | 0.05169 | 1.21377 | 0.08611 | 1.15776 | 0.08191 |
| 0.25 | 1.7115 | +0.0251 | 1.15461 | 0.05675 | 1.21904 | 0.09289 | 1.16208 | 0.08812 |
| 0.26 | 1.7359 | +0.0275 | 1.15772 | 0.06207 | 1.22417 | 0.09994 | 1.16629 | 0.09460 |
| 0.27 | 1.7604 | +0.0297 | 1.16074 | 0.06764 | 1.22915 | 0.10719 | 1.17039 | 0.10130 |
| 0.28 | 1.7849 | +0.0320 | 1.16367 | 0.07347 | 1.23398 | 0.11471 | 1.17439 | 0.10822 |
| 0.29 | 1.8095 | +0.0343 | 1.16651 | 0.07948 | 1.23868 | 0.12239 | 1.17829 | 0.11532 |
| 0.30 | 1.8336 | +0.0364 | 1.16926 | 0.08573 | 1.24324 | 0.13026 | 1.18208 | 0.12260 |
| 0.31 | 1.8579 | +0.0384 | 1.17190 | 0.09213 | 1.24770 | 0.13831 | 1.18579 | 0.13006 |
| 0.32 | 1.8822 | +0.0402 | 1.17450 | 0.09868 | 1.25203 | 0.14649 | 1.18940 | 0.13764 |
| 0.33 | 1.9062 | +0.0421 | 1.17700 | 0.10540 | 1.25625 | 0.15485 | 1.19292 | 0.14536 |
| 0.34 | 1.9302 | +0.0436 | 1.17942 | 0.11224 | 1.26034 | 0.16330 | 1.19637 | 0.15318 |
| 0.35 | 1.9543 | +0.0450 | 1.18180 | 0.11923 | 1.26433 | 0.17186 | 1.19973 | 0.16111 |
| 0.36 | 1.9783 | +0.0463 | 1.18410 | 0.12632 | 1.26822 | 0.18053 | 1.20300 | 0.16912 |
| 0.37 | 2.0020 | +0.0473 | 1.18633 | 0.13350 | 1.27201 | 0.18925 | 1.20619 | 0.17720 |
| 0.38 | 2.0259 | +0.0481 | 1.18852 | 0.14074 | 1.27571 | 0.19806 | 1.20931 | 0.18534 |
| 0.39 | 2.0498 | +0.0486 | 1.19061 | 0.14809 | 1.27931 | 0.20687 | 1.21235 | 0.19354 |
| 0.40 | 2.0735 | +0.0490 | 1.19268 | 0.15551 | 1.28282 | 0.21576 | 1.21532 | 0.20177 |
| 0.41 | 2.0973 | +0.0492 | 1.19467 | 0.16295 | 1.28626 | 0.22466 | 1.21822 | 0.21003 |
| 0.42 | 2.1210 | +0.0490 | 1.19661 | 0.17040 | 1.28958 | 0.23356 | 1.22105 | 0.21831 |
| 0.43 | 2.1445 | +0.0488 | 1.19852 | 0.17789 | 1.29284 | 0.24251 | 1.22383 | 0.22660 |
| 0.44 | 2.1681 | +0.0482 | 1.20035 | 0.18540 | 1.29602 | 0.25142 | 1.22653 | 0.23489 |
| 0.45 | 2.1917 | +0.0474 | 1.20217 | 0.19295 | 1.29915 | 0.26037 | 1.22918 | 0.24318 |
| 0.46 | 2.2153 | +0.0462 | 1.20393 | 0.20050 | 1.30217 | 0.26926 | 1.23177 | 0.25144 |
| 0.47 | 2.2389 | +0.0450 | 1.20563 | 0.20800 | 1.30514 | 0.27816 | 1.23429 | 0.25969 |
| 0.48 | 2.2623 | +0.0435 | 1.20731 | 0.21559 | 1.30803 | 0.28701 | 1.23677 | 0.26792 |
| 0.49 | 2.2857 | +0.0417 | 1.20892 | 0.22310 | 1.31095 | 0.29585 | 1.23919 | 0.27612 |
| 0.50 | 2.3092 | +0.0396 | 1.21049 | 0.23061 | 1.31363 | 0.30462 | 1.24158 | 0.28428 |

TABLE 1 (continued)

$$\tau_1 = 1.00$$

| $\mu$ | $X_1$ | $Y_1$ | $X_r$ | $Y_r$ | $X^{(1)}$ | $Y^{(1)}$ | $X^{(2)}$ | $Y^{(2)}$ |
|---|---|---|---|---|---|---|---|---|
| 0.51 | 2.3326 | +0.0372 | 1.21204 | 0.23809 | 1.31630 | 0.31337 | 1.24388 | 0.29240 |
| 0.52 | 2.3558 | +0.0346 | 1.21355 | 0.24554 | 1.31893 | 0.32206 | 1.24615 | 0.30048 |
| 0.53 | 2.3791 | +0.0317 | 1.21501 | 0.25295 | 1.32153 | 0.33070 | 1.24836 | 0.30852 |
| 0.54 | 2.4023 | +0.0287 | 1.21647 | 0.26035 | 1.32406 | 0.33930 | 1.25054 | 0.31650 |
| 0.55 | 2.4255 | +0.0254 | 1.21786 | 0.26768 | 1.32653 | 0.34781 | 1.25265 | 0.32444 |
| 0.56 | 2.4488 | +0.0218 | 1.21922 | 0.27498 | 1.32894 | 0.35629 | 1.25474 | 0.33232 |
| 0.57 | 2.4721 | +0.0179 | 1.22056 | 0.28224 | 1.33132 | 0.36467 | 1.25678 | 0.34014 |
| 0.58 | 2.4952 | +0.0139 | 1.22186 | 0.28942 | 1.33363 | 0.37301 | 1.25877 | 0.34790 |
| 0.59 | 2.5185 | +0.0096 | 1.22314 | 0.29662 | 1.33588 | 0.38128 | 1.26072 | 0.35560 |
| 0.60 | 2.5417 | +0.0049 | 1.22442 | 0.30372 | 1.33810 | 0.38949 | 1.26264 | 0.36323 |
| 0.61 | 2.5646 | +0.0002 | 1.22567 | 0.31074 | 1.34026 | 0.39764 | 1.26452 | 0.37081 |
| 0.62 | 2.5880 | -0.0049 | 1.22686 | 0.31775 | 1.34239 | 0.40569 | 1.26636 | 0.37831 |
| 0.63 | 2.6110 | -0.0101 | 1.22803 | 0.32468 | 1.34447 | 0.41366 | 1.26815 | 0.38576 |
| 0.64 | 2.6341 | -0.0155 | 1.22917 | 0.33157 | 1.34651 | 0.42158 | 1.26993 | 0.39314 |
| 0.65 | 2.6572 | -0.0213 | 1.23031 | 0.33842 | 1.34851 | 0.42941 | 1.27167 | 0.40045 |
| 0.66 | 2.6804 | -0.0273 | 1.23142 | 0.34516 | 1.35047 | 0.43717 | 1.27336 | 0.40768 |
| 0.67 | 2.7034 | -0.0336 | 1.23251 | 0.35187 | 1.35240 | 0.44486 | 1.27503 | 0.41485 |
| 0.68 | 2.7265 | -0.0399 | 1.23356 | 0.35854 | 1.35428 | 0.45248 | 1.27668 | 0.42196 |
| 0.69 | 2.7495 | -0.0465 | 1.23461 | 0.36512 | 1.35613 | 0.45999 | 1.27828 | 0.42898 |
| 0.70 | 2.7725 | -0.0533 | 1.23564 | 0.37166 | 1.35795 | 0.46745 | 1.27986 | 0.43596 |
| 0.71 | 2.7956 | -0.0604 | 1.23664 | 0.37811 | 1.35973 | 0.47483 | 1.28140 | 0.44285 |
| 0.72 | 2.8186 | -0.0676 | 1.23763 | 0.38452 | 1.36147 | 0.48213 | 1.28293 | 0.44968 |
| 0.73 | 2.8415 | -0.0751 | 1.23861 | 0.39085 | 1.36319 | 0.48936 | 1.28442 | 0.45644 |
| 0.74 | 2.8646 | -0.0827 | 1.23955 | 0.39712 | 1.36489 | 0.49651 | 1.28569 | 0.46312 |
| 0.75 | 2.8875 | -0.0906 | 1.24049 | 0.40334 | 1.36653 | 0.50359 | 1.28733 | 0.46975 |
| 0.76 | 2.9105 | -0.0986 | 1.24140 | 0.40950 | 1.36815 | 0.51061 | 1.28875 | 0.47630 |
| 0.77 | 2.9335 | -0.1069 | 1.24229 | 0.41562 | 1.36974 | 0.51753 | 1.29013 | 0.48278 |
| 0.78 | 2.9565 | -0.1153 | 1.24317 | 0.42165 | 1.37131 | 0.52440 | 1.29149 | 0.48919 |
| 0.79 | 2.9795 | -0.1240 | 1.24404 | 0.42760 | 1.37284 | 0.53117 | 1.29283 | 0.49553 |
| 0.80 | 3.0024 | -0.1328 | 1.24489 | 0.43349 | 1.37437 | 0.53789 | 1.29417 | 0.50181 |
| 0.81 | 3.0253 | -0.1418 | 1.24573 | 0.43934 | 1.37585 | 0.54452 | 1.29545 | 0.50801 |
| 0.82 | 3.0480 | -0.1508 | 1.24656 | 0.44515 | 1.37731 | 0.55107 | 1.29673 | 0.51416 |
| 0.83 | 3.0710 | -0.1602 | 1.24735 | 0.45085 | 1.37875 | 0.55758 | 1.29797 | 0.52024 |
| 0.84 | 3.0939 | -0.1696 | 1.24815 | 0.45652 | 1.38015 | 0.56399 | 1.29921 | 0.52625 |
| 0.85 | 3.1169 | -0.1794 | 1.24892 | 0.46212 | 1.38153 | 0.57035 | 1.30042 | 0.53219 |
| 0.86 | 3.1395 | -0.1892 | 1.24969 | 0.46767 | 1.38289 | 0.57664 | 1.30161 | 0.53807 |
| 0.87 | 3.1626 | -0.1992 | 1.25042 | 0.47316 | 1.38424 | 0.58286 | 1.30279 | 0.54390 |
| 0.88 | 3.1851 | -0.2093 | 1.25117 | 0.47858 | 1.38555 | 0.58899 | 1.30394 | 0.54966 |
| 0.89 | 3.2080 | -0.2196 | 1.25189 | 0.48396 | 1.38685 | 0.59508 | 1.30507 | 0.55535 |
| 0.90 | 3.2309 | -0.2301 | 1.25262 | 0.48929 | 1.38813 | 0.60108 | 1.30620 | 0.56099 |
| 0.91 | 3.2538 | -0.2407 | 1.25331 | 0.49454 | 1.38940 | 0.60703 | 1.30730 | 0.56656 |
| 0.92 | 3.2766 | -0.2515 | 1.25400 | 0.49974 | 1.39062 | 0.61292 | 1.30838 | 0.57207 |
| 0.93 | 3.2995 | -0.2624 | 1.25467 | 0.50491 | 1.39184 | 0.61874 | 1.30946 | 0.57753 |
| 0.94 | 3.3223 | -0.2734 | 1.25533 | 0.51000 | 1.39305 | 0.62450 | 1.31052 | 0.58293 |
| 0.95 | 3.3452 | -0.2847 | 1.25600 | 0.51503 | 1.39425 | 0.63017 | 1.31155 | 0.58825 |
| 0.96 | 3.3680 | -0.2960 | 1.25664 | 0.52004 | 1.39541 | 0.63581 | 1.31259 | 0.59353 |
| 0.97 | 3.3907 | -0.3075 | 1.25727 | 0.52498 | 1.39657 | 0.64138 | 1.31362 | 0.59875 |
| 0.98 | 3.4134 | -0.3191 | 1.25791 | 0.52988 | 1.39770 | 0.64689 | 1.31462 | 0.60392 |
| 0.99 | 3.4361 | -0.3308 | 1.25854 | 0.53472 | 1.39884 | 0.65234 | 1.31559 | 0.60902 |
| 1.00 | 3.4588 | -0.3427 | 1.25915 | 0.53952 | 1.39994 | 0.65773 | 1.31656 | 0.61406 |

## TABLE 2

THE MOMENTS $\alpha_n$, $\beta_n$, $A_n$ AND $B_n$ OF ORDER $n$ OF $X_1(\mu)$, $Y_1(\mu)$, $X_r(\mu)$ AND $Y_r(\mu)$

AND OTHER AUXILIARY CONSTANTS

### (i) The Moments of $X_1(\mu)$ and $Y_1(\mu)$

| $\tau_1$ | $\alpha_0$ | $\alpha_1$ | $\alpha_2$ | $\alpha_3$ | $\beta_0$ | $\beta_1$ | $\beta_2$ | $\beta_3$ |
|---|---|---|---|---|---|---|---|---|
| 0.05 | 2.31548 | 1.36836 | 0.98217 | 0.76806 | -0.25732 | -0.29170 | -0.25732 | -0.22236 |
| 0.10 | 2.30945 | 1.36152 | 0.97610 | 0.76279 | -0.21904 | -0.24507 | -0.21904 | -0.19101 |
| 0.15 | 2.30593 | 1.35764 | 0.97262 | 0.75975 | -0.19143 | -0.21222 | -0.19143 | -0.16811 |
| 0.20 | 2.30365 | 1.35517 | 0.97039 | 0.75780 | -0.16975 | -0.18684 | -0.16975 | -0.14995 |
| 0.25 | 2.30219 | 1.35357 | 0.96894 | 0.75650 | -0.15203 | -0.16636 | -0.15203 | -0.13495 |
| 0.50 | 2.2983 | 1.3496 | 0.9654 | 0.7534 | -0.0946 | -0.1015 | -0.0946 | -0.0855 |
| 1.00 | 2.292 | 1.345 | 0.962 | 0.750 | -0.0430 | -0.0453 | -0.0430 | -0.0397 |

### (ii) The Moments of $X_r(\mu)$ and $Y_r(\mu)$

| $\tau_1$ | $A_0$ | $A_1$ | $A_2$ | $A_3$ | $A_4$ | $B_0$ | $B_1$ | $B_2$ | $B_3$ | $B_4$ |
|---|---|---|---|---|---|---|---|---|---|---|
| 0.05 | 1.05438 | 0.52855 | 0.35254 | 0.26445 | 0.21158 | 0.88070 | 0.48306 | 0.32848 | 0.24825 | 0.19938 |
| 0.10 | 1.08370 | 0.54500 | 0.36381 | 0.27300 | 0.21846 | 0.80150 | 0.45978 | 0.31738 | 0.24145 | 0.19461 |
| 0.15 | 1.10451 | 0.55713 | 0.37223 | 0.27942 | 0.22364 | 0.73653 | 0.43627 | 0.30493 | 0.23335 | 0.18872 |
| 0.20 | 1.12030 | 0.56660 | 0.37888 | 0.28453 | 0.22778 | 0.68061 | 0.41341 | 0.29207 | 0.22472 | 0.18230 |
| 0.25 | 1.13273 | 0.57423 | 0.38430 | 0.28870 | 0.23117 | 0.63124 | 0.39145 | 0.27919 | 0.21587 | 0.17563 |
| 0.50 | 1.16813 | 0.59692 | 0.40076 | 0.30155 | 0.24168 | 0.44655 | 0.29682 | 0.21937 | 0.17307 | 0.14259 |
| 1.00 | 1.19003 | 0.61195 | 0.41208 | 0.31058 | 0.24918 | 0.23813 | 0.17019 | 0.13137 | 0.10654 | 0.08941 |

### (iii) The Constants Which Occur in the Expressions for $\psi$, $\phi$, $\chi$, $\zeta$, $\xi$, $\eta$, $\sigma$ and $\theta$ ; and $\bar{s}$

| $\tau_1$ | $\nu_1$ | $\nu_2$ | $\nu_3$ | $\nu_4$ | $u_3$ | $u_4$ | $Q$ | $\bar{s}$ |
|---|---|---|---|---|---|---|---|---|
| 0.05 | -0.20070 | -0.22308 | +0.90345 | -1.34589 | -0.29656 | -0.31992 | 20.999 | 0.0448 |
| 0.10 | -0.20515 | -0.24690 | +0.78333 | -1.26655 | -0.30167 | -0.34630 | 10.775 | 0.0843 |
| 0.15 | -0.20747 | -0.26665 | +0.69768 | -1.21356 | -0.30395 | -0.36834 | 7.3193 | 0.1194 |
| 0.20 | -0.20869 | -0.28386 | +0.63077 | -1.17446 | -0.30466 | -0.38764 | 5.5678 | 0.1510 |
| 0.25 | -0.20923 | -0.29922 | +0.57606 | -1.14409 | -0.30439 | -0.40492 | 4.5048 | 0.1800 |
| 0.50 | -0.2069 | -0.3587 | +0.3973 | -1.0559 | -0.2956 | -0.4724 | 2.3224 | 0.2963 |
| 1.00 | -0.1926 | -0.4326 | +0.2281 | -0.9882 | -0.2664 | -0.5581 | 1.1723 | 0.4484 |

671

TABLE 3

THE FUNCTIONS $\psi$ , $\phi$ , $\chi$ , $\zeta$ , $\xi$ , $\eta$ , $\sigma$ AND $\theta$ IN TERMS OF WHICH $S$ AND $T$ ARE EXPRESSED

$\tau_1 = 0.05$

| μ | $\psi$ | $\phi$ | $\chi$ | $\zeta$ | $\xi$ | $\eta$ | $\sigma$ | $\theta$ |
|---|---|---|---|---|---|---|---|---|
| 0 | 0 | 1.00000 | 1.00000 | 0 | 0 | 0 | 0 | 0 |
| 0.02 | 0.00444 | 1.05101 | 1.03200 | 0.00200 | 0.00391 | 0.11822 | 0.10622 | 0.00193 |
| 0.04 | 0.00787 | 1.07147 | 1.04627 | 0.00309 | 0.00660 | 0.34656 | 0.32630 | 0.00301 |
| 0.06 | 0.01113 | 1.08037 | 1.05360 | 0.00366 | 0.00897 | 0.50697 | 0.48301 | 0.00360 |
| 0.08 | 0.01477 | 1.08391 | 1.05799 | 0.00400 | 0.01169 | 0.61389 | 0.58695 | 0.00395 |
| 0.10 | 0.01904 | 1.08434 | 1.06090 | 0.00423 | 0.01500 | 0.68780 | 0.66379 | 0.00418 |
| 0.12 | 0.02402 | 1.08260 | 1.06295 | 0.00439 | 0.01901 | 0.74069 | 0.71904 | 0.00436 |
| 0.14 | 0.02977 | 1.07918 | 1.06449 | 0.00451 | 0.02378 | 0.77941 | 0.76137 | 0.00448 |
| 0.16 | 0.03632 | 1.07434 | 1.06568 | 0.00460 | 0.02933 | 0.80810 | 0.79481 | 0.00458 |
| 0.18 | 0.04369 | 1.06826 | 1.06663 | 0.00468 | 0.03570 | 0.82933 | 0.82184 | 0.00465 |
| 0.20 | 0.05187 | 1.06103 | 1.06740 | 0.00473 | 0.04290 | 0.84480 | 0.84415 | 0.00473 |
| 0.22 | 0.06088 | 1.05271 | 1.06804 | 0.00480 | 0.05091 | 0.85572 | 0.86286 | 0.00477 |
| 0.24 | 0.07073 | 1.04335 | 1.06859 | 0.00483 | 0.05976 | 0.86288 | 0.87878 | 0.00482 |
| 0.26 | 0.08143 | 1.03300 | 1.06905 | 0.00487 | 0.06946 | 0.86686 | 0.89248 | 0.00485 |
| 0.28 | 0.09296 | 1.02164 | 1.06946 | 0.00491 | 0.07998 | 0.86814 | 0.90440 | 0.00489 |
| 0.30 | 0.10532 | 1.00932 | 1.06980 | 0.00494 | 0.09135 | 0.86704 | 0.91486 | 0.00491 |
| 0.32 | 0.11854 | 0.99606 | 1.07012 | 0.00496 | 0.10357 | 0.86380 | 0.92411 | 0.00494 |
| 0.34 | 0.13261 | 0.98186 | 1.07039 | 0.00497 | 0.11662 | 0.85865 | 0.93236 | 0.00497 |
| 0.36 | 0.14752 | 0.96672 | 1.07063 | 0.00500 | 0.13053 | 0.85173 | 0.93976 | 0.00498 |
| 0.38 | 0.16328 | 0.95065 | 1.07086 | 0.00502 | 0.14529 | 0.84316 | 0.94641 | 0.00500 |
| 0.40 | 0.17989 | 0.93369 | 1.07106 | 0.00502 | 0.16088 | 0.83307 | 0.95246 | 0.00502 |
| 0.42 | 0.19734 | 0.91577 | 1.07124 | 0.00505 | 0.17733 | 0.82157 | 0.95796 | 0.00503 |
| 0.44 | 0.21564 | 0.89695 | 1.07140 | 0.00506 | 0.19462 | 0.80867 | 0.96298 | 0.00505 |
| 0.46 | 0.23480 | 0.87724 | 1.07155 | 0.00506 | 0.21278 | 0.79445 | 0.96760 | 0.00506 |
| 0.48 | 0.25480 | 0.85661 | 1.07169 | 0.00509 | 0.23177 | 0.77896 | 0.97185 | 0.00507 |
| 0.50 | 0.27565 | 0.83507 | 1.07183 | 0.00509 | 0.25161 | 0.76227 | 0.97576 | 0.00508 |
| 0.52 | 0.29735 | 0.81264 | 1.07194 | 0.00511 | 0.27231 | 0.74439 | 0.97941 | 0.00508 |
| 0.54 | 0.31991 | 0.78930 | 1.07206 | 0.00512 | 0.29386 | 0.72535 | 0.98279 | 0.00510 |
| 0.56 | 0.34330 | 0.76505 | 1.07215 | 0.00512 | 0.31625 | 0.70521 | 0.98594 | 0.00510 |
| 0.58 | 0.36756 | 0.73991 | 1.07225 | 0.00512 | 0.33949 | 0.68395 | 0.98888 | 0.00512 |
| 0.60 | 0.39266 | 0.71386 | 1.07234 | 0.00514 | 0.36358 | 0.66164 | 0.99163 | 0.00512 |
| 0.62 | 0.41861 | 0.68694 | 1.07242 | 0.00514 | 0.38852 | 0.63822 | 0.99422 | 0.00513 |
| 0.64 | 0.44542 | 0.65909 | 1.07250 | 0.00515 | 0.41432 | 0.61378 | 0.99665 | 0.00513 |
| 0.66 | 0.47306 | 0.63038 | 1.07258 | 0.00516 | 0.44097 | 0.58828 | 0.99894 | 0.00514 |
| 0.68 | 0.50158 | 0.60077 | 1.07265 | 0.00516 | 0.46847 | 0.56178 | 1.00110 | 0.00515 |
| 0.70 | 0.53093 | 0.57025 | 1.07271 | 0.00517 | 0.49681 | 0.53426 | 1.00312 | 0.00515 |
| 0.72 | 0.56113 | 0.53883 | 1.07277 | 0.00518 | 0.52601 | 0.50575 | 1.00505 | 0.00515 |
| 0.74 | 0.59219 | 0.50653 | 1.07284 | 0.00518 | 0.55606 | 0.47622 | 1.00689 | 0.00516 |
| 0.76 | 0.62409 | 0.47333 | 1.07289 | 0.00518 | 0.58696 | 0.44574 | 1.00862 | 0.00517 |
| 0.78 | 0.65685 | 0.43924 | 1.07295 | 0.00519 | 0.61871 | 0.41427 | 1.01026 | 0.00517 |
| 0.80 | 0.69046 | 0.40425 | 1.07300 | 0.00519 | 0.65131 | 0.38183 | 1.01184 | 0.00518 |
| 0.82 | 0.72492 | 0.36838 | 1.07304 | 0.00519 | 0.68476 | 0.34842 | 1.01333 | 0.00518 |
| 0.84 | 0.76023 | 0.33161 | 1.07308 | 0.00520 | 0.71907 | 0.31405 | 1.01476 | 0.00518 |
| 0.86 | 0.79640 | 0.29395 | 1.07313 | 0.00520 | 0.75422 | 0.27873 | 1.01612 | 0.00519 |
| 0.88 | 0.83341 | 0.25540 | 1.07318 | 0.00520 | 0.79022 | 0.24247 | 1.01742 | 0.00519 |
| 0.90 | 0.87128 | 0.21595 | 1.07322 | 0.00521 | 0.82708 | 0.20525 | 1.01867 | 0.00519 |
| 0.92 | 0.90999 | 0.17562 | 1.07325 | 0.00521 | 0.86479 | 0.16708 | 1.01986 | 0.00520 |
| 0.94 | 0.94955 | 0.13439 | 1.07329 | 0.00521 | 0.90334 | 0.12799 | 1.02099 | 0.00520 |
| 0.96 | 0.98997 | 0.09228 | 1.07331 | 0.00521 | 0.94275 | 0.08795 | 1.02209 | 0.00521 |
| 0.98 | 1.03125 | 0.04926 | 1.07337 | 0.00522 | 0.98302 | 0.04698 | 1.02314 | 0.00521 |
| 1.00 | 1.07336 | 0.00535 | 1.07339 | 0.00522 | 1.02412 | 0.00509 | 1.02415 | 0.00521 |

TABLE 3 (continued)

$$\tau_1 = 0.10$$

| μ | ψ | φ | χ | ζ | ξ | η | σ | θ |
|---|---|---|---|---|---|---|---|---|
| 0 | 0 | 1.00000 | 1.00000 | 0 | 0 | 0 | 0 | 0 |
| 0.02 | 0.00528 | 1.05571 | 1.03468 | 0.00241 | 0.00437 | 0.08455 | 0.02693 | 0.00215 |
| 0.04 | 0.01053 | 1.08710 | 1.05666 | 0.00456 | 0.00844 | 0.13971 | 0.12282 | 0.00407 |
| 0.06 | 0.01545 | 1.10610 | 1.07123 | 0.00573 | 0.01193 | 0.26947 | 0.24539 | 0.00546 |
| 0.08 | 0.02041 | 1.11736 | 1.08120 | 0.00668 | 0.01527 | 0.38317 | 0.35453 | 0.00645 |
| 0.10 | 0.02572 | 1.12358 | 1.08836 | 0.00738 | 0.01884 | 0.47527 | 0.44446 | 0.00717 |
| 0.12 | 0.03157 | 1.12621 | 1.09370 | 0.00791 | 0.02288 | 0.54884 | 0.51774 | 0.00770 |
| 0.14 | 0.03807 | 1.12616 | 1.09785 | 0.00831 | 0.02752 | 0.60759 | 0.57782 | 0.00813 |
| 0.16 | 0.04527 | 1.12395 | 1.10115 | 0.00864 | 0.03284 | 0.65469 | 0.62769 | 0.00847 |
| 0.18 | 0.05326 | 1.11994 | 1.10385 | 0.00890 | 0.03889 | 0.69247 | 0.66959 | 0.00875 |
| 0.20 | 0.06203 | 1.11428 | 1.10609 | 0.00913 | 0.04573 | 0.72280 | 0.70521 | 0.00899 |
| 0.22 | 0.07161 | 1.10718 | 1.10797 | 0.00931 | 0.05336 | 0.74691 | 0.73585 | 0.00918 |
| 0.24 | 0.08203 | 1.09876 | 1.10957 | 0.00947 | 0.06181 | 0.76590 | 0.76241 | 0.00935 |
| 0.26 | 0.09330 | 1.08905 | 1.11096 | 0.00961 | 0.07110 | 0.78048 | 0.78567 | 0.00950 |
| 0.28 | 0.10541 | 1.07817 | 1.11217 | 0.00973 | 0.08122 | 0.79125 | 0.80620 | 0.00962 |
| 0.30 | 0.11836 | 1.06610 | 1.11324 | 0.00984 | 0.09219 | 0.79873 | 0.82446 | 0.00973 |
| 0.32 | 0.13219 | 1.05296 | 1.11419 | 0.00993 | 0.10403 | 0.80322 | 0.84076 | 0.00983 |
| 0.34 | 0.14688 | 1.03870 | 1.11504 | 0.01002 | 0.11671 | 0.80507 | 0.85544 | 0.00992 |
| 0.36 | 0.16245 | 1.02337 | 1.11580 | 0.01008 | 0.13027 | 0.80450 | 0.86872 | 0.01001 |
| 0.38 | 0.17888 | 1.00700 | 1.11651 | 0.01016 | 0.14469 | 0.80173 | 0.88076 | 0.01008 |
| 0.40 | 0.19617 | 0.98957 | 1.11713 | 0.01023 | 0.15997 | 0.79691 | 0.89176 | 0.01013 |
| 0.42 | 0.21435 | 0.97114 | 1.11768 | 0.01027 | 0.17614 | 0.79018 | 0.90183 | 0.01020 |
| 0.44 | 0.23339 | 0.95168 | 1.11819 | 0.01033 | 0.19318 | 0.78167 | 0.91110 | 0.01025 |
| 0.46 | 0.25333 | 0.93122 | 1.11868 | 0.01037 | 0.21110 | 0.77147 | 0.91962 | 0.01031 |
| 0.48 | 0.27414 | 0.90975 | 1.11912 | 0.01042 | 0.22988 | 0.75963 | 0.92753 | 0.01035 |
| 0.50 | 0.29582 | 0.88731 | 1.11954 | 0.01045 | 0.24955 | 0.74628 | 0.93484 | 0.01040 |
| 0.52 | 0.31839 | 0.86388 | 1.11992 | 0.01050 | 0.27010 | 0.73143 | 0.94168 | 0.01043 |
| 0.54 | 0.34183 | 0.83944 | 1.12028 | 0.01053 | 0.29151 | 0.71518 | 0.94804 | 0.01047 |
| 0.56 | 0.36615 | 0.81403 | 1.12060 | 0.01057 | 0.31381 | 0.69754 | 0.95397 | 0.01050 |
| 0.58 | 0.39135 | 0.78764 | 1.12091 | 0.01059 | 0.33700 | 0.67859 | 0.95954 | 0.01054 |
| 0.60 | 0.41744 | 0.76028 | 1.12121 | 0.01062 | 0.36106 | 0.65830 | 0.96477 | 0.01058 |
| 0.62 | 0.44441 | 0.73196 | 1.12147 | 0.01065 | 0.38601 | 0.63676 | 0.96970 | 0.01060 |
| 0.64 | 0.47226 | 0.70265 | 1.12172 | 0.01068 | 0.41183 | 0.61399 | 0.97433 | 0.01063 |
| 0.66 | 0.50099 | 0.67239 | 1.12197 | 0.01071 | 0.43853 | 0.58998 | 0.97870 | 0.01065 |
| 0.68 | 0.53061 | 0.64116 | 1.12220 | 0.01072 | 0.46612 | 0.56478 | 0.98284 | 0.01067 |
| 0.70 | 0.56111 | 0.60896 | 1.12242 | 0.01074 | 0.49459 | 0.53842 | 0.98676 | 0.01070 |
| 0.72 | 0.59248 | 0.57581 | 1.12262 | 0.01077 | 0.52394 | 0.51090 | 0.99047 | 0.01072 |
| 0.74 | 0.62475 | 0.54167 | 1.12280 | 0.01079 | 0.55418 | 0.48223 | 0.99400 | 0.01074 |
| 0.76 | 0.65789 | 0.50660 | 1.12299 | 0.01081 | 0.58530 | 0.45244 | 0.99735 | 0.01076 |
| 0.78 | 0.69193 | 0.47056 | 1.12317 | 0.01082 | 0.61731 | 0.42153 | 1.00054 | 0.01077 |
| 0.80 | 0.72684 | 0.43357 | 1.12332 | 0.01084 | 0.65019 | 0.38952 | 1.00359 | 0.01079 |
| 0.82 | 0.76264 | 0.39561 | 1.12348 | 0.01085 | 0.68396 | 0.35643 | 1.00649 | 0.01081 |
| 0.84 | 0.79932 | 0.35670 | 1.12364 | 0.01087 | 0.71860 | 0.32225 | 1.00926 | 0.01083 |
| 0.86 | 0.83689 | 0.31683 | 1.12377 | 0.01089 | 0.75414 | 0.28700 | 1.01191 | 0.01084 |
| 0.88 | 0.87533 | 0.27600 | 1.12391 | 0.01090 | 0.79057 | 0.25069 | 1.01446 | 0.01086 |
| 0.90 | 0.91467 | 0.23423 | 1.12405 | 0.01090 | 0.82786 | 0.21331 | 1.01687 | 0.01088 |
| 0.92 | 0.95489 | 0.19150 | 1.12416 | 0.01092 | 0.86606 | 0.17488 | 1.01921 | 0.01089 |
| 0.94 | 0.99600 | 0.14781 | 1.12429 | 0.01094 | 0.90512 | 0.13543 | 1.02144 | 0.01090 |
| 0.96 | 1.03798 | 0.10318 | 1.12441 | 0.01095 | 0.94507 | 0.09491 | 1.02359 | 0.01091 |
| 0.98 | 1.08086 | 0.05760 | 1.12452 | 0.01096 | 0.98592 | 0.05336 | 1.02566 | 0.01093 |
| 1.00 | 1.12461 | 0.01104 | 1.12463 | 0.01097 | 1.02765 | 0.01080 | 1.02766 | 0.01094 |

TABLE 3 (continued)

$$\tau_1 = 0.15$$

| μ | Ψ | φ | χ | ζ | ξ | η | σ | θ |
|---|---|---|---|---|---|---|---|---|
| 0 | 0 | 1.00000 | 1.00000 | 0 | 0 | 0 | 0 | 0 |
| 0.02 | 0.00575 | 1.05753 | 1.03575 | 0.00263 | 0.00446 | 0.02288 | 0.01786 | 0.00219 |
| 0.04 | 0.01194 | 1.09266 | 1.06045 | 0.00503 | 0.00913 | 0.07277 | 0.06037 | 0.00444 |
| 0.06 | 0.01812 | 1.11733 | 1.07914 | 0.00701 | 0.01347 | 0.13708 | 0.13722 | 0.00637 |
| 0.08 | 0.02430 | 1.13416 | 1.09326 | 0.00853 | 0.01753 | 0.24961 | 0.22375 | 0.00793 |
| 0.10 | 0.03071 | 1.14523 | 1.10409 | 0.00973 | 0.02161 | 0.33585 | 0.30590 | 0.00917 |
| 0.12 | 0.03751 | 1.15195 | 1.11258 | 0.01068 | 0.02592 | 0.41157 | 0.37932 | 0.01016 |
| 0.14 | 0.04485 | 1.15529 | 1.11940 | 0.01144 | 0.03065 | 0.47644 | 0.44360 | 0.01095 |
| 0.16 | 0.05282 | 1.15588 | 1.12496 | 0.01207 | 0.03598 | 0.53142 | 0.49956 | 0.01162 |
| 0.18 | 0.06149 | 1.15412 | 1.12961 | 0.01260 | 0.04185 | 0.57780 | 0.54834 | 0.01217 |
| 0.20 | 0.07091 | 1.15037 | 1.13351 | 0.01303 | 0.04845 | 0.61672 | 0.59105 | 0.01264 |
| 0.22 | 0.08110 | 1.14485 | 1.13684 | 0.01341 | 0.05581 | 0.64924 | 0.62864 | 0.01305 |
| 0.24 | 0.09212 | 1.13770 | 1.13974 | 0.01375 | 0.06394 | 0.67619 | 0.66191 | 0.01340 |
| 0.26 | 0.10395 | 1.12903 | 1.14224 | 0.01403 | 0.07287 | 0.69834 | 0.69152 | 0.01371 |
| 0.28 | 0.11662 | 1.11895 | 1.14445 | 0.01428 | 0.08262 | 0.71618 | 0.71801 | 0.01398 |
| 0.30 | 0.13015 | 1.10753 | 1.14643 | 0.01451 | 0.09321 | 0.73025 | 0.74185 | 0.01421 |
| 0.32 | 0.14455 | 1.09486 | 1.14818 | 0.01471 | 0.10465 | 0.74086 | 0.76339 | 0.01443 |
| 0.34 | 0.15981 | 1.08090 | 1.14974 | 0.01489 | 0.11695 | 0.74844 | 0.78295 | 0.01462 |
| 0.36 | 0.17596 | 1.06574 | 1.15117 | 0.01506 | 0.13011 | 0.75319 | 0.80079 | 0.01480 |
| 0.38 | 0.19298 | 1.04943 | 1.15244 | 0.01521 | 0.14415 | 0.75534 | 0.81711 | 0.01496 |
| 0.40 | 0.21090 | 1.03195 | 1.15362 | 0.01533 | 0.15906 | 0.75512 | 0.83209 | 0.01511 |
| 0.42 | 0.22969 | 1.01333 | 1.15468 | 0.01546 | 0.17486 | 0.75266 | 0.84590 | 0.01524 |
| 0.44 | 0.24939 | 0.99361 | 1.15569 | 0.01557 | 0.19155 | 0.74811 | 0.85865 | 0.01537 |
| 0.46 | 0.26998 | 0.97280 | 1.15660 | 0.01568 | 0.20912 | 0.74158 | 0.87048 | 0.01547 |
| 0.48 | 0.29147 | 0.95090 | 1.15743 | 0.01577 | 0.22759 | 0.73317 | 0.88148 | 0.01558 |
| 0.50 | 0.31385 | 0.92791 | 1.15821 | 0.01586 | 0.24695 | 0.72300 | 0.89172 | 0.01567 |
| 0.52 | 0.33713 | 0.90384 | 1.15893 | 0.01595 | 0.26720 | 0.71113 | 0.90129 | 0.01577 |
| 0.54 | 0.36133 | 0.87872 | 1.15961 | 0.01603 | 0.28836 | 0.69761 | 0.91023 | 0.01585 |
| 0.56 | 0.38641 | 0.85255 | 1.16024 | 0.01610 | 0.31041 | 0.68251 | 0.91864 | 0.01592 |
| 0.58 | 0.41241 | 0.82533 | 1.16083 | 0.01616 | 0.33337 | 0.66590 | 0.92652 | 0.01601 |
| 0.60 | 0.43930 | 0.79707 | 1.16139 | 0.01623 | 0.35722 | 0.64780 | 0.93395 | 0.01607 |
| 0.62 | 0.46710 | 0.76776 | 1.16191 | 0.01629 | 0.38198 | 0.62828 | 0.94096 | 0.01614 |
| 0.64 | 0.49582 | 0.73742 | 1.16240 | 0.01634 | 0.40765 | 0.60735 | 0.94757 | 0.01620 |
| 0.66 | 0.52543 | 0.70605 | 1.16286 | 0.01640 | 0.43422 | 0.58505 | 0.95383 | 0.01625 |
| 0.68 | 0.55595 | 0.67365 | 1.16330 | 0.01644 | 0.46169 | 0.56140 | 0.95978 | 0.01631 |
| 0.70 | 0.58739 | 0.64024 | 1.16373 | 0.01650 | 0.49008 | 0.53644 | 0.96540 | 0.01637 |
| 0.72 | 0.61971 | 0.60579 | 1.16412 | 0.01654 | 0.51935 | 0.51020 | 0.97075 | 0.01642 |
| 0.74 | 0.65296 | 0.57033 | 1.16450 | 0.01659 | 0.54955 | 0.48268 | 0.97583 | 0.01645 |
| 0.76 | 0.68711 | 0.53383 | 1.16485 | 0.01663 | 0.58064 | 0.45394 | 0.98069 | 0.01651 |
| 0.78 | 0.72218 | 0.49633 | 1.16520 | 0.01667 | 0.61266 | 0.42396 | 0.98530 | 0.01655 |
| 0.80 | 0.75815 | 0.45780 | 1.16551 | 0.01670 | 0.64558 | 0.39276 | 0.98971 | 0.01659 |
| 0.82 | 0.79503 | 0.41826 | 1.16583 | 0.01674 | 0.67939 | 0.36037 | 0.99393 | 0.01663 |
| 0.84 | 0.83283 | 0.37772 | 1.16612 | 0.01677 | 0.71413 | 0.32677 | 0.99797 | 0.01666 |
| 0.86 | 0.87152 | 0.33616 | 1.16641 | 0.01680 | 0.74977 | 0.29204 | 1.00183 | 0.01671 |
| 0.88 | 0.91114 | 0.29359 | 1.16667 | 0.01684 | 0.78632 | 0.25613 | 1.00552 | 0.01674 |
| 0.90 | 0.95167 | 0.25003 | 1.16693 | 0.01687 | 0.82379 | 0.21905 | 1.00906 | 0.01677 |
| 0.92 | 0.99310 | 0.20544 | 1.16717 | 0.01688 | 0.86216 | 0.18085 | 1.01249 | 0.01681 |
| 0.94 | 1.03544 | 0.15985 | 1.16741 | 0.01692 | 0.90144 | 0.14150 | 1.01576 | 0.01684 |
| 0.96 | 1.07870 | 0.11326 | 1.16763 | 0.01694 | 0.94164 | 0.10102 | 1.01890 | 0.01687 |
| 0.98 | 1.12287 | 0.06565 | 1.16785 | 0.01697 | 0.98274 | 0.05946 | 1.02193 | 0.01689 |
| 1.00 | 1.16795 | 0.01704 | 1.16806 | 0.01699 | 1.02475 | 0.01675 | 1.02485 | 0.01692 |

TABLE 3 (continued)

$$\tau_1 = 0.20$$

| μ | ψ | φ | χ | ζ | ζ | η | σ | θ |
|---|---|---|---|---|---|---|---|---|
| 0 | 0 | 1.00000 | 1.00000 | 0 | 0 | 0 | 0 | 0 |
| 0.02 | 0.00613 | 1.05827 | 1.03643 | 0.00281 | 0.00449 | 0.01896 | 0.01548 | 0.00221 |
| 0.04 | 0.01289 | 1.09532 | 1.06234 | 0.00550 | 0.00939 | 0.04870 | 0.03985 | 0.00456 |
| 0.06 | 0.01995 | 1.12312 | 1.08333 | 0.00787 | 0.01426 | 0.10235 | 0.08704 | 0.00683 |
| 0.08 | 0.02715 | 1.14367 | 1.10026 | 0.00986 | 0.01893 | 0.17168 | 0.15030 | 0.00881 |
| 0.10 | 0.03454 | 1.15837 | 1.11393 | 0.01152 | 0.02352 | 0.24457 | 0.21825 | 0.01048 |
| 0.12 | 0.04227 | 1.16842 | 1.12507 | 0.01288 | 0.02820 | 0.31418 | 0.28440 | 0.01190 |
| 0.14 | 0.05046 | 1.17468 | 1.13427 | 0.01403 | 0.03316 | 0.37767 | 0.34596 | 0.01308 |
| 0.16 | 0.05921 | 1.17780 | 1.14195 | 0.01498 | 0.03853 | 0.43419 | 0.40207 | 0.01410 |
| 0.18 | 0.06862 | 1.17828 | 1.14848 | 0.01580 | 0.04444 | 0.48384 | 0.45274 | 0.01495 |
| 0.20 | 0.07871 | 1.17645 | 1.15405 | 0.01651 | 0.05094 | 0.52703 | 0.49838 | 0.01570 |
| 0.22 | 0.08954 | 1.17254 | 1.15890 | 0.01711 | 0.05813 | 0.56438 | 0.53945 | 0.01636 |
| 0.24 | 0.10116 | 1.16677 | 1.16310 | 0.01765 | 0.06602 | 0.59639 | 0.57652 | 0.01693 |
| 0.26 | 0.11358 | 1.15925 | 1.16681 | 0.01812 | 0.07468 | 0.62362 | 0.61005 | 0.01744 |
| 0.28 | 0.12683 | 1.15016 | 1.17011 | 0.01853 | 0.08411 | 0.64649 | 0.64048 | 0.01789 |
| 0.30 | 0.14092 | 1.13955 | 1.17304 | 0.01891 | 0.09437 | 0.66544 | 0.66817 | 0.01829 |
| 0.32 | 0.15588 | 1.12750 | 1.17568 | 0.01925 | 0.10544 | 0.68082 | 0.69846 | 0.01864 |
| 0.34 | 0.17170 | 1.11405 | 1.17806 | 0.01956 | 0.11736 | 0.69292 | 0.71664 | 0.01898 |
| 0.36 | 0.18840 | 1.09929 | 1.18021 | 0.01982 | 0.13013 | 0.70200 | 0.73793 | 0.01927 |
| 0.38 | 0.20598 | 1.08323 | 1.18218 | 0.02007 | 0.14378 | 0.70829 | 0.75755 | 0.01956 |
| 0.40 | 0.22447 | 1.06591 | 1.18399 | 0.02031 | 0.15831 | 0.71196 | 0.77568 | 0.01980 |
| 0.42 | 0.24384 | 1.04735 | 1.18568 | 0.02053 | 0.17371 | 0.71318 | 0.79250 | 0.02004 |
| 0.44 | 0.26413 | 1.02759 | 1.18715 | 0.02071 | 0.19001 | 0.71212 | 0.80811 | 0.02026 |
| 0.46 | 0.28532 | 1.00663 | 1.18856 | 0.02089 | 0.20720 | 0.70890 | 0.82267 | 0.02044 |
| 0.48 | 0.30742 | 0.98451 | 1.18986 | 0.02107 | 0.22530 | 0.70362 | 0.83625 | 0.02062 |
| 0.50 | 0.33043 | 0.96123 | 1.19106 | 0.02122 | 0.24428 | 0.69639 | 0.84895 | 0.02079 |
| 0.52 | 0.35436 | 0.93681 | 1.19219 | 0.02136 | 0.26419 | 0.68725 | 0.86087 | 0.02095 |
| 0.54 | 0.37920 | 0.91126 | 1.19324 | 0.02149 | 0.28501 | 0.67635 | 0.87206 | 0.02110 |
| 0.56 | 0.40497 | 0.88457 | 1.19424 | 0.02163 | 0.30672 | 0.66368 | 0.88258 | 0.02125 |
| 0.58 | 0.43166 | 0.85675 | 1.19517 | 0.02174 | 0.32936 | 0.64939 | 0.89250 | 0.02137 |
| 0.60 | 0.45926 | 0.82785 | 1.19605 | 0.02186 | 0.35292 | 0.63344 | 0.90188 | 0.02149 |
| 0.62 | 0.48779 | 0.79783 | 1.19686 | 0.02196 | 0.37739 | 0.61593 | 0.91073 | 0.02161 |
| 0.64 | 0.51724 | 0.76672 | 1.19763 | 0.02206 | 0.40279 | 0.59690 | 0.91913 | 0.02172 |
| 0.66 | 0.54763 | 0.73452 | 1.19835 | 0.02216 | 0.42911 | 0.57636 | 0.92707 | 0.02181 |
| 0.68 | 0.57893 | 0.70122 | 1.19906 | 0.02225 | 0.45633 | 0.55440 | 0.93462 | 0.02192 |
| 0.70 | 0.61117 | 0.66686 | 1.19971 | 0.02233 | 0.48450 | 0.53096 | 0.94181 | 0.02202 |
| 0.72 | 0.64434 | 0.63141 | 1.20033 | 0.02241 | 0.51359 | 0.50614 | 0.94864 | 0.02210 |
| 0.74 | 0.67843 | 0.59488 | 1.20093 | 0.02248 | 0.54361 | 0.47996 | 0.95516 | 0.02218 |
| 0.76 | 0.71346 | 0.55729 | 1.20148 | 0.02255 | 0.57455 | 0.45244 | 0.96137 | 0.02226 |
| 0.78 | 0.74942 | 0.51863 | 1.20202 | 0.02263 | 0.60642 | 0.42357 | 0.96732 | 0.02233 |
| 0.80 | 0.78629 | 0.47889 | 1.20253 | 0.02270 | 0.63921 | 0.39339 | 0.97298 | 0.02241 |
| 0.82 | 0.82411 | 0.43808 | 1.20302 | 0.02276 | 0.67294 | 0.36194 | 0.97841 | 0.02247 |
| 0.84 | 0.86285 | 0.39623 | 1.20349 | 0.02281 | 0.70759 | 0.32918 | 0.98362 | 0.02255 |
| 0.86 | 0.90252 | 0.35330 | 1.20394 | 0.02287 | 0.74317 | 0.29518 | 0.98861 | 0.02261 |
| 0.88 | 0.94314 | 0.30931 | 1.20437 | 0.02293 | 0.77969 | 0.25993 | 0.99340 | 0.02267 |
| 0.90 | 0.98469 | 0.26427 | 1.20477 | 0.02298 | 0.81714 | 0.22345 | 0.99799 | 0.02273 |
| 0.92 | 1.02715 | 0.21817 | 1.20516 | 0.02303 | 0.85551 | 0.18576 | 1.00240 | 0.02278 |
| 0.94 | 1.07058 | 0.17102 | 1.20556 | 0.02308 | 0.89482 | 0.14682 | 1.00666 | 0.02283 |
| 0.96 | 1.11493 | 0.12283 | 1.20590 | 0.02313 | 0.93508 | 0.10669 | 1.01075 | 0.02289 |
| 0.98 | 1.16021 | 0.07355 | 1.20624 | 0.02318 | 0.97624 | 0.06538 | 1.01469 | 0.02293 |
| 1.00 | 1.20641 | 0.02324 | 1.20659 | 0.02322 | 1.01834 | 0.02289 | 1.01848 | 0.02298 |

TABLE 3 (continued)

$$\tau_i = 0.25$$

| μ | ψ | φ | χ | ζ | ζ | η | σ | θ |
|---|---|---|---|---|---|---|---|---|
| 0 | 0 | 1.00000 | 1.00000 | 0 | 0 | 0 | 0 | 0 |
| 0.02 | 0.00646 | 1.05885 | 1.03686 | 0.00298 | 0.00451 | 0.01667 | 0.01412 | 0.00221 |
| 0.04 | 0.01367 | 1.09705 | 1.06853 | 0.00586 | 0.00949 | 0.03828 | 0.03204 | 0.00460 |
| 0.06 | 0.02136 | 1.12663 | 1.08591 | 0.00853 | 0.01465 | 0.07437 | 0.06290 | 0.00703 |
| 0.08 | 0.02933 | 1.14951 | 1.10472 | 0.01088 | 0.01976 | 0.12535 | 0.10844 | 0.00930 |
| 0.10 | 0.03758 | 1.16680 | 1.12050 | 0.01294 | 0.02480 | 0.18433 | 0.16243 | 0.01133 |
| 0.12 | 0.04616 | 1.17949 | 1.13376 | 0.01468 | 0.02986 | 0.24495 | 0.21911 | 0.01311 |
| 0.14 | 0.05517 | 1.18821 | 1.14498 | 0.01619 | 0.03511 | 0.30345 | 0.27490 | 0.01466 |
| 0.16 | 0.06470 | 1.19360 | 1.15454 | 0.01748 | 0.04066 | 0.35789 | 0.32796 | 0.01602 |
| 0.18 | 0.07481 | 1.19604 | 1.16279 | 0.01859 | 0.04664 | 0.40753 | 0.37755 | 0.01721 |
| 0.20 | 0.08559 | 1.19595 | 1.16994 | 0.01956 | 0.05314 | 0.45205 | 0.42340 | 0.01823 |
| 0.22 | 0.09707 | 1.19361 | 1.17623 | 0.02043 | 0.06023 | 0.49168 | 0.46563 | 0.01916 |
| 0.24 | 0.10931 | 1.18920 | 1.18173 | 0.02118 | 0.06797 | 0.52653 | 0.50445 | 0.01997 |
| 0.26 | 0.12232 | 1.18288 | 1.18663 | 0.02187 | 0.07641 | 0.55696 | 0.54010 | 0.02070 |
| 0.28 | 0.13614 | 1.17482 | 1.19102 | 0.02247 | 0.08561 | 0.58324 | 0.57286 | 0.02135 |
| 0.30 | 0.15081 | 1.16507 | 1.19495 | 0.02300 | 0.09556 | 0.60573 | 0.60306 | 0.02195 |
| 0.32 | 0.16631 | 1.15377 | 1.19851 | 0.02351 | 0.10633 | 0.62465 | 0.63091 | 0.02248 |
| 0.34 | 0.18269 | 1.14092 | 1.20174 | 0.02395 | 0.11791 | 0.64027 | 0.65665 | 0.02297 |
| 0.36 | 0.19993 | 1.12663 | 1.20468 | 0.02437 | 0.13032 | 0.65281 | 0.68050 | 0.02341 |
| 0.38 | 0.21806 | 1.11097 | 1.20736 | 0.02474 | 0.14360 | 0.66243 | 0.70264 | 0.02382 |
| 0.40 | 0.23709 | 1.09394 | 1.20983 | 0.02508 | 0.15774 | 0.66934 | 0.72323 | 0.02421 |
| 0.42 | 0.25702 | 1.07556 | 1.21210 | 0.02540 | 0.17275 | 0.67375 | 0.74241 | 0.02455 |
| 0.44 | 0.27785 | 1.05591 | 1.21420 | 0.02570 | 0.18865 | 0.67572 | 0.76035 | 0.02487 |
| 0.46 | 0.29961 | 1.03496 | 1.21614 | 0.02596 | 0.20544 | 0.67540 | 0.77711 | 0.02517 |
| 0.48 | 0.32228 | 1.01278 | 1.21794 | 0.02622 | 0.22314 | 0.67290 | 0.79284 | 0.02545 |
| 0.50 | 0.34588 | 0.98937 | 1.21963 | 0.02646 | 0.24174 | 0.66832 | 0.80762 | 0.02571 |
| 0.52 | 0.37038 | 0.96472 | 1.22120 | 0.02667 | 0.26126 | 0.66174 | 0.82153 | 0.02596 |
| 0.54 | 0.39584 | 0.93890 | 1.22267 | 0.02689 | 0.28169 | 0.65325 | 0.83463 | 0.02618 |
| 0.56 | 0.42222 | 0.91188 | 1.22406 | 0.02707 | 0.30303 | 0.64289 | 0.84699 | 0.02639 |
| 0.58 | 0.44953 | 0.88365 | 1.22535 | 0.02725 | 0.32530 | 0.63079 | 0.85869 | 0.02659 |
| 0.60 | 0.47778 | 0.85428 | 1.22656 | 0.02743 | 0.34850 | 0.61692 | 0.86975 | 0.02678 |
| 0.62 | 0.50697 | 0.82374 | 1.22772 | 0.02759 | 0.37263 | 0.60139 | 0.88023 | 0.02696 |
| 0.64 | 0.53710 | 0.79206 | 1.22880 | 0.02774 | 0.39769 | 0.58419 | 0.89020 | 0.02713 |
| 0.66 | 0.56816 | 0.75920 | 1.22984 | 0.02789 | 0.42368 | 0.56545 | 0.89966 | 0.02728 |
| 0.68 | 0.60018 | 0.72523 | 1.23082 | 0.02803 | 0.45060 | 0.54512 | 0.90867 | 0.02744 |
| 0.70 | 0.63315 | 0.69012 | 1.23173 | 0.02816 | 0.47847 | 0.52329 | 0.91725 | 0.02759 |
| 0.72 | 0.66705 | 0.65389 | 1.23261 | 0.02827 | 0.50728 | 0.49996 | 0.92543 | 0.02772 |
| 0.74 | 0.70189 | 0.61651 | 1.23346 | 0.02839 | 0.53702 | 0.47517 | 0.93324 | 0.02786 |
| 0.76 | 0.73769 | 0.57802 | 1.23424 | 0.02850 | 0.56771 | 0.44892 | 0.94072 | 0.02798 |
| 0.78 | 0.77443 | 0.53842 | 1.23500 | 0.02861 | 0.59935 | 0.42127 | 0.94787 | 0.02809 |
| 0.80 | 0.81213 | 0.49769 | 1.23573 | 0.02872 | 0.63193 | 0.39224 | 0.95470 | 0.02821 |
| 0.82 | 0.85078 | 0.45585 | 1.23643 | 0.02881 | 0.66545 | 0.36180 | 0.96126 | 0.02832 |
| 0.84 | 0.89037 | 0.41289 | 1.23710 | 0.02891 | 0.69991 | 0.33005 | 0.96754 | 0.02843 |
| 0.86 | 0.93092 | 0.36884 | 1.23773 | 0.02899 | 0.73532 | 0.29697 | 0.97358 | 0.02853 |
| 0.88 | 0.97242 | 0.32368 | 1.23834 | 0.02908 | 0.77169 | 0.26254 | 0.97939 | 0.02863 |
| 0.90 | 1.01487 | 0.27741 | 1.23892 | 0.02916 | 0.80901 | 0.22681 | 0.98496 | 0.02872 |
| 0.92 | 1.05827 | 0.23005 | 1.23949 | 0.02924 | 0.84726 | 0.18978 | 0.99032 | 0.02880 |
| 0.94 | 1.10263 | 0.18159 | 1.24001 | 0.02931 | 0.88648 | 0.15146 | 0.99549 | 0.02889 |
| 0.96 | 1.14793 | 0.13202 | 1.24053 | 0.02939 | 0.92664 | 0.11189 | 1.00046 | 0.02896 |
| 0.98 | 1.19419 | 0.08135 | 1.24103 | 0.02946 | 0.96774 | 0.07107 | 1.00527 | 0.02905 |
| 1.00 | 1.24140 | 0.02961 | 1.24151 | 0.02952 | 1.00981 | 0.02897 | 1.00990 | 0.02913 |

TABLE 3 (continued)

$$\tau_1 = 0.50$$

| μ | ψ | φ | χ | ς | ζ | η | σ | θ |
|---|---|---|---|---|---|---|---|---|
| 0 | 0 | 1.00000 | 1.00000 | 0 | 0 | 0 | 0 | 0 |
| 0.02 | 0.00777 | 1.06082 | 1.03857 | 0.00361 | 0.00447 | 0.00975 | 0.01029 | 0.00219 |
| 0.04 | 0.01647 | 1.10127 | 1.06785 | 0.00719 | 0.00947 | 0.02169 | 0.02183 | 0.00457 |
| 0.06 | 0.02600 | 1.13438 | 1.09252 | 0.01069 | 0.01491 | 0.03521 | 0.03462 | 0.00711 |
| 0.08 | 0.03625 | 1.16168 | 1.11511 | 0.01408 | 0.02070 | 0.05194 | 0.04981 | 0.00979 |
| 0.10 | 0.04715 | 1.18437 | 1.13565 | 0.01733 | 0.02675 | 0.07296 | 0.06898 | 0.01252 |
| 0.12 | 0.05866 | 1.20334 | 1.15440 | 0.02039 | 0.03299 | 0.09861 | 0.09247 | 0.01524 |
| 0.14 | 0.07074 | 1.21865 | 1.17150 | 0.02325 | 0.03938 | 0.12812 | 0.11981 | 0.01790 |
| 0.16 | 0.08336 | 1.23052 | 1.18710 | 0.02591 | 0.04591 | 0.16027 | 0.15009 | 0.02044 |
| 0.18 | 0.09658 | 1.23939 | 1.20130 | 0.02836 | 0.05265 | 0.19390 | 0.18227 | 0.02285 |
| 0.20 | 0.11040 | 1.24541 | 1.21427 | 0.03063 | 0.05965 | 0.22791 | 0.21554 | 0.02512 |
| 0.22 | 0.12487 | 1.24882 | 1.22611 | 0.03273 | 0.06695 | 0.26151 | 0.24916 | 0.02726 |
| 0.24 | 0.14001 | 1.24970 | 1.23696 | 0.03466 | 0.07464 | 0.29402 | 0.28264 | 0.02926 |
| 0.26 | 0.15586 | 1.24828 | 1.24692 | 0.03644 | 0.08277 | 0.32503 | 0.31560 | 0.03112 |
| 0.28 | 0.17244 | 1.24462 | 1.25608 | 0.03809 | 0.09140 | 0.35427 | 0.34774 | 0.03287 |
| 0.30 | 0.18979 | 1.23887 | 1.26454 | 0.03963 | 0.10057 | 0.38148 | 0.37894 | 0.03449 |
| 0.32 | 0.20793 | 1.23110 | 1.27239 | 0.04104 | 0.11035 | 0.40653 | 0.40910 | 0.03603 |
| 0.34 | 0.22688 | 1.22145 | 1.27965 | 0.04236 | 0.12076 | 0.42933 | 0.43813 | 0.03745 |
| 0.36 | 0.24668 | 1.20997 | 1.28638 | 0.04359 | 0.13186 | 0.44983 | 0.46601 | 0.03879 |
| 0.38 | 0.26734 | 1.19672 | 1.29267 | 0.04474 | 0.14367 | 0.46801 | 0.49278 | 0.04005 |
| 0.40 | 0.28887 | 1.18175 | 1.29855 | 0.04581 | 0.15622 | 0.48390 | 0.51843 | 0.04123 |
| 0.42 | 0.31128 | 1.16510 | 1.30399 | 0.04681 | 0.16954 | 0.49753 | 0.54299 | 0.04234 |
| 0.44 | 0.33461 | 1.14686 | 1.30912 | 0.04777 | 0.18367 | 0.50889 | 0.56648 | 0.04338 |
| 0.46 | 0.35884 | 1.12699 | 1.31392 | 0.04865 | 0.19860 | 0.51800 | 0.58899 | 0.04436 |
| 0.48 | 0.38399 | 1.10557 | 1.31844 | 0.04948 | 0.21437 | 0.52495 | 0.61051 | 0.04529 |
| 0.50 | 0.41009 | 1.08266 | 1.32272 | 0.05026 | 0.23100 | 0.52970 | 0.63113 | 0.04617 |
| 0.52 | 0.43711 | 1.05823 | 1.32672 | 0.05101 | 0.24848 | 0.53245 | 0.65086 | 0.04700 |
| 0.54 | 0.46509 | 1.03232 | 1.33050 | 0.05172 | 0.26684 | 0.53315 | 0.66977 | 0.04779 |
| 0.56 | 0.49402 | 1.00496 | 1.33409 | 0.05238 | 0.28610 | 0.53179 | 0.68789 | 0.04854 |
| 0.58 | 0.52392 | 0.97616 | 1.33749 | 0.05302 | 0.30626 | 0.52851 | 0.70526 | 0.04925 |
| 0.60 | 0.55479 | 0.94600 | 1.34072 | 0.05361 | 0.32735 | 0.52323 | 0.72192 | 0.04992 |
| 0.62 | 0.58663 | 0.91442 | 1.34377 | 0.05418 | 0.34936 | 0.51606 | 0.73790 | 0.05058 |
| 0.64 | 0.61945 | 0.88144 | 1.34670 | 0.05473 | 0.37229 | 0.50707 | 0.75326 | 0.05119 |
| 0.66 | 0.65325 | 0.84708 | 1.34947 | 0.05524 | 0.39616 | 0.49629 | 0.76802 | 0.05177 |
| 0.68 | 0.68803 | 0.81138 | 1.35212 | 0.05573 | 0.42097 | 0.48364 | 0.78221 | 0.05234 |
| 0.70 | 0.72381 | 0.77433 | 1.35463 | 0.05621 | 0.44675 | 0.46929 | 0.79584 | 0.05287 |
| 0.72 | 0.76058 | 0.73592 | 1.35703 | 0.05665 | 0.47347 | 0.45322 | 0.80898 | 0.05339 |
| 0.74 | 0.79836 | 0.69621 | 1.35936 | 0.05708 | 0.50118 | 0.43543 | 0.82164 | 0.05387 |
| 0.76 | 0.83714 | 0.65517 | 1.36154 | 0.05750 | 0.52985 | 0.41593 | 0.83381 | 0.05435 |
| 0.78 | 0.87690 | 0.61279 | 1.36367 | 0.05789 | 0.55947 | 0.39482 | 0.84555 | 0.05480 |
| 0.80 | 0.91768 | 0.56912 | 1.36568 | 0.05828 | 0.59008 | 0.37206 | 0.85689 | 0.05523 |
| 0.82 | 0.95947 | 0.52414 | 1.36763 | 0.05864 | 0.62167 | 0.34767 | 0.86781 | 0.05564 |
| 0.84 | 1.00227 | 0.47788 | 1.36948 | 0.05898 | 0.65426 | 0.32169 | 0.87837 | 0.05605 |
| 0.86 | 1.04610 | 0.43032 | 1.37127 | 0.05933 | 0.68785 | 0.29411 | 0.88857 | 0.05643 |
| 0.88 | 1.09092 | 0.38145 | 1.37301 | 0.05965 | 0.72240 | 0.26502 | 0.89843 | 0.05682 |
| 0.90 | 1.13675 | 0.33131 | 1.37467 | 0.05995 | 0.75796 | 0.23436 | 0.90797 | 0.05718 |
| 0.92 | 1.18360 | 0.27988 | 1.37630 | 0.06026 | 0.79450 | 0.20222 | 0.91719 | 0.05752 |
| 0.94 | 1.23148 | 0.22717 | 1.37783 | 0.06055 | 0.83204 | 0.16852 | 0.92614 | 0.05786 |
| 0.96 | 1.28037 | 0.17320 | 1.37933 | 0.06082 | 0.87060 | 0.13333 | 0.93478 | 0.05819 |
| 0.98 | 1.33028 | 0.11793 | 1.38078 | 0.06109 | 0.91015 | 0.09669 | 0.94818 | 0.05850 |
| 1.00 | 1.38121 | 0.06143 | 1.38218 | 0.06136 | 0.95071 | 0.05852 | 0.95133 | 0.05881 |

TABLE 3 (continued)

$$\tau_1 = 1.00$$

| $\mu$ | $\psi$ | $\phi$ | $\chi$ | $\zeta$ | $\xi$ | $\eta$ | $\sigma$ | $\theta$ |
|---|---|---|---|---|---|---|---|---|
| 0 | 0 | 1.0000 | 1.00000 | 0 | 0 | 0 | 0 | 0 |
| 0.02 | 0.0094 | 1.0623 | 1.04062 | 0.00439 | 0.0042 | 0.0056 | 0.00728 | 0.00203 |
| 0.04 | 0.0199 | 1.1044 | 1.07152 | 0.00883 | 0.0089 | 0.0117 | 0.01528 | 0.00424 |
| 0.06 | 0.0314 | 1.1387 | 1.09898 | 0.01325 | 0.0140 | 0.0187 | 0.02385 | 0.00662 |
| 0.08 | 0.0439 | 1.1682 | 1.12420 | 0.01764 | 0.0195 | 0.0257 | 0.03295 | 0.00913 |
| 0.10 | 0.0572 | 1.1934 | 1.14767 | 0.02198 | 0.0254 | 0.0337 | 0.04278 | 0.01180 |
| 0.12 | 0.0715 | 1.2153 | 1.16978 | 0.02624 | 0.0317 | 0.0424 | 0.05327 | 0.01459 |
| 0.14 | 0.0865 | 1.2337 | 1.19067 | 0.03042 | 0.0383 | 0.0523 | 0.06470 | 0.01749 |
| 0.16 | 0.1026 | 1.2496 | 1.21053 | 0.03450 | 0.0452 | 0.0636 | 0.07729 | 0.02049 |
| 0.18 | 0.1192 | 1.2624 | 1.22939 | 0.03848 | 0.0524 | 0.0763 | 0.09113 | 0.02353 |
| 0.20 | 0.1367 | 1.2724 | 1.24734 | 0.04233 | 0.0598 | 0.0899 | 0.10623 | 0.02663 |
| 0.22 | 0.1550 | 1.2801 | 1.26438 | 0.04607 | 0.0676 | 0.1048 | 0.12253 | 0.02974 |
| 0.24 | 0.1740 | 1.2854 | 1.28062 | 0.04967 | 0.0756 | 0.1206 | 0.14003 | 0.03285 |
| 0.26 | 0.1938 | 1.2883 | 1.29609 | 0.05314 | 0.0838 | 0.1376 | 0.15858 | 0.03592 |
| 0.28 | 0.2145 | 1.2889 | 1.31081 | 0.05648 | 0.0924 | 0.1549 | 0.17802 | 0.03896 |
| 0.30 | 0.2359 | 1.2875 | 1.32484 | 0.05968 | 0.1013 | 0.1727 | 0.19818 | 0.04195 |
| 0.32 | 0.2581 | 1.2841 | 1.33818 | 0.06276 | 0.1104 | 0.1903 | 0.21884 | 0.04487 |
| 0.34 | 0.2810 | 1.2782 | 1.35087 | 0.06571 | 0.1200 | 0.2079 | 0.23996 | 0.04773 |
| 0.36 | 0.3049 | 1.2707 | 1.36302 | 0.06855 | 0.1300 | 0.2253 | 0.26138 | 0.05050 |
| 0.38 | 0.3295 | 1.2610 | 1.37460 | 0.07127 | 0.1404 | 0.2418 | 0.28295 | 0.05321 |
| 0.40 | 0.3550 | 1.2494 | 1.38562 | 0.07387 | 0.1513 | 0.2576 | 0.30463 | 0.05583 |
| 0.42 | 0.3814 | 1.2360 | 1.39613 | 0.07637 | 0.1627 | 0.2726 | 0.32626 | 0.05839 |
| 0.44 | 0.4086 | 1.2205 | 1.40619 | 0.07875 | 0.1746 | 0.2868 | 0.34777 | 0.06086 |
| 0.46 | 0.4368 | 1.2034 | 1.41583 | 0.08105 | 0.1871 | 0.2997 | 0.36917 | 0.06327 |
| 0.48 | 0.4658 | 1.1844 | 1.42504 | 0.08325 | 0.2002 | 0.3119 | 0.39036 | 0.06558 |
| 0.50 | 0.4957 | 1.1637 | 1.43379 | 0.08536 | 0.2138 | 0.3226 | 0.41128 | 0.06782 |
| 0.52 | 0.5265 | 1.1411 | 1.44222 | 0.08739 | 0.2282 | 0.3318 | 0.43191 | 0.06999 |
| 0.54 | 0.5582 | 1.1169 | 1.45029 | 0.08934 | 0.2432 | 0.3399 | 0.45223 | 0.07208 |
| 0.56 | 0.5909 | 1.0909 | 1.45799 | 0.09121 | 0.2589 | 0.3467 | 0.47220 | 0.07411 |
| 0.58 | 0.6245 | 1.0633 | 1.46539 | 0.09300 | 0.2753 | 0.3520 | 0.49180 | 0.07608 |
| 0.60 | 0.6592 | 1.0340 | 1.47253 | 0.09475 | 0.2925 | 0.3557 | 0.51108 | 0.07797 |
| 0.62 | 0.6948 | 1.0030 | 1.47939 | 0.09640 | 0.3104 | 0.3581 | 0.52996 | 0.07981 |
| 0.64 | 0.7312 | 0.9705 | 1.48596 | 0.09801 | 0.3291 | 0.3593 | 0.54845 | 0.08160 |
| 0.66 | 0.7689 | 0.9363 | 1.49231 | 0.09956 | 0.3486 | 0.3585 | 0.56655 | 0.08332 |
| 0.68 | 0.8073 | 0.9006 | 1.49839 | 0.10105 | 0.3689 | 0.3562 | 0.58430 | 0.08498 |
| 0.70 | 0.8469 | 0.8631 | 1.50429 | 0.10250 | 0.3900 | 0.3525 | 0.60167 | 0.08659 |
| 0.72 | 0.8874 | 0.8242 | 1.50997 | 0.10388 | 0.4121 | 0.3472 | 0.61864 | 0.08815 |
| 0.74 | 0.9289 | 0.7838 | 1.51544 | 0.10522 | 0.4349 | 0.3404 | 0.63524 | 0.08968 |
| 0.76 | 0.9714 | 0.7417 | 1.52072 | 0.10652 | 0.4585 | 0.3320 | 0.65147 | 0.09115 |
| 0.78 | 1.0150 | 0.6981 | 1.52580 | 0.10777 | 0.4832 | 0.3220 | 0.66735 | 0.09256 |
| 0.80 | 1.0597 | 0.6530 | 1.53075 | 0.10899 | 0.5086 | 0.3102 | 0.68285 | 0.09396 |
| 0.82 | 1.1051 | 0.6063 | 1.53553 | 0.11016 | 0.5350 | 0.2972 | 0.69803 | 0.09530 |
| 0.84 | 1.1519 | 0.5581 | 1.54012 | 0.11129 | 0.5624 | 0.2825 | 0.71283 | 0.09661 |
| 0.86 | 1.1994 | 0.5084 | 1.54456 | 0.11239 | 0.5906 | 0.2659 | 0.72731 | 0.09788 |
| 0.88 | 1.2481 | 0.4572 | 1.54887 | 0.11344 | 0.6197 | 0.2481 | 0.74146 | 0.09912 |
| 0.90 | 1.2979 | 0.4046 | 1.55305 | 0.11447 | 0.6498 | 0.2285 | 0.75531 | 0.10031 |
| 0.92 | 1.3488 | 0.3504 | 1.55707 | 0.11546 | 0.6808 | 0.2075 | 0.76882 | 0.10148 |
| 0.94 | 1.4006 | 0.2947 | 1.56098 | 0.11644 | 0.7128 | 0.1850 | 0.78202 | 0.10260 |
| 0.96 | 1.4537 | 0.2376 | 1.56477 | 0.11738 | 0.7459 | 0.1608 | 0.79495 | 0.10369 |
| 0.98 | 1.5074 | 0.1790 | 1.56846 | 0.11829 | 0.7797 | 0.1350 | 0.80759 | 0.10477 |
| 1.00 | 1.5624 | 0.1190 | 1.57207 | 0.11918 | 0.8146 | 0.1076 | 0.81995 | 0.10581 |

TABLE 4

THE FUNCTIONS $\gamma_l(\mu)$ AND $\gamma_n(\mu)$ IN TERMS OF WHICH THE EFFECT OF

A GROUND REFLECTING ACCORDING TO LAMBERT'S LAW IS EXPRESSED

| $\mu$ | $\tau_1 = 0.05$ | | $\tau_1 = 0.10$ | | $\tau_1 = 0.15$ | | $\tau_1 = 0.20$ | |
|---|---|---|---|---|---|---|---|---|
| | $\gamma_l$ | $\gamma_n$ | $\gamma_l$ | $\gamma_n$ | $\gamma_l$ | $\gamma_n$ | $\gamma_l$ | $\gamma_n$ |
| 0 | 0.44337 | 0.46289 | 0.41003 | 0.43831 | 0.38434 | 0.41818 | 0.36317 | 0.40088 |
| 0.02 | 0.52227 | 0.52860 | 0.45127 | 0.46718 | 0.41951 | 0.44245 | 0.39547 | 0.42351 |
| 0.04 | 0.63542 | 0.63805 | 0.51134 | 0.52044 | 0.45673 | 0.47234 | 0.42440 | 0.44550 |
| 0.06 | 0.71309 | 0.71449 | 0.57603 | 0.58168 | 0.50302 | 0.51380 | 0.45877 | 0.47452 |
| 0.08 | 0.76501 | 0.76588 | 0.63090 | 0.63468 | 0.54933 | 0.55708 | 0.49617 | 0.50810 |
| 0.10 | 0.80147 | 0.80207 | 0.67511 | 0.67781 | 0.59111 | 0.59689 | 0.53278 | 0.54204 |
| 0.12 | 0.82831 | 0.82874 | 0.71068 | 0.71272 | 0.62745 | 0.63189 | 0.56668 | 0.57403 |
| 0.14 | 0.84883 | 0.84916 | 0.73964 | 0.74121 | 0.65871 | 0.66222 | 0.59728 | 0.60323 |
| 0.16 | 0.86501 | 0.86528 | 0.76353 | 0.76478 | 0.68561 | 0.68844 | 0.62462 | 0.62953 |
| 0.18 | 0.87809 | 0.87830 | 0.78353 | 0.78455 | 0.70886 | 0.71120 | 0.64895 | 0.65307 |
| 0.20 | 0.88886 | 0.88903 | 0.80049 | 0.80134 | 0.72909 | 0.73104 | 0.67063 | 0.67413 |
| 0.22 | 0.89789 | 0.89804 | 0.81502 | 0.81574 | 0.74680 | 0.74846 | 0.68999 | 0.69300 |
| 0.24 | 0.90557 | 0.90569 | 0.82761 | 0.82823 | 0.76241 | 0.76384 | 0.70734 | 0.70995 |
| 0.26 | 0.91218 | 0.91229 | 0.83862 | 0.83914 | 0.77627 | 0.77751 | 0.72295 | 0.72522 |
| 0.28 | 0.91792 | 0.91803 | 0.84832 | 0.84877 | 0.78863 | 0.78971 | 0.73705 | 0.73906 |
| 0.30 | 0.92296 | 0.92305 | 0.85693 | 0.85734 | 0.79974 | 0.80068 | 0.74984 | 0.75162 |
| 0.32 | 0.92742 | 0.92750 | 0.86462 | 0.86497 | 0.80975 | 0.81059 | 0.76148 | 0.76307 |
| 0.34 | 0.93140 | 0.93146 | 0.87154 | 0.87185 | 0.81882 | 0.81956 | 0.77211 | 0.77354 |
| 0.36 | 0.93496 | 0.93502 | 0.87778 | 0.87807 | 0.82709 | 0.82774 | 0.78187 | 0.78315 |
| 0.38 | 0.93816 | 0.93822 | 0.88345 | 0.88372 | 0.83463 | 0.83522 | 0.79084 | 0.79200 |
| 0.40 | 0.94107 | 0.94112 | 0.88862 | 0.88886 | 0.84156 | 0.84209 | 0.79911 | 0.80017 |
| 0.42 | 0.94372 | 0.94376 | 0.89335 | 0.89357 | 0.84793 | 0.84841 | 0.80676 | 0.80772 |
| 0.44 | 0.94613 | 0.94618 | 0.89770 | 0.89790 | 0.85381 | 0.85425 | 0.81387 | 0.81474 |
| 0.46 | 0.94836 | 0.94840 | 0.90171 | 0.90188 | 0.85926 | 0.85966 | 0.82047 | 0.82128 |
| 0.48 | 0.95040 | 0.95043 | 0.90541 | 0.90559 | 0.86432 | 0.86468 | 0.82663 | 0.82737 |
| 0.50 | 0.95228 | 0.95233 | 0.90886 | 0.90901 | 0.86903 | 0.86937 | 0.83239 | 0.83307 |
| 0.52 | 0.95403 | 0.95407 | 0.91205 | 0.91219 | 0.87342 | 0.87373 | 0.83778 | 0.83841 |
| 0.54 | 0.95566 | 0.95570 | 0.91504 | 0.91518 | 0.87754 | 0.87782 | 0.84284 | 0.84341 |
| 0.56 | 0.95717 | 0.95720 | 0.91782 | 0.91795 | 0.88140 | 0.88165 | 0.84758 | 0.84813 |
| 0.58 | 0.95859 | 0.95862 | 0.92044 | 0.92055 | 0.88502 | 0.88525 | 0.85206 | 0.85257 |
| 0.60 | 0.95991 | 0.95994 | 0.92288 | 0.92300 | 0.88842 | 0.88864 | 0.85628 | 0.85676 |
| 0.62 | 0.96116 | 0.96119 | 0.92519 | 0.92529 | 0.89163 | 0.89184 | 0.86027 | 0.86072 |
| 0.64 | 0.96232 | 0.96235 | 0.92736 | 0.92746 | 0.89467 | 0.89486 | 0.86404 | 0.86447 |
| 0.66 | 0.96342 | 0.96345 | 0.92940 | 0.92950 | 0.89753 | 0.89771 | 0.86762 | 0.86802 |
| 0.68 | 0.96446 | 0.96448 | 0.93134 | 0.93143 | 0.90025 | 0.90043 | 0.87102 | 0.87139 |
| 0.70 | 0.96544 | 0.96546 | 0.93317 | 0.93326 | 0.90283 | 0.90299 | 0.87424 | 0.87460 |
| 0.72 | 0.96636 | 0.96639 | 0.93491 | 0.93499 | 0.90527 | 0.90543 | 0.87732 | 0.87765 |
| 0.74 | 0.96724 | 0.96727 | 0.93656 | 0.93665 | 0.90760 | 0.90775 | 0.88024 | 0.88055 |
| 0.76 | 0.96808 | 0.96810 | 0.93813 | 0.93821 | 0.90981 | 0.90996 | 0.88303 | 0.88333 |
| 0.78 | 0.96887 | 0.96889 | 0.93963 | 0.93969 | 0.91192 | 0.91206 | 0.88570 | 0.88597 |
| 0.80 | 0.96962 | 0.96964 | 0.94105 | 0.94112 | 0.91393 | 0.91407 | 0.88824 | 0.88850 |
| 0.82 | 0.97034 | 0.97036 | 0.94240 | 0.94247 | 0.91585 | 0.91599 | 0.89067 | 0.89092 |
| 0.84 | 0.97102 | 0.97105 | 0.94370 | 0.94376 | 0.91769 | 0.91783 | 0.89300 | 0.89325 |
| 0.86 | 0.97168 | 0.97169 | 0.94494 | 0.94500 | 0.91945 | 0.91959 | 0.89523 | 0.89547 |
| 0.88 | 0.97231 | 0.97233 | 0.94613 | 0.94619 | 0.92115 | 0.92127 | 0.89737 | 0.89760 |
| 0.90 | 0.97290 | 0.97293 | 0.94726 | 0.94731 | 0.92277 | 0.92289 | 0.89944 | 0.89965 |
| 0.92 | 0.97347 | 0.97350 | 0.94835 | 0.94841 | 0.92432 | 0.92444 | 0.90141 | 0.90161 |
| 0.94 | 0.97402 | 0.97404 | 0.94940 | 0.94945 | 0.92581 | 0.92594 | 0.90331 | 0.90352 |
| 0.96 | 0.97455 | 0.97457 | 0.95040 | 0.95045 | 0.92725 | 0.92736 | 0.90515 | 0.90533 |
| 0.98 | 0.97506 | 0.97507 | 0.95137 | 0.95142 | 0.92863 | 0.92875 | 0.90691 | 0.90707 |
| 1.00 | 0.97554 | 0.97556 | 0.95229 | 0.95235 | 0.92996 | 0.93007 | 0.90860 | 0.90877 |

TABLE 4 (continued)

| μ | $\tau_1 = 0.25$ | | $\tau_1 = 0.50$ | | $\tau_1 = 1.00$ | |
|---|---|---|---|---|---|---|
| | $\gamma_l$ | $\gamma_r$ | $\gamma_l$ | $\gamma_r$ | $\gamma_l$ | $\gamma_r$ |
| 0 | 0.34514 | 0.38558 | 0.28122 | 0.32744 | 0.21053 | 0.25592 |
| 0.02 | 0.37544 | 0.40703 | 0.30508 | 0.34507 | 0.22828 | 0.26954 |
| 0.04 | 0.40078 | 0.42604 | 0.32429 | 0.35995 | 0.24284 | 0.28090 |
| 0.06 | 0.42840 | 0.44833 | 0.34230 | 0.37409 | 0.25525 | 0.29156 |
| 0.08 | 0.45894 | 0.47474 | 0.35994 | 0.38816 | 0.26754 | 0.30181 |
| 0.10 | 0.49039 | 0.50305 | 0.37779 | 0.40284 | 0.27948 | 0.31181 |
| 0.12 | 0.52094 | 0.53124 | 0.39614 | 0.41830 | 0.29116 | 0.32165 |
| 0.14 | 0.54965 | 0.55813 | 0.41479 | 0.43440 | 0.30270 | 0.33142 |
| 0.16 | 0.57610 | 0.58319 | 0.43345 | 0.45089 | 0.31426 | 0.34122 |
| 0.18 | 0.60028 | 0.60628 | 0.45193 | 0.46745 | 0.32575 | 0.35107 |
| 0.20 | 0.62228 | 0.62741 | 0.47001 | 0.48387 | 0.33714 | 0.36101 |
| 0.22 | 0.64231 | 0.64673 | 0.48755 | 0.49995 | 0.34853 | 0.37099 |
| 0.24 | 0.66052 | 0.66437 | 0.50441 | 0.51556 | 0.35992 | 0.38104 |
| 0.26 | 0.67712 | 0.68049 | 0.52058 | 0.53065 | 0.37125 | 0.39114 |
| 0.28 | 0.69228 | 0.69525 | 0.53603 | 0.54513 | 0.38251 | 0.40123 |
| 0.30 | 0.70617 | 0.70880 | 0.55075 | 0.55900 | 0.39369 | 0.41129 |
| 0.32 | 0.71893 | 0.72126 | 0.56475 | 0.57228 | 0.40472 | 0.42126 |
| 0.34 | 0.73066 | 0.73276 | 0.57806 | 0.58492 | 0.41554 | 0.43114 |
| 0.36 | 0.74150 | 0.74338 | 0.59071 | 0.59698 | 0.42624 | 0.44092 |
| 0.38 | 0.75152 | 0.75323 | 0.60274 | 0.60849 | 0.43664 | 0.45055 |
| 0.40 | 0.76082 | 0.76236 | 0.61415 | 0.61944 | 0.44685 | 0.46004 |
| 0.42 | 0.76947 | 0.77087 | 0.62500 | 0.62986 | 0.45685 | 0.46933 |
| 0.44 | 0.77753 | 0.77880 | 0.63533 | 0.63979 | 0.46660 | 0.47844 |
| 0.46 | 0.78505 | 0.78622 | 0.64513 | 0.64926 | 0.47611 | 0.48737 |
| 0.48 | 0.79210 | 0.79316 | 0.65446 | 0.65829 | 0.48544 | 0.49611 |
| 0.50 | 0.79871 | 0.79967 | 0.66334 | 0.66690 | 0.49449 | 0.50463 |
| 0.52 | 0.80491 | 0.80580 | 0.67182 | 0.67512 | 0.50325 | 0.51295 |
| 0.54 | 0.81075 | 0.81157 | 0.67990 | 0.68297 | 0.51180 | 0.52107 |
| 0.56 | 0.81626 | 0.81701 | 0.68760 | 0.69048 | 0.52014 | 0.52898 |
| 0.58 | 0.82145 | 0.82215 | 0.69497 | 0.69765 | 0.52824 | 0.53669 |
| 0.60 | 0.82637 | 0.82701 | 0.70202 | 0.70452 | 0.53614 | 0.54422 |
| 0.62 | 0.83102 | 0.83162 | 0.70875 | 0.71111 | 0.54382 | 0.55155 |
| 0.64 | 0.83544 | 0.83598 | 0.71521 | 0.71741 | 0.55129 | 0.55867 |
| 0.66 | 0.83963 | 0.84015 | 0.72139 | 0.72347 | 0.55856 | 0.56563 |
| 0.68 | 0.84362 | 0.84409 | 0.72731 | 0.72927 | 0.56561 | 0.57239 |
| 0.70 | 0.84742 | 0.84786 | 0.73301 | 0.73484 | 0.57247 | 0.57899 |
| 0.72 | 0.85104 | 0.85143 | 0.73848 | 0.74020 | 0.57917 | 0.58541 |
| 0.74 | 0.85449 | 0.85485 | 0.74375 | 0.74538 | 0.58567 | 0.59167 |
| 0.76 | 0.85778 | 0.85812 | 0.74881 | 0.75032 | 0.59199 | 0.59776 |
| 0.78 | 0.86093 | 0.86125 | 0.75368 | 0.75510 | 0.59816 | 0.60369 |
| 0.80 | 0.86395 | 0.86424 | 0.75837 | 0.75969 | 0.60414 | 0.60947 |
| 0.82 | 0.86684 | 0.86712 | 0.76289 | 0.76413 | 0.60995 | 0.61511 |
| 0.84 | 0.86961 | 0.86985 | 0.76726 | 0.76841 | 0.61565 | 0.62060 |
| 0.86 | 0.87227 | 0.87251 | 0.77147 | 0.77254 | 0.62113 | 0.62594 |
| 0.88 | 0.87482 | 0.87504 | 0.77554 | 0.77654 | 0.62650 | 0.63115 |
| 0.90 | 0.87728 | 0.87746 | 0.77947 | 0.78039 | 0.63176 | 0.63623 |
| 0.92 | 0.87964 | 0.87981 | 0.78327 | 0.78413 | 0.63687 | 0.64118 |
| 0.94 | 0.88191 | 0.88207 | 0.78695 | 0.78774 | 0.64188 | 0.64600 |
| 0.96 | 0.88410 | 0.88424 | 0.79050 | 0.79123 | 0.64675 | 0.65072 |
| 0.98 | 0.88622 | 0.88633 | 0.79395 | 0.79461 | 0.65144 | 0.65531 |
| 1.00 | 0.88825 | 0.88835 | 0.79728 | 0.79790 | 0.65603 | 0.65981 |

TABLE 5

THE ILLUMINATION AND POLARIZATION OF THE SKY FOR VARIOUS ZENITH DISTANCES

OF THE SUN AND FOR AN ATMOSPHERE OF OPTICAL THICKNESS $\tau_1 = 0.15$

| $\mu$ | $\mu_0 = 0; \varphi = 0°$ | | $\mu_0 = 0; \varphi = 10°$ | | | $\mu_0 = 0; \varphi = 20°$ | | | $\mu_0 = 0; \varphi = 30°$ | | | $\mu_0 = 0; \varphi = 40°$ | | |
|---|---|---|---|---|---|---|---|---|---|---|---|---|---|---|
| | $I_1+I_r$ | $\delta$ | $I_1+I_r$ | $\chi°$ | $\delta$ | $I_1+I_r$ | $\chi°$ | $\delta$ | $I_1+I_r$ | $\chi°$ | $\delta$ | $I_1+I_r$ | $\chi°$ | $\delta$ |
| 0 | .. | .. | .. | .. | .. | .. | .. | .. | .. | .. | .. | .. | .. | .. |
| 0.04 | 0.6465 | -15 | 0.6398 | - 0.8 | -16 | 0.6205 | - 1.3 | -20 | 0.5911 | - 1.4 | -26 | 0.5549 | - 1.3 | -34 |
| 0.08 | 1.1218 | - 8 | 1.1079 | - 3.3 | -10 | 1.0679 | - 4.5 | -14 | 1.0065 | - 4.3 | -21 | 0.9313 | - 3.6 | -31 |
| 0.12 | 1.2438 | - 6 | 1.2278 | - 6.5 | - 8 | 1.1817 | - 8.1 | -13 | 1.1110 | - 7.2 | -20 | 1.0244 | - 5.9 | -30 |
| 0.16 | 1.2140 | - 5 | 1.1982 | -10.2 | - 7 | 1.1527 | -11.7 | -12 | 1.0829 | -10.2 | -20 | 0.9973 | - 8.1 | -30 |
| 0.20 | 1.1369 | - 4 | 1.1220 | -14.7 | - 6 | 1.0793 | -15.6 | -12 | 1.0139 | -13.1 | -20 | 0.9336 | -10.4 | -31 |
| 0.24 | 1.0492 | - 3 | 1.0356 | -20.2 | - 6 | 0.9963 | -19.7 | -12 | 0.9361 | -16.1 | -20 | 0.8623 | -12.6 | -31 |
| 0.28 | 0.9637 | - 2 | 0.9513 | -26.9 | - 6 | 0.9155 | -23.9 | -12 | 0.8606 | -19.2 | -20 | 0.7934 | -14.9 | -31 |
| 0.32 | 0.8844 | 0 | 0.8731 | -34.8 | - 6 | 0.8406 | -28.2 | -12 | 0.7908 | -22.2 | -21 | 0.7298 | -17.1 | -32 |
| 0.36 | 0.8121 | + 1 | 0.8019 | -42.9 | - 6 | 0.7725 | -32.6 | -13 | 0.7274 | -25.2 | -22 | 0.6721 | -19.4 | -33 |
| 0.40 | 0.7466 | + 3 | 0.7374 | +39.7 | + 7 | 0.7108 | -36.8 | -14 | 0.6701 | -28.2 | -23 | 0.6201 | -21.6 | -34 |
| 0.44 | 0.6872 | + 5 | 0.6789 | +33.6 | + 8 | 0.6549 | -40.9 | -15 | 0.6182 | -31.2 | -24 | 0.5732 | -23.9 | -35 |
| 0.48 | 0.6332 | + 7 | 0.6257 | +28.8 | +10 | 0.6042 | -44.8 | -16 | 0.5712 | -34.1 | -25 | 0.5307 | -26.1 | -36 |
| 0.52 | 0.5838 | +10 | 0.5771 | +25.1 | +12 | 0.5579 | +41.8 | +18 | 0.5283 | -36.9 | -27 | 0.4921 | -28.3 | -38 |
| 0.56 | 0.5385 | +13 | 0.5326 | +22.2 | +15 | 0.5154 | +38.6 | +21 | 0.4891 | -39.6 | -29 | 0.4568 | -30.4 | -40 |
| 0.60 | 0.4968 | +16 | 0.4915 | +19.9 | +18 | 0.4763 | +35.8 | +23 | 0.4530 | -42.1 | -32 | 0.4244 | -32.5 | -42 |
| 0.64 | 0.4580 | +19 | 0.4534 | +18.0 | +21 | 0.4401 | +33.3 | +27 | 0.4196 | -44.5 | -34 | 0.3945 | -34.6 | -44 |
| 0.68 | 0.4220 | +24 | 0.4179 | +16.5 | +25 | 0.4064 | +31.1 | +30 | 0.3886 | +43.2 | +38 | 0.3669 | -36.6 | -47 |
| 0.72 | 0.3882 | +28 | 0.3848 | +15.2 | +30 | 0.3749 | +29.1 | +34 | 0.3597 | +41.1 | +41 | 0.3411 | -38.5 | -51 |
| 0.76 | 0.3566 | +34 | 0.3537 | +14.1 | +35 | 0.3454 | +27.3 | +39 | 0.3326 | +39.1 | +46 | 0.3170 | -40.4 | -54 |
| 0.80 | 0.3267 | +40 | 0.3244 | +13.2 | +41 | 0.3176 | +25.8 | +45 | 0.3071 | +37.3 | +51 | 0.2944 | -42.2 | -59 |
| 0.84 | 0.2985 | +47 | 0.2966 | +12.4 | +48 | 0.2913 | +24.4 | +52 | 0.2831 | +35.6 | +57 | 0.2731 | -43.9 | -63 |
| 0.88 | 0.2717 | +56 | 0.2703 | +11.7 | +56 | 0.2664 | +23.1 | +59 | 0.2603 | +34.1 | +63 | 0.2529 | +44.4 | +69 |
| 0.92 | 0.2461 | +65 | 0.2452 | +11.1 | +66 | 0.2427 | +22.0 | +68 | 0.2387 | +32.6 | +71 | 0.2338 | +42.9 | +75 |
| 0.96 | 0.2218 | +77 | 0.2213 | +10.5 | +77 | 0.2200 | +20.9 | +79 | 0.2181 | +31.2 | +80 | 0.2157 | +41.4 | +83 |
| 1.00 | 0.1984 | +91 | 0.1984 | +10.0 | +91 | 0.1984 | +20.0 | +91 | 0.1984 | +30.0 | +91 | 0.1984 | +40.0 | +91 |
| 0.96 | 0.2218 | +77 | 0.2213 | +10.5 | +77 | 0.2200 | +20.9 | +79 | 0.2181 | +31.2 | +80 | 0.2157 | +41.4 | +83 |
| 0.92 | 0.2461 | +65 | 0.2452 | +11.1 | +66 | 0.2427 | +22.0 | +68 | 0.2387 | +32.6 | +71 | 0.2338 | +42.9 | +75 |
| 0.88 | 0.2717 | +56 | 0.2703 | +11.7 | +56 | 0.2664 | +23.1 | +59 | 0.2603 | +34.1 | +63 | 0.2529 | +44.4 | +69 |
| 0.84 | 0.2985 | +47 | 0.2966 | +12.4 | +48 | 0.2913 | +24.4 | +52 | 0.2831 | +35.6 | +57 | 0.2731 | -43.9 | -68 |
| 0.80 | 0.3267 | +40 | 0.3244 | +13.2 | +41 | 0.3176 | +25.8 | +45 | 0.3071 | +37.3 | +51 | 0.2944 | -42.2 | -59 |
| 0.76 | 0.3566 | +34 | 0.3537 | +14.1 | +35 | 0.3454 | +27.3 | +39 | 0.3326 | +39.1 | +46 | 0.3170 | -40.4 | -54 |
| 0.72 | 0.3882 | +28 | 0.3848 | +15.2 | +30 | 0.3749 | +29.1 | +34 | 0.3597 | +41.1 | +41 | 0.3411 | -38.5 | -51 |
| 0.68 | 0.4220 | +24 | 0.4179 | +16.5 | +25 | 0.4064 | +31.1 | +30 | 0.3886 | +43.2 | +38 | 0.3669 | -34.6 | -47 |
| 0.64 | 0.4580 | +19 | 0.4534 | +18.0 | +21 | 0.4401 | +33.3 | +27 | 0.4196 | -44.5 | -34 | 0.3945 | -34.6 | -44 |
| 0.60 | 0.4968 | +16 | 0.4915 | +19.9 | +18 | 0.4763 | +35.8 | +23 | 0.4530 | -42.1 | -32 | 0.4244 | -32.5 | -42 |
| 0.56 | 0.5385 | +13 | 0.5326 | +22.2 | +15 | 0.5154 | +38.6 | +21 | 0.4891 | -39.6 | -29 | 0.4568 | -30.4 | -40 |
| 0.52 | 0.5838 | +10 | 0.5771 | +25.1 | +12 | 0.5579 | +41.8 | +18 | 0.5283 | -36.9 | -27 | 0.4921 | -28.3 | -38 |
| 0.48 | 0.6332 | + 7 | 0.6257 | +28.8 | +10 | 0.6042 | -44.8 | -16 | 0.5712 | -34.1 | -25 | 0.5307 | -26.1 | -36 |
| 0.44 | 0.6872 | + 5 | 0.6789 | +33.6 | + 8 | 0.6549 | -40.9 | -15 | 0.6182 | -31.2 | -24 | 0.5732 | -23.9 | -35 |
| 0.40 | 0.7466 | + 3 | 0.7374 | +39.7 | + 7 | 0.7108 | -36.8 | -14 | 0.6701 | -28.2 | -23 | 0.6201 | -21.6 | -34 |
| 0.36 | 0.8121 | + 1 | 0.8019 | -42.9 | - 6 | 0.7725 | -32.6 | -13 | 0.7274 | -25.2 | -22 | 0.6721 | -19.4 | -33 |
| 0.32 | 0.8844 | 0 | 0.8731 | -34.8 | - 6 | 0.8406 | -28.2 | -12 | 0.7908 | -22.2 | -21 | 0.7298 | -17.1 | -32 |
| 0.28 | 0.9637 | - 2 | 0.9513 | -26.9 | - 6 | 0.9155 | -23.9 | -12 | 0.8606 | -19.2 | -20 | 0.7934 | -14.9 | -31 |
| 0.24 | 1.0492 | - 3 | 1.0356 | -20.2 | - 6 | 0.9963 | -19.7 | -12 | 0.9361 | -16.1 | -20 | 0.8623 | -12.6 | -31 |
| 0.20 | 1.1369 | - 4 | 1.1220 | -14.7 | - 6 | 1.0793 | -15.6 | -12 | 1.0139 | -13.1 | -20 | 0.9336 | -10.4 | -31 |
| 0.16 | 1.2140 | - 5 | 1.1982 | -10.2 | - 7 | 1.1527 | -11.7 | -12 | 1.0829 | -10.2 | -20 | 0.9973 | - 8.1 | -30 |
| 0.12 | 1.2438 | - 6 | 1.2278 | - 6.5 | - 8 | 1.1817 | - 8.1 | -13 | 1.1110 | - 7.2 | -20 | 1.0244 | - 5.9 | -30 |
| 0.08 | 1.1218 | - 8 | 1.1079 | - 3.3 | -10 | 1.0679 | - 4.5 | -14 | 1.0065 | - 4.3 | -21 | 0.9313 | - 3.6 | -31 |
| 0.04 | 0.6465 | -15 | 0.6398 | - 0.8 | -16 | 0.6205 | - 1.3 | -20 | 0.5911 | - 1.4 | -26 | 0.5549 | - 1.3 | -34 |
| 0 | .. | .. | .. | .. | .. | .. | .. | .. | .. | .. | .. | .. | .. | .. |

TABLE 5 (continued)

| μ | $\mu_0 = 0;\ \varphi = 50°$ | | | $\mu_0 = 0;\ \varphi = 60°$ | | | $\mu_0 = 0;\ \varphi = 70°$ | | | $\mu_0 = 0;\ \varphi = 80°$ | | | $\mu_0 = 0;\ \varphi = 90°$ | |
|---|---|---|---|---|---|---|---|---|---|---|---|---|---|---|
| | $I_1+I_r$ | $\chi°$ | $\delta$ | $I_1+I_r$ | $\chi°$ | $\delta$ | $I_1+I_r$ | $\chi°$ | $\delta$ | $I_1+I_r$ | $\chi°$ | $\delta$ | $I_1+I_r$ | $\delta$ |
| 0 | .. | .. | .. | .. | .. | .. | .. | .. | .. | .. | .. | .. | .. | .. |
| 0.04 | 0.5164 | - 1.1 | -44 | 0.4802 | - 0.8 | -55 | 0.4507 | - 0.6 | -65 | 0.4315 | - 0.3 | -73 | 0.4248 | -75 |
| 0.08 | 0.8512 | - 2.8 | -44 | 0.7760 | - 2.1 | -58 | 0.7147 | - 1.3 | -71 | 0.6746 | - 0.7 | -81 | 0.6607 | -85 |
| 0.12 | 0.9321 | - 4.5 | -41 | 0.8454 | - 3.3 | -58 | 0.7748 | - 2.1 | -73 | 0.7287 | - 1.0 | -84 | 0.7127 | -88 |
| 0.16 | 0.9063 | - 6.2 | -44 | 0.8207 | - 4.4 | -59 | 0.7509 | - 2.9 | -74 | 0.7054 | - 1.4 | -85 | 0.6896 | -89 |
| 0.20 | 0.8482 | - 7.8 | -44 | 0.7679 | - 5.6 | -59 | 0.7025 | - 3.6 | -74 | 0.6598 | - 1.8 | -86 | 0.6450 | -90 |
| 0.24 | 0.7838 | - 9.5 | -44 | 0.7100 | - 6.8 | -60 | 0.6498 | - 4.4 | -75 | 0.6106 | - 2.2 | -86 | 0.5969 | -90 |
| 0.28 | 0.7218 | -11.2 | -45 | 0.6545 | - 8.0 | -60 | 0.5997 | - 5.2 | -75 | 0.5639 | - 2.5 | -86 | 0.5514 | -91 |
| 0.32 | 0.6648 | -12.9 | -46 | 0.6037 | - 9.2 | -61 | 0.5539 | - 5.9 | -75 | 0.5215 | - 2.9 | -87 | 0.5102 | -91 |
| 0.36 | 0.6133 | -14.5 | -46 | 0.5580 | -10.4 | -61 | 0.5129 | - 6.7 | -76 | 0.4835 | - 3.3 | -87 | 0.4733 | -91 |
| 0.40 | 0.5670 | -16.2 | -47 | 0.5170 | -11.6 | -62 | 0.4763 | - 7.5 | -76 | 0.4497 | - 3.7 | -87 | 0.4405 | -91 |
| 0.44 | 0.5253 | -17.9 | -48 | 0.4803 | -12.8 | -63 | 0.4436 | - 8.2 | -77 | 0.4196 | - 4.0 | -87 | 0.4113 | -91 |
| 0.48 | 0.4877 | -19.6 | -49 | 0.4472 | -14.0 | -64 | 0.4142 | - 9.0 | -77 | 0.3927 | - 4.4 | -87 | 0.3852 | -91 |
| 0.52 | 0.4536 | -21.2 | -51 | 0.4174 | -15.2 | -65 | 0.3878 | - 9.8 | -78 | 0.3685 | - 4.8 | -88 | 0.3619 | -91 |
| 0.56 | 0.4225 | -22.9 | -52 | 0.3902 | -16.4 | -66 | 0.3639 | -10.6 | -79 | 0.3468 | - 5.2 | -88 | 0.3408 | -91 |
| 0.60 | 0.3940 | -24.6 | -54 | 0.3655 | -17.7 | -67 | 0.3422 | -11.4 | -79 | 0.3270 | - 5.6 | -88 | 0.3217 | -91 |
| 0.64 | 0.3679 | -26.2 | -56 | 0.3428 | -18.9 | -69 | 0.3224 | -12.3 | -80 | 0.3090 | - 6.0 | -88 | 0.3044 | -91 |
| 0.68 | 0.3437 | -27.9 | -58 | 0.3219 | -20.1 | -70 | 0.3042 | -13.1 | -81 | 0.2926 | - 6.4 | -88 | 0.2886 | -91 |
| 0.72 | 0.3212 | -29.5 | -61 | 0.3026 | -21.4 | -72 | 0.2874 | -13.9 | -82 | 0.2775 | - 6.9 | -89 | 0.2741 | -91 |
| 0.76 | 0.3003 | -31.1 | -64 | 0.2847 | -22.6 | -74 | 0.2719 | -14.8 | -83 | 0.2636 | - 7.3 | -89 | 0.2607 | -91 |
| 0.80 | 0.2807 | -32.7 | -67 | 0.2680 | -23.9 | -76 | 0.2575 | -15.6 | -84 | 0.2507 | - 7.7 | -89 | 0.2484 | -91 |
| 0.84 | 0.2624 | -34.2 | -71 | 0.2523 | -25.1 | -78 | 0.2441 | -16.5 | -85 | 0.2388 | - 8.2 | -89 | 0.2369 | -91 |
| 0.88 | 0.2450 | -35.7 | -75 | 0.2376 | -26.3 | -81 | 0.2316 | -17.3 | -86 | 0.2277 | - 8.6 | -90 | 0.2263 | -91 |
| 0.92 | 0.2287 | -37.2 | -80 | 0.2238 | -27.6 | -84 | 0.2199 | -18.2 | -88 | 0.2173 | - 9.1 | -90 | 0.2164 | -91 |
| 0.96 | 0.2132 | -38.6 | -85 | 0.2108 | -28.8 | -87 | 0.2088 | -19.1 | -89 | 0.2076 | - 9.5 | -91 | 0.2071 | -91 |
| 1.00 | 0.1984 | -40.0 | -91 | 0.1984 | -30.0 | -91 | 0.1984 | -20.0 | -91 | 0.1984 | -10.0 | -91 | 0.1984 | -91 |
| 0.96 | 0.2132 | -38.6 | -85 | 0.2108 | -28.8 | -87 | 0.2088 | -19.1 | -89 | 0.2076 | - 9.5 | -91 | 0.2071 | -91 |
| 0.92 | 0.2287 | -37.2 | -80 | 0.2238 | -27.6 | -84 | 0.2199 | -18.2 | -88 | 0.2173 | - 9.1 | -90 | 0.2164 | -91 |
| 0.88 | 0.2450 | -35.7 | -75 | 0.2376 | -26.3 | -81 | 0.2316 | -17.3 | -86 | 0.2277 | - 8.6 | -90 | 0.2263 | -91 |
| 0.84 | 0.2624 | -34.2 | -71 | 0.2523 | -25.1 | -78 | 0.2441 | -16.5 | -85 | 0.2388 | - 8.2 | -89 | 0.2369 | -91 |
| 0.80 | 0.2807 | -32.7 | -67 | 0.2680 | -23.9 | -76 | 0.2575 | -15.6 | -84 | 0.2507 | - 7.7 | -89 | 0.2484 | -91 |
| 0.76 | 0.3003 | -31.1 | -64 | 0.2847 | -22.6 | -74 | 0.2719 | -14.8 | -83 | 0.2636 | - 7.3 | -89 | 0.2607 | -91 |
| 0.72 | 0.3212 | -29.5 | -61 | 0.3026 | -21.4 | -72 | 0.2874 | -13.9 | -82 | 0.2775 | - 6.9 | -89 | 0.2741 | -91 |
| 0.68 | 0.3437 | -27.9 | -58 | 0.3219 | -20.1 | -70 | 0.3042 | -13.1 | -81 | 0.2926 | - 6.4 | -88 | 0.2886 | -91 |
| 0.64 | 0.3679 | -26.2 | -56 | 0.3428 | -18.9 | -69 | 0.3224 | -12.3 | -80 | 0.3090 | - 6.0 | -88 | 0.3044 | -91 |
| 0.60 | 0.3940 | -24.6 | -54 | 0.3655 | -17.7 | -67 | 0.3422 | -11.4 | -79 | 0.3270 | - 5.6 | -88 | 0.3217 | -91 |
| 0.56 | 0.4225 | -22.9 | -52 | 0.3902 | -16.4 | -66 | 0.3639 | -10.6 | -79 | 0.3468 | - 5.2 | -88 | 0.3408 | -91 |
| 0.52 | 0.4536 | -21.2 | -51 | 0.4174 | -15.2 | -65 | 0.3878 | - 9.8 | -78 | 0.3685 | - 4.8 | -88 | 0.3619 | -91 |
| 0.48 | 0.4877 | -19.6 | -49 | 0.4472 | -14.0 | -64 | 0.4142 | - 9.0 | -77 | 0.3927 | - 4.4 | -87 | 0.3852 | -91 |
| 0.44 | 0.5253 | -17.9 | -48 | 0.4803 | -12.8 | -63 | 0.4436 | - 8.2 | -77 | 0.4196 | - 4.0 | -87 | 0.4113 | -91 |
| 0.40 | 0.5670 | -16.2 | -47 | 0.5170 | -11.6 | -62 | 0.4763 | - 7.5 | -76 | 0.4497 | - 3.7 | -87 | 0.4405 | -91 |
| 0.36 | 0.6133 | -14.5 | -46 | 0.5580 | -10.4 | -61 | 0.5129 | - 6.7 | -76 | 0.4835 | - 3.3 | -87 | 0.4733 | -91 |
| 0.32 | 0.6648 | -12.9 | -46 | 0.6037 | - 9.2 | -61 | 0.5539 | - 5.9 | -75 | 0.5215 | - 2.9 | -87 | 0.5102 | -91 |
| 0.28 | 0.7218 | -11.2 | -45 | 0.6545 | - 8.0 | -60 | 0.5997 | - 5.2 | -75 | 0.5639 | - 2.5 | -86 | 0.5514 | -91 |
| 0.24 | 0.7838 | - 9.5 | -44 | 0.7100 | - 6.8 | -60 | 0.6498 | - 4.4 | -75 | 0.6106 | - 2.2 | -86 | 0.5969 | -90 |
| 0.20 | 0.8482 | - 7.8 | -44 | 0.7679 | - 5.6 | -59 | 0.7025 | - 3.6 | -74 | 0.6598 | - 1.8 | -86 | 0.6450 | -90 |
| 0.16 | 0.9063 | - 6.2 | -44 | 0.8207 | - 4.4 | -59 | 0.7509 | - 2.9 | -74 | 0.7054 | - 1.4 | -85 | 0.6896 | -89 |
| 0.12 | 0.9321 | - 4.5 | -41 | 0.8454 | - 3.3 | -58 | 0.7748 | - 2.1 | -73 | 0.7287 | - 1.0 | -84 | 0.7127 | -88 |
| 0.08 | 0.8512 | - 2.8 | -44 | 0.7760 | - 2.1 | -58 | 0.7147 | - 1.3 | -71 | 0.6746 | - 0.7 | -81 | 0.6607 | -85 |
| 0.04 | 0.5164 | - 1.1 | -44 | 0.4802 | - 0.8 | -55 | 0.4507 | - 0.6 | -65 | 0.4315 | - 0.3 | -73 | 0.4248 | -75 |
| 0 | .. | .. | .. | .. | .. | .. | .. | .. | .. | .. | .. | .. | .. | .. |

TABLE 5 (continued)

| μ | $\mu_0 = 0.08$; $\varphi = 0°$ | | $\mu_0 = 0.08$; $\varphi = 10°$ | | | $\mu_0 = 0.08$; $\varphi = 20°$ | | | $\mu_0 = 0.08$; $\varphi = 30°$ | | |
|---|---|---|---|---|---|---|---|---|---|---|---|
| | $I_1+I_r$ | $\delta$ | $I_1+I_r$ | $\chi°$ | $\delta$ | $I_1+I_r$ | $\chi°$ | $\delta$ | $I_1+I_r$ | $\chi°$ | $\delta$ |
| 0 | 0.0897 | − 8 | 0.0886 | + 3.7 | −10 | 0.0854 | + 5.3 | −14 | 0.0805 | + 5.4 | −21 |
| 0.04 | 0.1359 | − 8 | 0.1342 | + 2.2 | − 9 | 0.1291 | + 3.2 | −13 | 0.1214 | + 3.4 | −21 |
| 0.08 | . . | . . | . . | . . | . . | . . | . . | . . | . . | . . | . . |
| 0.12 | 0.1302 | − 6 | 0.1285 | − 1.9 | − 8 | 0.1236 | − 2.1 | −12 | 0.1161 | − 1.4 | −20 |
| 0.16 | 0.1152 | − 6 | 0.1137 | − 4.3 | − 8 | 0.1094 | − 5.0 | −12 | 0.1027 | − 4.0 | −19 |
| 0.20 | 0.1019 | − 6 | 0.1006 | − 7.1 | − 7 | 0.0967 | − 8.1 | −12 | 0.0909 | − 6.6 | −19 |
| 0.24 | 0.0906 | − 5 | 0.0894 | −10.3 | − 7 | 0.0861 | −11.5 | −12 | 0.0809 | − 9.4 | −19 |
| 0.28 | 0.0811 | − 4 | 0.0801 | −14.3 | − 6 | 0.0771 | −15.0 | −12 | 0.0725 | −12.2 | −20 |
| 0.32 | 0.0731 | − 3 | 0.0722 | −19.3 | − 6 | 0.0695 | −18.8 | −12 | 0.0654 | −15.1 | −20 |
| 0.36 | 0.0663 | − 2 | 0.0654 | −25.6 | − 6 | 0.0630 | −22.9 | −12 | 0.0594 | −18.1 | −20 |
| 0.40 | 0.0603 | − 1 | 0.0596 | −33.2 | − 6 | 0.0574 | −27.2 | −12 | 0.0542 | −21.1 | −21 |
| 0.44 | 0.0551 | + 1 | 0.0544 | −41.5 | − 6 | 0.0525 | −31.7 | −13 | 0.0496 | −24.3 | −21 |
| 0.48 | 0.0504 | + 3 | 0.0498 | +40.6 | + 7 | 0.0481 | −36.2 | −14 | 0.0455 | −27.4 | −22 |
| 0.52 | 0.0463 | + 5 | 0.0458 | +34.0 | + 8 | 0.0442 | −40.5 | −15 | 0.0419 | −30.6 | −24 |
| 0.56 | 0.0425 | + 7 | 0.0421 | +28.8 | +10 | 0.0407 | −44.6 | −16 | 0.0386 | −33.7 | −25 |
| 0.60 | 0.0391 | +10 | 0.0387 | +24.8 | +12 | 0.0375 | +41.6 | +18 | 0.0357 | −36.7 | −27 |
| 0.64 | 0.0360 | +13 | 0.0356 | +21.7 | +15 | 0.0346 | +38.2 | +21 | 0.0330 | −39.7 | −29 |
| 0.68 | 0.0331 | +17 | 0.0328 | +19.3 | +19 | 0.0319 | +35.1 | +24 | 0.0304 | −42.5 | −32 |
| 0.72 | 0.0304 | +21 | 0.0301 | +17.3 | +23 | 0.0293 | +32.4 | +28 | 0.0281 | +44.9 | +35 |
| 0.76 | 0.0279 | +26 | 0.0276 | +15.8 | +28 | 0.0270 | +30.0 | +32 | 0.0259 | +42.4 | +39 |
| 0.80 | 0.0255 | +32 | 0.0253 | +14.4 | +33 | 0.0247 | +27.9 | +37 | 0.0239 | +40.0 | +43 |
| 0.84 | 0.0232 | +38 | 0.0231 | +13.3 | +40 | 0.0226 | +26.1 | +43 | 0.0219 | +37.8 | +48 |
| 0.88 | 0.0210 | +46 | 0.0209 | +12.4 | +47 | 0.0206 | +24.4 | +50 | 0.0201 | +35.8 | +55 |
| 0.92 | 0.0190 | +56 | 0.0189 | +11.5 | +57 | 0.0187 | +22.9 | +59 | 0.0183 | +33.9 | +63 |
| 0.96 | 0.0169 | +69 | 0.0168 | +10.8 | +69 | 0.0167 | +21.5 | +71 | 0.0165 | +32.1 | +73 |
| 1.00 | 0.0145 | +89 | 0.0145 | +10.0 | +89 | 0.0145 | +20.0 | +89 | 0.0145 | +30.0 | +89 |
| 0.96 | 0.0157 | +82 | 0.0156 | +10.3 | +82 | 0.0156 | +20.5 | +83 | 0.0155 | +30.6 | +85 |
| 0.92 | 0.0172 | +72 | 0.0172 | +10.7 | +72 | 0.0170 | +21.3 | +74 | 0.0168 | +31.6 | +77 |
| 0.88 | 0.0190 | +62 | 0.0189 | +11.2 | +63 | 0.0187 | +22.2 | +66 | 0.0183 | +32.8 | +70 |
| 0.84 | 0.0209 | +54 | 0.0208 | +11.8 | +55 | 0.0204 | +23.3 | +58 | 0.0199 | +34.1 | +63 |
| 0.80 | 0.0229 | +47 | 0.0228 | +12.5 | +48 | 0.0223 | +24.5 | +52 | 0.0217 | +35.6 | +57 |
| 0.76 | 0.0251 | +40 | 0.0249 | +13.3 | +41 | 0.0244 | +25.8 | +46 | 0.0235 | +37.1 | +52 |
| 0.72 | 0.0275 | +34 | 0.0272 | +14.2 | +36 | 0.0266 | +27.3 | +40 | 0.0256 | +38.8 | +47 |
| 0.68 | 0.0300 | +29 | 0.0297 | +15.2 | +31 | 0.0290 | +29.0 | +36 | 0.0278 | +40.6 | +43 |
| 0.64 | 0.0328 | +24 | 0.0325 | +16.5 | +26 | 0.0316 | +30.8 | +31 | 0.0302 | +42.6 | +39 |
| 0.60 | 0.0358 | +20 | 0.0355 | +18.0 | +22 | 0.0344 | +32.9 | +28 | 0.0328 | +44.7 | +36 |
| 0.56 | 0.0392 | +17 | 0.0387 | +19.8 | +19 | 0.0375 | +35.3 | +25 | 0.0357 | −43.1 | −33 |
| 0.52 | 0.0429 | +13 | 0.0424 | +22.0 | +16 | 0.0410 | +37.9 | +22 | 0.0389 | −40.8 | −31 |
| 0.48 | 0.0470 | +10 | 0.0464 | +24.8 | +13 | 0.0449 | +40.9 | +20 | 0.0425 | −38.3 | −29 |
| 0.44 | 0.0516 | + 8 | 0.0510 | +28.3 | +11 | 0.0492 | +44.1 | +18 | 0.0465 | −35.7 | −27 |
| 0.40 | 0.0568 | + 5 | 0.0561 | +32.8 | + 9 | 0.0541 | −42.3 | −16 | 0.0511 | −33.1 | −26 |
| 0.36 | 0.0628 | + 3 | 0.0620 | +38.3 | + 8 | 0.0598 | −38.6 | −15 | 0.0564 | −30.3 | −24 |
| 0.32 | 0.0697 | + 2 | 0.0688 | −44.9 | − 7 | 0.0663 | −34.6 | −14 | 0.0625 | −27.6 | −23 |
| 0.28 | 0.0778 | 0 | 0.0768 | −37.5 | − 6 | 0.0740 | −30.6 | −13 | 0.0696 | −24.8 | −22 |
| 0.24 | 0.0874 | − 1 | 0.0863 | −30.1 | − 6 | 0.0831 | −26.6 | −13 | 0.0781 | −22.0 | −22 |
| 0.20 | 0.0989 | − 3 | 0.0976 | −23.5 | − 6 | 0.0939 | −22.7 | −12 | 0.0882 | −19.2 | −21 |
| 0.16 | 0.1125 | − 4 | 0.1110 | −18.0 | − 7 | 0.1068 | −19.0 | −12 | 0.1003 | −16.4 | −21 |
| 0.12 | 0.1279 | − 5 | 0.1263 | −13.5 | − 7 | 0.1214 | −15.4 | −12 | 0.1141 | −13.7 | −20 |
| 0.08 | . . | . . | . . | . . | . . | . . | . . | . . | . . | . . | . . |
| 0.04 | 0.1351 | − 7 | 0.1334 | − 6.3 | − 9 | 0.1284 | − 8.3 | −13 | 0.1207 | − 8.1 | −21 |
| 0 | 0.0897 | − 8 | 0.0886 | − 3.7 | −10 | 0.0854 | − 5.3 | −14 | 0.0805 | − 5.4 | −21 |

TABLE 5 (continued)

| μ | $\mu_o = 0.08; \varphi = 40°$ | | | $\mu_o = 0.08; \varphi = 50°$ | | | $\mu_o = 0.08; \varphi = 60°$ | | | $\mu_o = 0.08; \varphi = 70°$ | | |
|---|---|---|---|---|---|---|---|---|---|---|---|---|
| | $I_1 + I_r$ | $\chi°$ | $\delta$ | $I_1 + I_r$ | $\chi°$ | $\delta$ | $I_1 + I_r$ | $\chi°$ | $\delta$ | $I_1 + I_r$ | $\chi°$ | $\delta$ |
| 0 | 0.0745 | + 5.2 | -31 | 0.0681 | + 4.8 | -43 | 0.0621 | + 4.5 | -57 | 0.0572 | + 4.3 | -71 |
| 0.04 | 0.1120 | + 3.4 | -31 | 0.1019 | + 3.5 | -44 | 0.0924 | + 3.5 | -58 | 0.0847 | + 3.7 | -73 |
| 0.08 | .. | .. | .. | | | | | | | | | |
| 0.12 | 0.1068 | - 0.5 | -30 | 0.0969 | + 0.4 | -43 | 0.0876 | + 1.4 | -58 | 0.0799 | + 2.3 | -74 |
| 0.16 | 0.0945 | - 2.5 | -30 | 0.0857 | - 1.1 | -43 | 0.0775 | + 0.2 | -58 | 0.0707 | + 1.5 | -74 |
| 0.20 | 0.0836 | - 4.7 | -30 | 0.0759 | - 2.7 | -43 | 0.0686 | - 0.9 | -58 | 0.0626 | + 0.8 | -74 |
| 0.24 | 0.0745 | - 6.8 | -30 | 0.0677 | - 4.4 | -43 | 0.0612 | - 2.1 | -58 | 0.0559 | 0 | -74 |
| 0.28 | 0.0668 | - 9.0 | -30 | 0.0608 | - 6.0 | -44 | 0.0551 | - 3.3 | -59 | 0.0503 | - 0.8 | -74 |
| 0.32 | 0.0604 | -11.2 | -31 | 0.0550 | - 7.7 | -44 | 0.0499 | - 4.5 | -59 | 0.0456 | - 1.5 | -74 |
| 0.36 | 0.0549 | -13.5 | -31 | 0.0501 | - 9.4 | -44 | 0.0455 | - 5.7 | -59 | 0.0417 | - 2.4 | -74 |
| 0.40 | 0.0501 | -15.8 | -32 | 0.0458 | -11.1 | -45 | 0.0417 | - 7.0 | -59 | 0.0383 | - 3.2 | -74 |
| 0.44 | 0.0460 | -18.1 | -32 | 0.0421 | -12.9 | -45 | 0.0384 | - 8.2 | -60 | 0.0353 | - 4.0 | -74 |
| 0.48 | 0.0423 | -20.5 | -33 | 0.0388 | -14.6 | -46 | 0.0355 | - 9.5 | -60 | 0.0327 | - 4.9 | -74 |
| 0.52 | 0.0390 | -22.9 | -34 | 0.0359 | -16.5 | -47 | 0.0329 | -10.8 | -61 | 0.0305 | - 5.8 | -74 |
| 0.56 | 0.0361 | -25.3 | -36 | 0.0333 | -18.3 | -48 | 0.0306 | -12.2 | -62 | 0.0284 | - 6.7 | -75 |
| 0.60 | 0.0334 | -27.7 | -37 | 0.0309 | -20.1 | -49 | 0.0286 | -13.6 | -62 | 0.0266 | - 7.6 | -75 |
| 0.64 | 0.0309 | -30.1 | -39 | 0.0288 | -22.0 | -51 | 0.0267 | -15.0 | -64 | 0.0249 | - 8.5 | -76 |
| 0.68 | 0.0287 | -32.4 | -41 | 0.0268 | -23.9 | -53 | 0.0250 | -16.4 | -65 | 0.0234 | - 9.5 | -76 |
| 0.72 | 0.0266 | -34.8 | -44 | 0.0250 | -25.8 | -55 | 0.0234 | -17.9 | -66 | 0.0221 | -10.5 | -77 |
| 0.76 | 0.0247 | -37.0 | -47 | 0.0233 | -27.7 | -57 | 0.0219 | -19.4 | -68 | 0.0208 | -11.6 | -78 |
| 0.80 | 0.0228 | -39.3 | -51 | 0.0217 | -29.7 | -60 | 0.0206 | -20.9 | -70 | 0.0196 | -12.7 | -79 |
| 0.84 | 0.0211 | -41.4 | -55 | 0.0202 | -31.6 | -64 | 0.0193 | -22.4 | -72 | 0.0185 | -13.8 | -80 |
| 0.88 | 0.0195 | -43.5 | -61 | 0.0188 | -33.5 | -68 | 0.0181 | -24.1 | -75 | 0.0175 | -15.0 | -81 |
| 0.92 | 0.0179 | +44.4 | +67 | 0.0174 | -35.5 | -73 | 0.0169 | -25.7 | -78 | 0.0165 | -16.3 | -83 |
| 0.96 | 0.0163 | +42.4 | +76 | 0.0161 | -37.5 | -79 | 0.0158 | -27.5 | -82 | 0.0156 | -17.8 | -85 |
| 1.00 | 0.0145 | +40.0 | +89 | 0.0145 | -40.0 | -89 | 0.0145 | -30.0 | -89 | 0.0145 | -20.0 | -89 |
| 0.96 | 0.0154 | +40.5 | +86 | 0.0153 | -39.6 | -88 | 0.0152 | -29.9 | -89 | 0.0151 | -20.4 | -90 |
| 0.92 | 0.0166 | +41.7 | +80 | 0.0163 | -38.6 | -84 | 0.0161 | -29.2 | -87 | 0.0159 | -19.9 | -89 |
| 0.88 | 0.0179 | +42.9 | +75 | 0.0174 | -37.5 | -80 | 0.0170 | -28.3 | -85 | 0.0168 | -19.4 | -88 |
| 0.84 | 0.0193 | +44.3 | +69 | 0.0187 | -36.2 | -76 | 0.0181 | -27.3 | -83 | 0.0177 | -18.8 | -88 |
| 0.80 | 0.0209 | -44.3 | -65 | 0.0200 | -35.0 | -73 | 0.0193 | -26.3 | -80 | 0.0187 | -18.2 | -87 |
| 0.76 | 0.0225 | -42.7 | -60 | 0.0215 | -33.6 | -69 | 0.0206 | -25.3 | -78 | 0.0198 | -17.6 | -86 |
| 0.72 | 0.0244 | -41.1 | -56 | 0.0231 | -32.3 | -66 | 0.0219 | -24.3 | -76 | 0.0211 | -16.9 | -85 |
| 0.68 | 0.0263 | -39.4 | -53 | 0.0248 | -30.9 | -64 | 0.0234 | -23.2 | -75 | 0.0224 | -16.2 | -84 |
| 0.64 | 0.0285 | -37.7 | -50 | 0.0267 | -29.4 | -61 | 0.0251 | -22.2 | -73 | 0.0238 | -15.6 | -83 |
| 0.60 | 0.0309 | -35.9 | -47 | 0.0288 | -28.0 | -59 | 0.0269 | -21.1 | -71 | 0.0255 | -14.9 | -82 |
| 0.56 | 0.0335 | -34.0 | -44 | 0.0311 | -26.5 | -57 | 0.0290 | -20.0 | -70 | 0.0273 | -14.2 | -81 |
| 0.52 | 0.0364 | -32.1 | -42 | 0.0337 | -25.0 | -55 | 0.0312 | -18.9 | -69 | 0.0293 | -13.5 | -81 |
| 0.48 | 0.0396 | -30.1 | -40 | 0.0366 | -23.5 | -53 | 0.0338 | -17.8 | -67 | 0.0315 | -12.8 | -80 |
| 0.44 | 0.0433 | -28.1 | -39 | 0.0398 | -22.0 | -52 | 0.0366 | -16.8 | -66 | 0.0341 | -12.1 | -79 |
| 0.40 | 0.0474 | -26.1 | -37 | 0.0435 | -20.5 | -51 | 0.0399 | -15.7 | -65 | 0.0371 | -11.5 | -79 |
| 0.36 | 0.0522 | -24.0 | -36 | 0.0478 | -18.9 | -49 | 0.0437 | -14.6 | -64 | 0.0405 | -10.8 | -78 |
| 0.32 | 0.0578 | -22.0 | -35 | 0.0528 | -17.4 | -48 | 0.0482 | -13.5 | -63 | 0.0445 | -10.1 | -78 |
| 0.28 | 0.0643 | -19.9 | -34 | 0.0587 | -15.8 | -48 | 0.0534 | -12.4 | -63 | 0.0492 | - 9.4 | -77 |
| 0.24 | 0.0720 | -17.8 | -33 | 0.0656 | -14.3 | -47 | 0.0596 | -11.3 | -62 | 0.0548 | - 8.7 | -77 |
| 0.20 | 0.0813 | -15.7 | -32 | 0.0740 | -12.7 | -46 | 0.0671 | -10.2 | -61 | 0.0616 | - 8.0 | -76 |
| 0.16 | 0.0924 | -13.7 | -32 | 0.0840 | -11.2 | -45 | 0.0761 | - 9.1 | -61 | 0.0698 | - 7.3 | -76 |
| 0.12 | 0.1050 | -11.6 | -31 | 0.0954 | - 9.6 | -45 | 0.0864 | - 8.0 | -60 | 0.0791 | - 6.5 | -75 |
| 0.08 | .. | .. | .. | .. | .. | .. | .. | .. | .. | .. | .. | .. |
| 0.04 | 0.1114 | - 7.3 | -31 | 0.1014 | - 6.4 | -44 | 0.0920 | - 5.7 | -59 | 0.0844 | - 5.0 | -73 |
| 0 | 0.0745 | - 5.2 | -31 | 0.0681 | - 4.8 | -43 | 0.0621 | - 4.5 | -57 | 0.0572 | - 4.3 | -71 |

TABLE 5 (continued)

| μ | μ_0 = 0.08; φ = 80° | | | μ_0 = 0.08; φ = 90° | | | μ_0 = 0.24; φ = 0° | | μ_0 = 0.24; φ = 10° | | |
|---|---|---|---|---|---|---|---|---|---|---|---|
| | $I_1+I_r$ | $\chi°$ | δ | $I_1+I_r$ | $\chi°$ | δ | $I_1+I_r$ | δ | $I_1+I_r$ | $\chi°$ | δ |
| 0 | 0.0540 | + 4.2 | -81 | 0.0529 | + 4.1 | -85 | 0.2518 | - 3 | 0.2485 | +21.7 | - 6 |
| 0.04 | 0.0796 | + 3.9 | -84 | 0.0778 | + 4.2 | -88 | 0.2922 | - 4 | 0.2884 | +15.2 | - 7 |
| 0.08 | .. | .. | .. | .. | .. | .. | 0.2718 | - 5 | 0.2683 | +11.9 | - 7 |
| 0.12 | 0.0749 | + 3.2 | -85 | 0.0730 | + 4.2 | -90 | 0.2363 | - 5 | 0.2332 | + 8.9 | - 7 |
| 0.16 | 0.0662 | + 2.8 | -85 | 0.0645 | + 4.2 | -90 | 0.2049 | - 6 | 0.2022 | + 6.1 | - 7 |
| 0.20 | 0.0586 | + 2.5 | -85 | 0.0571 | + 4.2 | -90 | 0.1794 | - 6 | 0.1771 | + 3.6 | - 7 |
| 0.24 | 0.0523 | + 2.1 | -85 | 0.0509 | + 4.2 | -91 | .. | .. | .. | .. | .. |
| 0.28 | 0.0471 | + 1.7 | -85 | 0.0458 | + 4.1 | -91 | 0.1422 | - 6 | 0.1404 | - 1.5 | - 7 |
| 0.32 | 0.0428 | + 1.3 | -85 | 0.0416 | + 4.1 | -91 | 0.1282 | - 5 | 0.1266 | - 4.3 | - 7 |
| 0.36 | 0.0391 | + 0.9 | -85 | 0.0380 | + 4.0 | -91 | 0.1164 | - 5 | 0.1150 | - 7.6 | - 6 |
| 0.40 | 0.0359 | + 0.4 | -85 | 0.0349 | + 4.0 | -91 | 0.1063 | - 4 | 0.1050 | -11.6 | - 6 |
| 0.44 | 0.0332 | 0 | -85 | 0.0322 | + 3.9 | -91 | 0.0974 | - 3 | 0.0968 | -16.7 | - 5 |
| 0.48 | 0.0308 | - 0.5 | -85 | 0.0299 | + 3.8 | -90 | 0.0896 | - 2 | 0.0886 | -23.4 | - 5 |
| 0.52 | 0.0287 | - 0.9 | -85 | 0.0279 | + 3.8 | -90 | 0.0826 | - 1 | 0.0817 | -31.9 | - 5 |
| 0.56 | 0.0269 | - 1.4 | -85 | 0.0261 | + 3.7 | -90 | 0.0763 | + 1 | 0.0755 | -41.4 | - 5 |
| 0.60 | 0.0252 | - 1.9 | -85 | 0.0245 | + 3.6 | -90 | 0.0705 | + 3 | 0.0698 | +39.5 | + 6 |
| 0.64 | 0.0237 | - 2.5 | -85 | 0.0231 | + 3.4 | -90 | 0.0652 | + 5 | 0.0646 | +32.2 | + 8 |
| 0.68 | 0.0223 | - 3.0 | -85 | 0.0218 | + 3.3 | -90 | 0.0603 | + 8 | 0.0597 | +26.7 | +10 |
| 0.72 | 0.0211 | - 3.6 | -85 | 0.0206 | + 3.2 | -90 | 0.0557 | +11 | 0.0552 | +22.7 | +13 |
| 0.76 | 0.0200 | - 4.2 | -86 | 0.0195 | + 3.0 | -90 | 0.0513 | +15 | 0.0508 | +19.6 | +16 |
| 0.80 | 0.0189 | - 4.9 | -86 | 0.0185 | + 2.8 | -90 | 0.0471 | +19 | 0.0467 | +17.3 | +21 |
| 0.84 | 0.0179 | - 5.6 | -86 | 0.0176 | + 2.5 | -90 | 0.0431 | +25 | 0.0428 | +15.4 | +26 |
| 0.88 | 0.0170 | - 6.3 | -86 | 0.0167 | + 2.2 | -89 | 0.0391 | +32 | 0.0389 | +13.9 | +33 |
| 0.92 | 0.0162 | - 7.2 | -87 | 0.0160 | + 1.9 | -89 | 0.0352 | +41 | 0.0350 | +12.6 | +42 |
| 0.96 | 0.0154 | - 8.2 | -87 | 0.0152 | + 1.3 | -89 | 0.0311 | +53 | 0.0310 | +11.4 | +54 |
| 1.00 | 0.0145 | -10.0 | -89 | 0.0145 | 0 | -89 | 0.0254 | +81 | 0.0254 | +10.0 | +81 |
| 0.96 | 0.0152 | -10.8 | -90 | 0.0152 | + 1.3 | -89 | 0.0252 | +89 | 0.0251 | + 9.8 | +90 |
| 0.92 | 0.0159 | -10.9 | -90 | 0.0160 | + 1.9 | -89 | 0.0269 | +84 | 0.0268 | +10.0 | +85 |
| 0.88 | 0.0167 | -10.8 | -90 | 0.0167 | + 2.2 | -89 | 0.0291 | +77 | 0.0290 | +10.2 | +78 |
| 0.84 | 0.0175 | -10.6 | -90 | 0.0176 | + 2.5 | -90 | 0.0316 | +70 | 0.0315 | +10.6 | +71 |
| 0.80 | 0.0185 | -10.4 | -90 | 0.0185 | + 2.8 | -90 | 0.0345 | +63 | 0.0343 | +11.0 | +64 |
| 0.76 | 0.0195 | -10.2 | -90 | 0.0195 | + 3.0 | -90 | 0.0377 | +56 | 0.0375 | +11.5 | +57 |
| 0.72 | 0.0206 | - 9.9 | -90 | 0.0206 | + 3.2 | -90 | 0.0413 | +50 | 0.0410 | +12.0 | +51 |
| 0.68 | 0.0218 | - 9.7 | -90 | 0.0218 | + 3.3 | -90 | 0.0452 | +44 | 0.0448 | +12.6 | +46 |
| 0.64 | 0.0231 | - 9.4 | -89 | 0.0231 | + 3.4 | -90 | 0.0495 | +39 | 0.0491 | +13.3 | +40 |
| 0.60 | 0.0246 | - 9.1 | -89 | 0.0245 | + 3.6 | -90 | 0.0543 | +34 | 0.0538 | +14.1 | +35 |
| 0.56 | 0.0263 | - 8.8 | -89 | 0.0261 | + 3.7 | -90 | 0.0596 | +29 | 0.0591 | +15.0 | +31 |
| 0.52 | 0.0281 | - 8.5 | -89 | 0.0279 | + 3.8 | -90 | 0.0656 | +25 | 0.0649 | +16.0 | +27 |
| 0.48 | 0.0302 | - 8.2 | -89 | 0.0299 | + 3.8 | -90 | 0.0724 | +21 | 0.0716 | +17.3 | +23 |
| 0.44 | 0.0326 | - 7.9 | -88 | 0.0322 | + 3.9 | -91 | 0.0800 | +18 | 0.0792 | +18.8 | +20 |
| 0.40 | 0.0353 | - 7.6 | -88 | 0.0349 | + 4.0 | -91 | 0.0888 | +15 | 0.0878 | +20.6 | +17 |
| 0.36 | 0.0385 | - 7.3 | -88 | 0.0380 | + 4.0 | -91 | 0.0991 | +12 | 0.0979 | +22.8 | +15 |
| 0.32 | 0.0422 | - 7.0 | -88 | 0.0416 | + 4.1 | -91 | 0.1111 | + 9 | 0.1097 | +25.5 | +12 |
| 0.28 | 0.0465 | - 6.7 | -87 | 0.0458 | + 4.1 | -91 | 0.1254 | + 7 | 0.1238 | +28.9 | +10 |
| 0.24 | 0.0518 | - 6.3 | -87 | 0.0509 | + 4.2 | -91 | .. | .. | .. | .. | .. |
| 0.20 | 0.0581 | - 6.0 | -87 | 0.0571 | + 4.2 | -90 | 0.1639 | + 3 | 0.1619 | +38.4 | + 8 |
| 0.16 | 0.0657 | - 5.7 | -87 | 0.0645 | + 4.2 | -90 | 0.1906 | + 2 | 0.1882 | +44.7 | + 7 |
| 0.12 | 0.0745 | - 5.3 | -86 | 0.0730 | + 4.2 | -90 | 0.2238 | 0 | 0.2210 | -38.4 | - 6 |
| 0.08 | .. | .. | .. | .. | .. | .. | 0.2622 | - 1 | 0.2589 | -31.2 | - 6 |
| 0.04 | 0.0795 | - 4.6 | -84 | 0.0778 | + 4.2 | -88 | 0.2870 | - 3 | 0.2833 | -24.7 | - 6 |
| 0 | 0.0540 | - 4.2 | -81 | 0.0529 | + 4.1 | -85 | 0.2518 | - 3 | 0.2485 | -21.7 | - 6 |

TABLE 5 (continued)

| μ | $\mu_o = 0.24;\ \varphi = 20°$ | | | $\mu_o = 0.24;\ \varphi = 30°$ | | | $\mu_o = 0.24;\ \varphi = 40°$ | | | $\mu_o = 0.24;\ \varphi = 50°$ | | |
|---|---|---|---|---|---|---|---|---|---|---|---|---|
| | $I_1+I_r$ | $\chi°$ | $\delta$ | $I_1+I_r$ | $\chi°$ | $\delta$ | $I_1+I_r$ | $\chi°$ | $\delta$ | $I_1+I_r$ | $\chi°$ | $\delta$ |
| 0 | 0.2391 | +22.2 | -12 | 0.2247 | +19.8 | -20 | 0.2070 | +17.5 | -31 | 0.1881 | +15.7 | -44 |
| 0.04 | 0.2775 | +17.6 | -12 | 0.2609 | +16.5 | -20 | 0.2404 | +15.1 | -30 | 0.2186 | +14.0 | -44 |
| 0.08 | 0.2582 | +14.5 | -12 | 0.2426 | +14.0 | -19 | 0.2235 | +13.1 | -30 | 0.2031 | +12.5 | -43 |
| 0.12 | 0.2244 | +11.4 | -12 | 0.2109 | +11.4 | -19 | 0.1942 | +11.1 | -29 | 0.1764 | +10.9 | -42 |
| 0.16 | 0.1946 | + 8.3 | -11 | 0.1829 | + 8.8 | -19 | 0.1685 | + 9.0 | -29 | 0.1530 | + 9.3 | -42 |
| 0.20 | 0.1705 | + 5.3 | -11 | 0.1603 | + 6.1 | -18 | 0.1477 | + 6.9 | -28 | 0.1342 | + 7.7 | -41 |
| 0.24 | $\cdot\ \cdot$ | $\cdot\ \cdot$ | $\cdot\ \cdot$ | $\cdot\ \cdot$ | $\cdot\ \cdot$ | $\cdot\ \cdot$ | $\cdot\ \cdot$ | $\cdot\ \cdot$ | $\cdot\ \cdot$ | $\cdot\ \cdot$ | $\cdot\ \cdot$ | $\cdot\ \cdot$ |
| 0.28 | 0.1352 | - 1.0 | -11 | 0.1273 | + 0.6 | -18 | 0.1174 | + 2.5 | -28 | 0.1068 | + 4.3 | -40 |
| 0.32 | 0.1220 | - 4.3 | -11 | 0.1149 | - 2.2 | -18 | 0.1061 | + 0.2 | -28 | 0.0966 | + 2.6 | -40 |
| 0.36 | 0.1109 | - 7.8 | -11 | 0.1045 | - 5.2 | -18 | 0.0966 | - 2.2 | -28 | 0.0881 | + 0.8 | -40 |
| 0.40 | 0.1013 | -11.6 | -11 | 0.0956 | - 8.3 | -18 | 0.0884 | - 4.6 | -28 | 0.0807 | - 1.1 | -40 |
| 0.44 | 0.0929 | -15.8 | -11 | 0.0878 | -11.6 | -18 | 0.0814 | - 7.2 | -28 | 0.0744 | - 3.0 | -40 |
| 0.48 | 0.0856 | -20.4 | -11 | 0.0809 | -15.0 | -18 | 0.0751 | - 9.8 | -28 | 0.0688 | - 5.0 | -40 |
| 0.52 | 0.0790 | -25.3 | -11 | 0.0748 | -18.5 | -18 | 0.0696 | -12.4 | -28 | 0.0639 | - 7.1 | -40 |
| 0.56 | 0.0730 | -30.5 | -11 | 0.0693 | -22.2 | -19 | 0.0646 | -15.2 | -29 | 0.0595 | - 9.2 | -40 |
| 0.60 | 0.0676 | -35.8 | -12 | 0.0643 | -26.0 | -20 | 0.0601 | -18.1 | -29 | 0.0555 | -11.4 | -41 |
| 0.64 | 0.0626 | -40.9 | -14 | 0.0597 | -29.8 | -21 | 0.0559 | -21.0 | -30 | 0.0518 | -13.7 | -42 |
| 0.68 | 0.0580 | +44.2 | +15 | 0.0554 | -33.6 | -23 | 0.0521 | -24.1 | -32 | 0.0484 | -16.0 | -42 |
| 0.72 | 0.0537 | +39.9 | +18 | 0.0514 | -37.3 | -25 | 0.0485 | -27.1 | -33 | 0.0453 | -18.4 | -44 |
| 0.76 | 0.0496 | +36.0 | +21 | 0.0476 | -40.9 | -27 | 0.0451 | -30.2 | -36 | 0.0423 | -21.0 | -45 |
| 0.80 | 0.0457 | +32.6 | +25 | 0.0440 | -44.4 | -31 | 0.0419 | -33.3 | -38 | 0.0396 | -23.6 | -47 |
| 0.84 | 0.0419 | +29.7 | +29 | 0.0405 | +42.4 | +35 | 0.0388 | -36.3 | -42 | 0.0369 | -26.2 | -50 |
| 0.88 | 0.0382 | +27.1 | +36 | 0.0372 | +39.4 | +40 | 0.0358 | -39.4 | -46 | 0.0343 | -29.0 | -54 |
| 0.92 | 0.0346 | +24.8 | +44 | 0.0338 | +36.5 | +47 | 0.0328 | -42.4 | -52 | 0.0317 | -31.9 | -58 |
| 0.96 | 0.0307 | +22.7 | +55 | 0.0303 | +33.7 | +58 | 0.0297 | +44.5 | +61 | 0.0291 | -35.1 | -65 |
| 1.00 | 0.0254 | +20.0 | +81 | 0.0254 | +29.9 | +81 | 0.0254 | +39.9 | +81 | 0.0254 | -40.1 | -81 |
| 0.96 | 0.0251 | +19.5 | +90 | 0.0251 | +29.1 | +90 | 0.0252 | +38.7 | +90 | 0.0252 | -41.8 | -90 |
| 0.92 | 0.0267 | +19.8 | +86 | 0.0266 | +29.5 | +87 | 0.0265 | +39.1 | +89 | 0.0264 | -41.6 | -90 |
| 0.88 | 0.0288 | +20.3 | +80 | 0.0285 | +30.2 | +82 | 0.0281 | +39.7 | +86 | 0.0278 | -41.2 | -88 |
| 0.84 | 0.0312 | +21.0 | +73 | 0.0306 | +30.9 | +77 | 0.0301 | +40.4 | +82 | 0.0295 | -40.6 | -86 |
| 0.80 | 0.0338 | +21.7 | +67 | 0.0331 | +31.8 | +72 | 0.0325 | +41.3 | +78 | 0.0315 | -39.9 | -83 |
| 0.76 | 0.0368 | +22.5 | +61 | 0.0359 | +32.8 | +67 | 0.0347 | +42.2 | +74 | 0.0336 | -39.1 | -81 |
| 0.72 | 0.0401 | +23.4 | +55 | 0.0389 | +33.8 | +62 | 0.0375 | +43.3 | +69 | 0.0360 | -38.2 | -78 |
| 0.68 | 0.0438 | +24.4 | +50 | 0.0423 | +35.0 | +57 | 0.0405 | +44.4 | +66 | 0.0387 | -37.3 | -75 |
| 0.64 | 0.0479 | +25.5 | +45 | 0.0460 | +36.3 | +53 | 0.0439 | -44.4 | -62 | 0.0417 | -36.3 | -72 |
| 0.60 | 0.0524 | +26.8 | +41 | 0.0502 | +37.7 | +49 | 0.0476 | -43.1 | -59 | 0.0450 | -35.2 | -69 |
| 0.56 | 0.0574 | +28.2 | +37 | 0.0549 | +39.1 | +45 | 0.0518 | -41.8 | -55 | 0.0488 | -34.2 | -67 |
| 0.52 | 0.0630 | +29.7 | +33 | 0.0601 | +40.7 | +41 | 0.0566 | -40.4 | -52 | 0.0530 | -33.0 | -65 |
| 0.48 | 0.0694 | +31.5 | +29 | 0.0660 | +42.5 | +38 | 0.0619 | -38.9 | -50 | 0.0578 | -31.9 | -62 |
| 0.44 | 0.0766 | +33.5 | +26 | 0.0727 | +44.3 | +36 | 0.0681 | -37.4 | -47 | 0.0632 | -30.7 | -60 |
| 0.40 | 0.0849 | +35.7 | +24 | 0.0805 | -43.7 | -33 | 0.0751 | -35.8 | -45 | 0.0695 | -29.4 | -58 |
| 0.36 | 0.0945 | +38.1 | +21 | 0.0895 | -41.6 | -31 | 0.0833 | -34.1 | -43 | 0.0769 | -28.2 | -56 |
| 0.32 | 0.1059 | +40.8 | +19 | 0.1001 | -39.5 | -29 | 0.0930 | -32.4 | -41 | 0.0856 | -26.9 | -55 |
| 0.28 | 0.1194 | +43.8 | +18 | 0.1127 | -37.1 | -27 | 0.1045 | -30.6 | -39 | 0.0960 | -25.5 | -53 |
| 0.24 | $\cdot\ \cdot$ | $\cdot\ \cdot$ | $\cdot\ \cdot$ | $\cdot\ \cdot$ | $\cdot\ \cdot$ | $\cdot\ \cdot$ | $\cdot\ \cdot$ | $\cdot\ \cdot$ | $\cdot\ \cdot$ | $\cdot\ \cdot$ | $\cdot\ \cdot$ | $\cdot\ \cdot$ |
| 0.20 | 0.1559 | -39.6 | -15 | 0.1469 | -32.3 | -24 | 0.1358 | -26.9 | -36 | 0.1242 | -22.8 | -50 |
| 0.16 | 0.1812 | -36.0 | -14 | 0.1706 | -29.7 | -23 | 0.1576 | -25.0 | -35 | 0.1439 | -21.4 | -49 |
| 0.12 | 0.2127 | -32.2 | -13 | 0.2001 | -27.1 | -22 | 0.1847 | -23.1 | -34 | 0.1684 | -20.0 | -48 |
| 0.08 | 0.2492 | -28.4 | -13 | 0.2343 | -24.5 | -22 | 0.2161 | -21.1 | -33 | 0.1969 | -18.5 | -46 |
| 0.04 | 0.2727 | -24.5 | -12 | 0.2564 | -21.7 | -21 | 0.2364 | -19.1 | -32 | 0.2152 | -17.0 | -45 |
| 0 | 0.2391 | -22.2 | -12 | 0.2247 | -19.8 | -20 | 0.2070 | -17.5 | -31 | 0.1881 | -15.7 | -44 |

TABLE 5 (continued)

| μ | μ₀ = 0.24; φ = 60° | | | μ₀ = 0.24; φ = 70° | | | μ₀ = 0.24; φ = 80° | | | μ₀ = 0.24; φ = 90° | | |
|---|---|---|---|---|---|---|---|---|---|---|---|---|
| | $I_1+I_r$ | $\chi°$ | $\delta$ | $I_1+I_r$ | $\chi°$ | $\delta$ | $I_1+I_r$ | $\chi°$ | $\delta$ | $I_1+I_r$ | $\chi°$ | $\delta$ |
| 0 | 0.1704 | +14.5 | -59 | 0.1560 | +13.7 | -74 | 0.1465 | +13.2 | -86 | 0.1433 | +13.0 | -90 |
| 0.04 | 0.1980 | +13.2 | -59 | 0.1811 | +12.8 | -74 | 0.1699 | +12.7 | -85 | 0.1657 | +12.9 | -90 |
| 0.08 | 0.1837 | +12.1 | -58 | 0.1678 | +12.1 | -73 | 0.1570 | +12.4 | -85 | 0.1527 | +12.9 | -90 |
| 0.12 | 0.1595 | +11.0 | -58 | 0.1454 | +11.4 | -73 | 0.1359 | +12.0 | -85 | 0.1318 | +12.9 | -90 |
| 0.16 | 0.1383 | + 9.9 | -57 | 0.1260 | +10.6 | -72 | 0.1175 | +11.6 | -84 | 0.1137 | +12.8 | -90 |
| 0.20 | 0.1212 | + 8.7 | -56 | 0.1104 | + 9.8 | -72 | 0.1028 | +11.2 | -84 | 0.0993 | +12.8 | -90 |
| 0.24 | .. | .. | .. | .. | .. | .. | .. | .. | .. | .. | .. | .. |
| 0.28 | 0.0966 | + 6.2 | -55 | 0.0879 | + 8.2 | -70 | 0.0817 | +10.3 | -83 | 0.0786 | +12.6 | -90 |
| 0.32 | 0.0875 | + 4.9 | -55 | 0.0796 | + 7.3 | -70 | 0.0740 | + 9.8 | -83 | 0.0710 | +12.4 | -90 |
| 0.36 | 0.0798 | + 3.6 | -54 | 0.0727 | + 6.4 | -69 | 0.0675 | + 9.3 | -82 | 0.0647 | +12.3 | -90 |
| 0.40 | 0.0733 | + 2.3 | -54 | 0.0668 | + 5.5 | -69 | 0.0620 | + 8.8 | -82 | 0.0593 | +12.1 | -89 |
| 0.44 | 0.0676 | + 0.8 | -54 | 0.0617 | + 4.5 | -68 | 0.0573 | + 8.2 | -81 | 0.0548 | +11.9 | -89 |
| 0.48 | 0.0627 | - 0.6 | -54 | 0.0573 | + 3.5 | -68 | 0.0533 | + 7.6 | -81 | 0.0509 | +11.7 | -89 |
| 0.52 | 0.0583 | - 2.1 | -54 | 0.0534 | + 2.5 | -68 | 0.0497 | + 7.0 | -80 | 0.0474 | +11.5 | -88 |
| 0.56 | 0.0544 | - 3.7 | -54 | 0.0500 | + 1.4 | -67 | 0.0465 | + 6.3 | -80 | 0.0444 | +11.2 | -88 |
| 0.60 | 0.0509 | - 5.4 | -54 | 0.0469 | + 0.2 | -67 | 0.0437 | + 5.6 | -79 | 0.0417 | +10.9 | -87 |
| 0.64 | 0.0477 | - 7.1 | -54 | 0.0441 | - 1.0 | -67 | 0.0412 | + 4.8 | -79 | 0.0393 | +10.5 | -87 |
| 0.68 | 0.0448 | - 8.9 | -55 | 0.0415 | - 2.3 | -67 | 0.0389 | + 4.0 | -78 | 0.0371 | +10.1 | -86 |
| 0.72 | 0.0421 | -10.8 | -55 | 0.0392 | - 3.7 | -67 | 0.0368 | + 3.1 | -78 | 0.0352 | + 9.7 | -86 |
| 0.76 | 0.0396 | -12.7 | -56 | 0.0370 | - 5.1 | -67 | 0.0349 | + 2.1 | -77 | 0.0334 | + 9.1 | -85 |
| 0.80 | 0.0372 | -14.8 | -57 | 0.0350 | - 6.7 | -68 | 0.0331 | + 1.0 | -77 | 0.0318 | + 8.5 | -85 |
| 0.84 | 0.0349 | -17.0 | -59 | 0.0331 | - 8.4 | -68 | 0.0315 | - 0.2 | -77 | 0.0303 | + 7.8 | -84 |
| 0.88 | 0.0327 | -19.4 | -61 | 0.0312 | -10.3 | -69 | 0.0299 | - 1.6 | -77 | 0.0289 | + 6.9 | -83 |
| 0.92 | 0.0306 | -22.0 | -64 | 0.0295 | -12.5 | -71 | 0.0285 | - 3.3 | -77 | 0.0276 | + 5.7 | -82 |
| 0.96 | 0.0284 | -24.9 | -69 | 0.0277 | -15.1 | -73 | 0.0270 | - 5.4 | -78 | 0.0265 | + 4.1 | -81 |
| 1.00 | 0.0254 | -30.0 | -81 | 0.0254 | -20.0 | -81 | 0.0254 | -10.0 | -81 | 0.0254 | 0 | -81 |
| 0.96 | 0.0254 | -32.3 | -89 | 0.0256 | -23.0 | -87 | 0.0260 | -13.6 | -85 | 0.0265 | + 4.1 | -81 |
| 0.92 | 0.0264 | -32.5 | -90 | 0.0266 | -23.5 | -89 | 0.0270 | -14.6 | -86 | 0.0276 | + 5.7 | -82 |
| 0.88 | 0.0277 | -32.3 | -90 | 0.0278 | -23.7 | -90 | 0.0282 | -15.3 | -88 | 0.0289 | + 6.9 | -83 |
| 0.84 | 0.0292 | -32.0 | -89 | 0.0292 | -23.7 | -90 | 0.0295 | -15.7 | -88 | 0.0303 | + 7.8 | -84 |
| 0.80 | 0.0309 | -31.5 | -88 | 0.0307 | -23.6 | -90 | 0.0310 | -16.0 | -89 | 0.0318 | + 8.5 | -85 |
| 0.76 | 0.0328 | -31.0 | -86 | 0.0324 | -23.4 | -90 | 0.0326 | -16.2 | -90 | 0.0334 | + 9.1 | -85 |
| 0.72 | 0.0349 | -30.4 | -85 | 0.0343 | -23.2 | -89 | 0.0343 | -16.3 | -90 | 0.0352 | + 9.7 | -86 |
| 0.68 | 0.0373 | -29.7 | -83 | 0.0364 | -22.8 | -89 | 0.0363 | -16.4 | -90 | 0.0371 | +10.1 | -86 |
| 0.64 | 0.0399 | -29.0 | -82 | 0.0387 | -22.5 | -88 | 0.0385 | -16.4 | -90 | 0.0393 | +10.5 | -87 |
| 0.60 | 0.0426 | -28.3 | -80 | 0.0413 | -22.1 | -87 | 0.0409 | -16.3 | -90 | 0.0417 | +10.9 | -87 |
| 0.56 | 0.0461 | -27.5 | -78 | 0.0443 | -21.7 | -87 | 0.0437 | -16.3 | -90 | 0.0444 | +11.2 | -88 |
| 0.52 | 0.0498 | -26.7 | -76 | 0.0476 | -21.2 | -86 | 0.0467 | -16.2 | -90 | 0.0474 | +11.5 | -88 |
| 0.48 | 0.0541 | -25.9 | -75 | 0.0514 | -20.7 | -85 | 0.0503 | -16.1 | -90 | 0.0509 | +11.7 | -89 |
| 0.44 | 0.0589 | -25.1 | -73 | 0.0558 | -20.2 | -84 | 0.0543 | -15.9 | -90 | 0.0548 | +11.9 | -89 |
| 0.40 | 0.0645 | -24.2 | -72 | 0.0608 | -19.7 | -83 | 0.0590 | -15.7 | -90 | 0.0593 | +12.1 | -89 |
| 0.36 | 0.0711 | -23.3 | -70 | 0.0667 | -19.2 | -82 | 0.0645 | -15.6 | -90 | 0.0647 | +12.3 | -90 |
| 0.32 | 0.0789 | -22.4 | -69 | 0.0738 | -18.6 | -82 | 0.0710 | -15.3 | -89 | 0.0710 | +12.4 | -90 |
| 0.28 | 0.0882 | -21.4 | -68 | 0.0822 | -18.0 | -81 | 0.0788 | -15.1 | -89 | 0.0786 | +12.6 | -90 |
| 0.24 | .. | .. | .. | .. | .. | .. | .. | .. | .. | .. | .. | .. |
| 0.20 | 0.1135 | -19.5 | -65 | 0.1051 | -16.8 | -79 | 0.1001 | -14.6 | -88 | 0.0993 | +12.8 | -90 |
| 0.16 | 0.1312 | -18.5 | -64 | 0.1211 | -16.2 | -78 | 0.1150 | -14.3 | -88 | 0.1137 | +12.8 | -90 |
| 0.12 | 0.1533 | -17.5 | -63 | 0.1412 | -15.6 | -77 | 0.1337 | -14.0 | -87 | 0.1318 | +12.9 | -90 |
| 0.08 | 0.1789 | -16.4 | -62 | 0.1645 | -14.9 | -76 | 0.1554 | -13.7 | -87 | 0.1527 | +12.9 | -90 |
| 0.04 | 0.1954 | -15.4 | -61 | 0.1793 | -14.2 | -75 | 0.1690 | -13.4 | -86 | 0.1657 | +12.9 | -90 |
| 0 | 0.1704 | -14.5 | -59 | 0.1560 | -13.7 | -74 | 0.1465 | -13.2 | -86 | 0.1433 | +13.0 | -90 |

TABLE 5 (continued)

| μ | μ₀ = 0.52; φ = 0° | | μ₀ = 0.52; φ = 10° | | | μ₀ = 0.52; φ = 20° | | | μ₀ = 0.52; φ = 30° | | |
|---|---|---|---|---|---|---|---|---|---|---|---|
| | $I_1 + I_r$ | δ | $I_1 + I_r$ | $\chi°$ | δ | $I_1 + I_r$ | $\chi°$ | δ | $I_1 + I_r$ | $\chi°$ | δ |
| 0 | 0.3036 | +10 | 0.3001 | -21.5 | +13 | 0.2901 | -35.4 | +18 | 0.2747 | -44.2 | +27 |
| 0.04 | 0.3308 | + 7 | 0.3270 | -25.7 | +10 | 0.3162 | -39.3 | +16 | 0.2995 | +42.8 | -25 |
| 0.08 | 0.3009 | + 5 | 0.2975 | -29.2 | + 8 | 0.2876 | -42.2 | +15 | 0.2724 | +40.6 | -23 |
| 0.12 | 0.2608 | + 4 | 0.2578 | -33.5 | + 7 | 0.2492 | +44.7 | -13 | 0.2360 | +38.3 | -22 |
| 0.16 | 0.2271 | + 2 | 0.2245 | -38.8 | + 6 | 0.2170 | +41.3 | -12 | 0.2055 | +35.8 | -20 |
| 0.20 | 0.2003 | + 1 | 0.1981 | +44.8 | - 5 | 0.1915 | +37.6 | -11 | 0.1813 | +33.2 | -19 |
| 0.24 | 0.1790 | 0 | 0.1770 | +37.8 | - 5 | 0.1711 | +33.7 | -11 | 0.1621 | +30.5 | -18 |
| 0.28 | 0.1617 | - 1 | 0.1599 | +30.7 | - 5 | 0.1546 | +29.7 | -10 | 0.1465 | +27.6 | -17 |
| 0.32 | 0.1474 | - 2 | 0.1457 | +24.2 | - 5 | 0.1410 | +25.7 | -10 | 0.1336 | +24.6 | -16 |
| 0.36 | 0.1353 | - 3 | 0.1338 | +18.6 | - 5 | 0.1295 | +21.6 | - 9 | 0.1228 | +21.5 | -16 |
| 0.40 | 0.1249 | - 3 | 0.1236 | +13.9 | - 5 | 0.1197 | +17.5 | - 9 | 0.1136 | +18.3 | -15 |
| 0.44 | 0.1159 | - 4 | 0.1147 | + 9.8 | - 5 | 0.1111 | +13.5 | - 9 | 0.1056 | +14.9 | -14 |
| 0.48 | 0.1080 | - 4 | 0.1068 | + 6.1 | - 5 | 0.1036 | + 9.3 | - 8 | 0.0985 | +11.4 | -14 |
| 0.52 | .. | .. | .. | .. | .. | .. | .. | .. | .. | .. | .. |
| 0.56 | 0.0944 | - 4 | 0.0984 | - 1.3 | - 5 | 0.0907 | + 0.6 | - 8 | 0.0865 | + 3.8 | -13 |
| 0.60 | 0.0884 | - 3 | 0.0876 | - 5.9 | - 4 | 0.0851 | - 4.4 | - 8 | 0.0813 | - 0.4 | -13 |
| 0.64 | 0.0829 | - 2 | 0.0822 | -12.0 | - 4 | 0.0799 | -10.1 | - 7 | 0.0765 | - 5.0 | -13 |
| 0.68 | 0.0778 | - 1 | 0.0771 | -21.1 | - 4 | 0.0751 | -16.7 | - 7 | 0.0719 | -10.0 | -13 |
| 0.72 | 0.0728 | 0 | 0.0722 | -34.2 | - 3 | 0.0705 | -24.3 | - 7 | 0.0677 | -15.6 | -13 |
| 0.76 | 0.0681 | + 2 | 0.0676 | +41.5 | + 4 | 0.0661 | -32.5 | - 8 | 0.0636 | -21.5 | -13 |
| 0.80 | 0.0636 | + 4 | 0.0631 | +30.5 | + 6 | 0.0618 | -40.8 | - 9 | 0.0596 | -27.8 | -14 |
| 0.84 | 0.0591 | + 7 | 0.0587 | +23.4 | + 8 | 0.0575 | +41.8 | +11 | 0.0557 | -34.2 | -16 |
| 0.88 | 0.0545 | +11 | 0.0542 | +18.8 | +12 | 0.0533 | +35.5 | +15 | 0.0518 | -40.5 | -19 |
| 0.92 | 0.0497 | +17 | 0.0495 | +15.5 | +18 | 0.0488 | +30.1 | +20 | 0.0477 | +43.5 | +23 |
| 0.96 | 0.0445 | +26 | 0.0443 | +12.9 | +27 | 0.0439 | +25.6 | +28 | 0.0432 | +37.8 | +30 |
| 1.00 | 0.0353 | +53 | 0.0353 | + 9.9 | +53 | 0.0353 | +19.9 | +53 | 0.0353 | +29.8 | +53 |
| 0.96 | 0.0311 | +80 | 0.0312 | + 8.8 | +80 | 0.0313 | +17.6 | +79 | 0.0316 | +26.4 | +77 |
| 0.92 | 0.0311 | +87 | 0.0312 | + 8.6 | +87 | 0.0313 | +17.3 | +86 | 0.0316 | +25.8 | +85 |
| 0.88 | 0.0320 | +90 | 0.0320 | + 8.6 | +90 | 0.0321 | +17.2 | +89 | 0.0323 | +25.6 | +89 |
| 0.84 | 0.0335 | +89 | 0.0335 | + 8.7 | +89 | 0.0335 | +17.2 | +89 | 0.0336 | +25.6 | +90 |
| 0.80 | 0.0354 | +87 | 0.0354 | + 8.7 | +87 | 0.0353 | +17.4 | +88 | 0.0352 | +25.7 | +89 |
| 0.76 | 0.0378 | +84 | 0.0377 | + 8.9 | +84 | 0.0375 | +17.6 | +86 | 0.0373 | +26.0 | +88 |
| 0.72 | 0.0406 | +79 | 0.0405 | + 9.0 | +80 | 0.0402 | +17.8 | +82 | 0.0398 | +26.3 | +85 |
| 0.68 | 0.0439 | +75 | 0.0437 | + 9.2 | +76 | 0.0433 | +18.2 | +78 | 0.0426 | +26.7 | +82 |
| 0.64 | 0.0477 | +70 | 0.0475 | + 9.5 | +71 | 0.0468 | +18.6 | +74 | 0.0460 | +27.1 | +78 |
| 0.60 | 0.0520 | +65 | 0.0517 | + 9.7 | +66 | 0.0509 | +19.0 | +69 | 0.0498 | +27.6 | +74 |
| 0.56 | 0.0570 | +59 | 0.0567 | +10.0 | +61 | 0.0557 | +19.5 | +65 | 0.0542 | +28.2 | +71 |
| 0.52 | .. | .. | .. | .. | .. | .. | .. | .. | .. | .. | .. |
| 0.48 | 0.0693 | +50 | 0.0687 | +10.7 | +51 | 0.0672 | +20.7 | +56 | 0.0650 | +29.6 | +63 |
| 0.44 | 0.0769 | +45 | 0.0762 | +11.1 | +47 | 0.0744 | +21.4 | +52 | 0.0718 | +30.4 | +59 |
| 0.40 | 0.0858 | +41 | 0.0850 | +11.6 | +43 | 0.0829 | +22.1 | +47 | 0.0797 | +31.2 | +55 |
| 0.36 | 0.0962 | +37 | 0.0953 | +12.1 | +38 | 0.0928 | +23.0 | +44 | 0.0890 | +32.1 | +51 |
| 0.32 | 0.1087 | +33 | 0.1076 | +12.7 | +35 | 0.1046 | +23.9 | +40 | 0.1001 | +33.2 | +48 |
| 0.28 | 0.1237 | +29 | 0.1225 | +13.4 | +31 | 0.1189 | +25.0 | +36 | 0.1136 | +34.3 | +45 |
| 0.24 | 0.1422 | +26 | 0.1407 | +14.1 | +28 | 0.1365 | +26.1 | +33 | 0.1302 | +35.5 | +42 |
| 0.20 | 0.1653 | +22 | 0.1635 | +15.0 | +24 | 0.1585 | +27.4 | +30 | 0.1509 | +36.8 | +39 |
| 0.16 | 0.1946 | +19 | 0.1926 | +16.1 | +21 | 0.1865 | +28.9 | +27 | 0.1774 | +38.2 | +36 |
| 0.12 | 0.2323 | +16 | 0.2298 | +17.4 | +19 | 0.2224 | +30.5 | +25 | 0.2113 | +39.8 | +33 |
| 0.08 | 0.2786 | +14 | 0.2755 | +18.9 | +16 | 0.2666 | +32.4 | +22 | 0.2530 | +41.4 | +31 |
| 0.04 | 0.3183 | +11 | 0.3147 | +20.7 | +14 | 0.3044 | +34.4 | +20 | 0.2887 | +43.2 | +29 |
| 0 | 0.3036 | +10 | 0.3001 | +21.5 | +13 | 0.2901 | +35.4 | +18 | 0.2747 | +44.2 | +27 |

TABLE 5 (continued)

| μ | $\mu_o = 0.52; \varphi = 40°$ | | | $\mu_o = 0.52; \varphi = 50°$ | | | $\mu_o = 0.52; \varphi = 60°$ | | | $\mu_o = 0.52; \varphi = 70°$ | | |
|---|---|---|---|---|---|---|---|---|---|---|---|---|
| | $I_1+I_r$ | $\chi°$ | $\delta$ | $I_1+I_r$ | $\chi°$ | $\delta$ | $I_1+I_r$ | $\chi°$ | $\delta$ | $I_1+I_r$ | $\chi°$ | $\delta$ |
| 0 | 0.2559 | +40.0 | -38 | 0.2359 | +36.0 | -50 | 0.2170 | +33.2 | -64 | 0.2017 | +31.4 | -77 |
| 0.04 | 0.2791 | +37.7 | -36 | 0.2572 | +34.3 | -48 | 0.2364 | +32.0 | -62 | 0.2193 | +30.5 | -75 |
| 0.08 | 0.2536 | +36.0 | -34 | 0.2334 | +32.9 | -46 | 0.2141 | +31.0 | -60 | 0.1979 | +29.9 | -74 |
| 0.12 | 0.2196 | +34.2 | -32 | 0.2019 | +31.6 | -44 | 0.1849 | +30.0 | -58 | 0.1705 | +29.2 | -72 |
| 0.16 | 0.1912 | +32.3 | -30 | 0.1757 | +30.2 | -43 | 0.1606 | +28.9 | -56 | 0.1477 | +28.5 | -70 |
| 0.20 | 0.1687 | +30.4 | -29 | 0.1549 | +28.6 | -41 | 0.1415 | +27.8 | -55 | 0.1299 | +27.7 | -69 |
| 0.24 | 0.1508 | +28.3 | -28 | 0.1385 | +27.1 | -39 | 0.1264 | +26.7 | -53 | 0.1158 | +26.9 | -67 |
| 0.28 | 0.1363 | +26.1 | -26 | 0.1252 | +25.4 | -38 | 0.1142 | +25.4 | -51 | 0.1045 | +26.1 | -65 |
| 0.32 | 0.1244 | +23.8 | -25 | 0.1143 | +23.6 | -37 | 0.1043 | +24.1 | -50 | 0.0953 | +25.1 | -64 |
| 0.36 | 0.1145 | +21.4 | -24 | 0.1052 | +21.8 | -35 | 0.0960 | +22.7 | -48 | 0.0877 | +24.2 | -62 |
| 0.40 | 0.1059 | +18.9 | -23 | 0.0975 | +19.9 | -34 | 0.0890 | +21.3 | -47 | 0.0812 | +23.1 | -60 |
| 0.44 | 0.0986 | +16.2 | -22 | 0.0908 | +17.8 | -33 | 0.0829 | +19.7 | -45 | 0.0757 | +22.0 | -59 |
| 0.48 | 0.0921 | +13.4 | -22 | 0.0849 | +15.6 | -32 | 0.0777 | +18.1 | -44 | 0.0710 | +20.8 | -57 |
| 0.52 | .. | .. | .. | .. | .. | .. | .. | .. | .. | | | |
| 0.56 | 0.0811 | + 7.3 | -20 | 0.0751 | +10.8 | -30 | 0.0689 | +14.4 | -41 | 0.0631 | +18.1 | -54 |
| 0.60 | 0.0764 | + 3.9 | -20 | 0.0709 | + 8.2 | -29 | 0.0652 | +12.4 | -40 | 0.0598 | +16.6 | -52 |
| 0.64 | 0.0720 | + 0.3 | -20 | 0.0670 | + 5.3 | -28 | 0.0618 | +10.2 | -39 | 0.0568 | +14.9 | -51 |
| 0.68 | 0.0679 | - 3.6 | -19 | 0.0634 | + 2.2 | -28 | 0.0587 | + 7.8 | -38 | 0.0541 | +13.1 | -50 |
| 0.72 | 0.0641 | - 7.9 | -19 | 0.0600 | - 1.1 | -28 | 0.0557 | + 5.2 | -37 | 0.0516 | +11.1 | -48 |
| 0.76 | 0.0604 | -12.5 | -20 | 0.0568 | - 4.7 | -28 | 0.0530 | + 2.3 | -37 | 0.0493 | + 8.9 | -47 |
| 0.80 | 0.0569 | -17.5 | -20 | 0.0537 | - 8.7 | -28 | 0.0504 | - 0.9 | -36 | 0.0471 | + 6.4 | -46 |
| 0.84 | 0.0534 | -22.8 | -22 | 0.0507 | -13.1 | -28 | 0.0478 | - 4.4 | -36 | 0.0450 | + 3.6 | -45 |
| 0.88 | 0.0499 | -28.5 | -24 | 0.0477 | -17.9 | -30 | 0.0453 | - 8.5 | -37 | 0.0429 | + 0.3 | -45 |
| 0.92 | 0.0463 | -34.3 | -27 | 0.0446 | -23.3 | -32 | 0.0428 | -13.2 | -38 | 0.0409 | - 3.7 | -44 |
| 0.96 | 0.0423 | -40.6 | -33 | 0.0412 | -29.5 | -37 | 0.0400 | -19.0 | -41 | 0.0387 | - 8.9 | -45 |
| 1.00 | 0.0353 | +39.8 | +53 | 0.0353 | -40.2 | -53 | 0.0353 | -30.2 | -53 | 0.0353 | -20.1 | -52 |
| 0.96 | 0.0320 | +35.2 | +75 | 0.0326 | +44.0 | +72 | 0.0333 | -37.2 | -69 | 0.0342 | -28.3 | -65 |
| 0.92 | 0.0320 | +34.3 | +83 | 0.0326 | +42.7 | +80 | 0.0335 | -39.0 | -76 | 0.0345 | -30.7 | -71 |
| 0.88 | 0.0327 | +33.9 | +87 | 0.0333 | +42.0 | +84 | 0.0341 | -40.0 | -80 | 0.0352 | -32.2 | -75 |
| 0.84 | 0.0338 | +33.7 | +89 | 0.0343 | +41.6 | +87 | 0.0351 | -40.7 | -84 | 0.0362 | -33.2 | -79 |
| 0.80 | 0.0353 | +33.8 | +90 | 0.0356 | +41.5 | +89 | 0.0363 | -41.1 | -86 | 0.0374 | -34.0 | -81 |
| 0.76 | 0.0372 | +33.9 | +89 | 0.0373 | +41.5 | +90 | 0.0378 | -41.3 | -88 | 0.0389 | -34.5 | -84 |
| 0.72 | 0.0394 | +34.2 | +88 | 0.0393 | +41.6 | +90 | 0.0396 | -41.4 | -89 | 0.0406 | -34.9 | -86 |
| 0.68 | 0.0420 | +34.6 | +86 | 0.0416 | +41.8 | +89 | 0.0417 | -41.4 | -90 | 0.0425 | -35.2 | -87 |
| 0.64 | 0.0450 | +35.0 | +83 | 0.0443 | +42.1 | +88 | 0.0442 | -41.3 | -90 | 0.0448 | -35.4 | -88 |
| 0.60 | 0.0485 | +35.5 | +80 | 0.0475 | +42.5 | +86 | 0.0470 | -41.2 | -89 | 0.0474 | -35.5 | -89 |
| 0.56 | 0.0525 | +36.0 | +77 | 0.0511 | +42.9 | +84 | 0.0503 | -41.0 | -88 | 0.0504 | -35.5 | -89 |
| 0.52 | .. | .. | .. | .. | .. | .. | .. | .. | .. | .. | .. | .. |
| 0.48 | 0.0624 | +37.3 | +71 | 0.0601 | +43.9 | +80 | 0.0583 | -40.4 | -86 | 0.0577 | -35.4 | -90 |
| 0.44 | 0.0687 | +38.0 | +68 | 0.0657 | +44.5 | +77 | 0.0634 | -40.0 | -85 | 0.0624 | -35.3 | -89 |
| 0.40 | 0.0759 | +38.8 | +65 | 0.0723 | -44.8 | -75 | 0.0694 | -39.5 | -83 | 0.0678 | -35.1 | -89 |
| 0.36 | 0.0845 | +39.7 | +61 | 0.0801 | -44.2 | -72 | 0.0765 | -39.1 | -82 | 0.0743 | -34.9 | -88 |
| 0.32 | 0.0948 | +40.6 | +58 | 0.0894 | -43.4 | -69 | 0.0849 | -38.6 | -80 | 0.0821 | -34.6 | -87 |
| 0.28 | 0.1072 | +41.6 | +55 | 0.1008 | -42.6 | -67 | 0.0953 | -38.0 | -78 | 0.0915 | -34.3 | -86 |
| 0.24 | 0.1226 | +42.7 | +52 | 0.1148 | -41.8 | -64 | 0.1080 | -37.4 | -76 | 0.1032 | -34.0 | -85 |
| 0.20 | 0.1418 | +43.8 | +49 | 0.1324 | -40.9 | -62 | 0.1240 | -36.7 | -74 | 0.1179 | -33.6 | -84 |
| 0.16 | 0.1663 | -44.8 | -47 | 0.1548 | -39.9 | -59 | 0.1444 | -36.0 | -72 | 0.1366 | -33.2 | -83 |
| 0.12 | 0.1978 | -43.7 | -44 | 0.1836 | -38.9 | -57 | 0.1707 | -35.3 | -70 | 0.1607 | -32.7 | -81 |
| 0.08 | 0.2364 | -42.3 | -42 | 0.2190 | -37.8 | -55 | 0.2029 | -34.5 | -68 | 0.1903 | -32.2 | -80 |
| 0.04 | 0.2695 | -40.9 | -40 | 0.2491 | -36.7 | -52 | 0.2302 | -33.7 | -66 | 0.2150 | -31.7 | -78 |
| 0 | 0.2559 | -40.0 | -38 | 0.2359 | -36.0 | -50 | 0.2170 | -33.2 | -64 | 0.2017 | -31.4 | -77 |

TABLE 5 (continued)

| $\mu$ | $\mu_o = 0.52; \varphi = 80°$ | | | $\mu_o = 0.52; \varphi = 90°$ | | | $\mu_o = 0.64; \varphi = 0°$ | | $\mu_o = 0.64; \varphi = 10°$ | | |
|---|---|---|---|---|---|---|---|---|---|---|---|
| | $I_1+I_r$ | $\chi°$ | $\delta$ | $I_1+I_r$ | $\chi°$ | $\delta$ | $I_1+I_r$ | $\delta$ | $I_1+I_r$ | $\chi°$ | $\delta$ |
| 0 | 0.1916 | +30.4 | -87 | 0.1882 | +30.1 | -90 | 0.2931 | +20 | 0.2902 | -13.9 | +22 |
| 0.04 | 0.2076 | +29.9 | -85 | 0.2029 | +29.8 | -89 | 0.3168 | +17 | 0.3136 | -15.5 | +19 |
| 0.08 | 0.1867 | +29.5 | -84 | 0.1815 | +29.8 | -89 | 0.2880 | +14 | 0.2850 | -16.8 | +16 |
| 0.12 | 0.1601 | +29.1 | -83 | 0.1549 | +29.7 | -89 | 0.2502 | +12 | 0.2477 | -18.3 | +14 |
| 0.16 | 0.1382 | +28.7 | -82 | 0.1331 | +29.6 | -89 | 0.2187 | +10 | 0.2164 | -20.2 | +12 |
| 0.20 | 0.1211 | +28.3 | -81 | 0.1161 | +29.4 | -88 | 0.1937 | + 8 | 0.1918 | -22.5 | +10 |
| 0.24 | 0.1077 | +27.8 | -79 | 0.1027 | +29.3 | -88 | 0.1739 | + 6 | 0.1721 | -25.5 | + 8 |
| 0.28 | 0.0969 | +27.3 | -78 | 0.0921 | +29.0 | -87 | 0.1579 | + 4 | 0.1563 | -29.3 | + 7 |
| 0.32 | 0.0882 | +26.7 | -76 | 0.0835 | +28.8 | -86 | 0.1447 | + 3 | 0.1432 | -34.3 | + 6 |
| 0.36 | 0.0810 | +26.1 | -75 | 0.0764 | +28.5 | -85 | 0.1336 | + 1 | 0.1323 | -40.6 | + 5 |
| 0.40 | 0.0749 | +25.4 | -73 | 0.0705 | +28.1 | -84 | 0.1241 | 0 | 0.1229 | +41.8 | - 4 |
| 0.44 | 0.0698 | +24.6 | -72 | 0.0654 | +27.7 | -82 | 0.1158 | - 1 | 0.1147 | +33.6 | - 4 |
| 0.48 | 0.0653 | +23.8 | -70 | 0.0611 | +27.2 | -81 | 0.1086 | - 2 | 0.1075 | +25.7 | - 4 |
| 0.52 | .. | .. | .. | .. | .. | .. | 0.1021 | - 2 | 0.1011 | +19.0 | - 4 |
| 0.56 | 0.0581 | +22.0 | -67 | 0.0542 | +26.1 | -78 | 0.0961 | - 3 | 0.0953 | +13.2 | - 4 |
| 0.60 | 0.0551 | +20.9 | -65 | 0.0514 | +25.4 | -76 | 0.0907 | - 3 | 0.0900 | + 8.3 | - 4 |
| 0.64 | 0.0524 | +19.7 | -63 | 0.0489 | +24.7 | -74 | .. | .. | .. | .. | .. |
| 0.68 | 0.0500 | +18.4 | -61 | 0.0466 | +23.8 | -72 | 0.0810 | - 3 | 0.0804 | - 2.0 | - 3 |
| 0.72 | 0.0478 | +17.0 | -59 | 0.0447 | +22.8 | -70 | 0.0765 | - 2 | 0.0759 | - 9.1 | - 3 |
| 0.76 | 0.0458 | +15.3 | -58 | 0.0429 | +21.6 | -68 | 0.0722 | - 1 | 0.0717 | -20.2 | - 3 |
| 0.80 | 0.0440 | +13.4 | -56 | 0.0413 | +20.2 | -66 | 0.0679 | 0 | 0.0675 | -38.0 | - 3 |
| 0.84 | 0.0423 | +11.2 | -54 | 0.0398 | +18.6 | -63 | 0.0637 | + 2 | 0.0633 | +34.8 | + 4 |
| 0.88 | 0.0406 | + 8.5 | -53 | 0.0385 | +16.5 | -61 | 0.0594 | + 5 | 0.0591 | +24.2 | + 6 |
| 0.92 | 0.0391 | + 5.3 | -51 | 0.0373 | +13.9 | -58 | 0.0548 | + 9 | 0.0546 | +18.1 | +10 |
| 0.96 | 0.0375 | + 0.8 | -50 | 0.0363 | +10.1 | -55 | 0.0496 | +16 | 0.0495 | +14.1 | +17 |
| 1.00 | 0.0353 | -10.1 | -52 | 0.0353 | 0 | -52 | 0.0400 | +39 | 0.0400 | + 9.9 | +39 |
| 0.96 | 0.0352 | -19.3 | -60 | 0.0363 | +10.1 | -55 | 0.0344 | +68 | 0.0345 | + 8.3 | +67 |
| 0.92 | 0.0358 | -22.4 | -65 | 0.0373 | +13.9 | -58 | 0.0337 | +78 | 0.0337 | + 8.0 | +78 |
| 0.88 | 0.0367 | -24.4 | -68 | 0.0385 | +16.5 | -61 | 0.0338 | +85 | 0.0339 | + 7.9 | +84 |
| 0.84 | 0.0378 | -25.9 | -72 | 0.0398 | +18.6 | -63 | 0.0346 | +88 | 0.0347 | + 7.8 | +88 |
| 0.80 | 0.0391 | -27.0 | -74 | 0.0413 | +20.2 | -66 | 0.0359 | +90 | 0.0360 | + 7.8 | +90 |
| 0.76 | 0.0406 | -28.0 | -77 | 0.0429 | +21.6 | -68 | 0.0377 | +90 | 0.0377 | + 7.8 | +90 |
| 0.72 | 0.0422 | -28.7 | -79 | 0.0447 | +22.8 | -70 | 0.0399 | +88 | 0.0398 | + 7.9 | +88 |
| 0.68 | 0.0441 | -29.3 | -81 | 0.0466 | +23.8 | -72 | 0.0425 | +86 | 0.0424 | + 7.9 | +86 |
| 0.64 | 0.0463 | -29.8 | -83 | 0.0489 | +24.7 | -74 | .. | .. | .. | .. | .. |
| 0.60 | 0.0488 | -30.2 | -85 | 0.0514 | +25.4 | -76 | 0.0493 | +79 | 0.0492 | + 8.2 | +80 |
| 0.56 | 0.0516 | -30.6 | -86 | 0.0542 | +26.1 | -78 | 0.0537 | +74 | 0.0535 | + 8.3 | +75 |
| 0.52 | .. | .. | .. | .. | .. | .. | 0.0587 | +70 | 0.0584 | + 8.5 | +71 |
| 0.48 | 0.0586 | -31.1 | -88 | 0.0611 | +27.2 | -81 | 0.0646 | +66 | 0.0642 | + 8.7 | +67 |
| 0.44 | 0.0630 | -31.2 | -89 | 0.0654 | +27.7 | -82 | 0.0714 | +61 | 0.0710 | + 8.9 | +62 |
| 0.40 | 0.0681 | -31.3 | -89 | 0.0705 | +28.1 | -84 | 0.0795 | +56 | 0.0790 | + 9.2 | +58 |
| 0.36 | 0.0742 | -31.4 | -90 | 0.0764 | +28.5 | -85 | 0.0891 | +52 | 0.0885 | + 9.5 | +53 |
| 0.32 | 0.0815 | -31.4 | -90 | 0.0835 | +28.8 | -86 | 0.1006 | +48 | 0.0999 | + 9.8 | +49 |
| 0.28 | 0.0903 | -31.3 | -90 | 0.0921 | +29.0 | -87 | 0.1147 | +44 | 0.1137 | +10.1 | +45 |
| 0.24 | 0.1013 | -31.3 | -90 | 0.1027 | +29.3 | -88 | 0.1320 | +39 | 0.1309 | +10.5 | +41 |
| 0.20 | 0.1150 | -31.2 | -89 | 0.1161 | +29.4 | -88 | 0.1538 | +36 | 0.1524 | +11.0 | +37 |
| 0.16 | 0.1326 | -31.0 | -89 | 0.1331 | +29.6 | -89 | 0.1817 | +32 | 0.1800 | +11.5 | +34 |
| 0.12 | 0.1552 | -30.9 | -88 | 0.1549 | +29.7 | -89 | 0.2176 | +29 | 0.2156 | +12.1 | +30 |
| 0.08 | 0.1828 | -30.7 | -88 | 0.1815 | +29.8 | -89 | 0.2624 | +25 | 0.2598 | +12.8 | +27 |
| 0.04 | 0.2055 | -30.4 | -87 | 0.2029 | +29.8 | -89 | 0.3024 | +22 | 0.2994 | +13.5 | +24 |
| 0 | 0.1916 | -30.4 | -87 | 0.1882 | +30.1 | -90 | 0.2931 | +20 | 0.2902 | +13.9 | +22 |

TABLE 5 (continued)

| μ | μ₀ = 0.64; φ = 20° | | | μ₀ = 0.64; φ = 30° | | | μ₀ = 0.64; φ = 40° | | | μ₀ = 0.64; φ = 50° | | |
|---|---|---|---|---|---|---|---|---|---|---|---|---|
| | $I_1+I_r$ | $\chi°$ | $\delta$ | $I_1+I_r$ | $\chi°$ | $\delta$ | $I_1+I_r$ | $\chi°$ | $\delta$ | $I_1+I_r$ | $\chi°$ | $\delta$ |
| 0 | 0.2816 | −25.4 | +27 | 0.2686 | −34.0 | +34 | 0.2525 | −40.2 | +44 | 0.2354 | −44.6 | +56 |
| 0.04 | 0.3044 | −27.5 | +24 | 0.2903 | −36.0 | +31 | 0.2729 | −41.9 | +41 | 0.2542 | +44.0 | −53 |
| 0.08 | 0.2766 | −29.2 | +21 | 0.2636 | −37.6 | +29 | 0.2476 | −43.3 | +39 | 0.2302 | +42.9 | −50 |
| 0.12 | 0.2403 | −31.1 | +19 | 0.2289 | −39.3 | +27 | 0.2148 | −44.7 | +36 | 0.1995 | +41.7 | −48 |
| 0.16 | 0.2100 | −33.2 | +17 | 0.2000 | −41.2 | +25 | 0.1875 | +43.8 | −34 | 0.1740 | +40.5 | −45 |
| 0.20 | 0.1860 | −35.6 | +15 | 0.1772 | −43.2 | +23 | 0.1661 | +42.1 | −32 | 0.1540 | +39.2 | −48 |
| 0.24 | 0.1670 | −38.4 | +14 | 0.1591 | +44.6 | −21 | 0.1491 | +40.3 | −30 | 0.1382 | +37.8 | −41 |
| 0.28 | 0.1517 | −41.5 | +12 | 0.1445 | +42.1 | −19 | 0.1355 | +38.4 | −28 | 0.1255 | +36.3 | −39 |
| 0.32 | 0.1390 | +44.9 | −11 | 0.1325 | +39.5 | −18 | 0.1242 | +36.4 | −26 | 0.1151 | +34.7 | −37 |
| 0.36 | 0.1284 | +41.0 | −10 | 0.1224 | +36.6 | −16 | 0.1149 | +34.2 | −25 | 0.1064 | +33.0 | −35 |
| 0.40 | 0.1193 | +36.7 | −9 | 0.1138 | +33.5 | −15 | 0.1069 | +31.8 | −23 | 0.0991 | +31.2 | −33 |
| 0.44 | 0.1115 | +32.1 | −8 | 0.1064 | +30.2 | −14 | 0.1000 | +29.2 | −22 | 0.0928 | +29.2 | −31 |
| 0.48 | 0.1046 | +27.3 | −8 | 0.0999 | +26.7 | −13 | 0.0940 | +26.5 | −21 | 0.0873 | +27.0 | −30 |
| 0.52 | 0.0984 | +22.4 | −7 | 0.0941 | +22.9 | −12 | 0.0886 | +23.5 | −19 | 0.0824 | +24.7 | −28 |
| 0.56 | 0.0928 | +17.3 | −7 | 0.0888 | +18.8 | −12 | 0.0838 | +20.3 | −18 | 0.0781 | +22.2 | −27 |
| 0.60 | 0.0877 | +12.2 | −6 | 0.0840 | +14.6 | −11 | 0.0794 | +16.9 | −17 | 0.0741 | +19.6 | −26 |
| 0.64 | .. | .. | .. | .. | .. | .. | .. | .. | .. | .. | .. | .. |
| 0.68 | 0.0785 | +0.5 | −6 | 0.0755 | +4.6 | −10 | 0.0716 | +9.0 | −16 | 0.0672 | +13.2 | −23 |
| 0.72 | 0.0742 | −6.5 | −6 | 0.0715 | −1.1 | −10 | 0.0681 | +4.3 | −15 | 0.0641 | +9.6 | −22 |
| 0.76 | 0.0702 | −14.9 | −6 | 0.0678 | −7.6 | −10 | 0.0647 | −0.8 | −15 | 0.0611 | +5.5 | −22 |
| 0.80 | 0.0662 | −25.2 | −6 | 0.0641 | −15.1 | −10 | 0.0614 | −6.6 | −15 | 0.0582 | +0.9 | −21 |
| 0.84 | 0.0622 | −36.3 | −7 | 0.0604 | −23.4 | −10 | 0.0581 | −13.1 | −15 | 0.0554 | −4.3 | −21 |
| 0.88 | 0.0582 | +43.2 | +8 | 0.0567 | −32.2 | −12 | 0.0548 | −20.4 | −16 | 0.0525 | −10.4 | −21 |
| 0.92 | 0.0539 | +34.7 | +12 | 0.0528 | −40.9 | −14 | 0.0513 | −28.4 | −18 | 0.0495 | −17.4 | −22 |
| 0.96 | 0.0490 | +27.7 | +18 | 0.0483 | +40.7 | +20 | 0.0473 | −37.1 | −22 | 0.0462 | −25.7 | −25 |
| 1.00 | 0.0400 | +19.8 | +39 | 0.0400 | +29.7 | +39 | 0.0400 | +39.7 | +39 | 0.0400 | −40.3 | −39 |
| 0.96 | 0.0347 | +16.6 | +66 | 0.0351 | +25.0 | +64 | 0.0357 | +33.3 | +62 | 0.0364 | +41.7 | +59 |
| 0.92 | 0.0340 | +16.0 | +77 | 0.0344 | +23.9 | +75 | 0.0351 | +31.8 | +72 | 0.0359 | +39.7 | +68 |
| 0.88 | 0.0341 | +15.7 | +83 | 0.0346 | +23.4 | +81 | 0.0352 | +31.0 | +78 | 0.0361 | +38.5 | +74 |
| 0.84 | 0.0349 | +15.5 | +87 | 0.0352 | +23.1 | +86 | 0.0358 | +30.5 | +83 | 0.0367 | +37.8 | +80 |
| 0.80 | 0.0361 | +15.5 | +89 | 0.0364 | +23.0 | +89 | 0.0368 | +30.3 | +87 | 0.0376 | +37.3 | +83 |
| 0.76 | 0.0377 | +15.5 | +90 | 0.0379 | +23.0 | +90 | 0.0382 | +30.2 | +89 | 0.0389 | +37.0 | +86 |
| 0.72 | 0.0398 | +15.6 | +89 | 0.0398 | +23.0 | +90 | 0.0400 | +30.1 | +90 | 0.0405 | +36.9 | +88 |
| 0.68 | 0.0423 | +15.7 | +87 | 0.0421 | +23.2 | +89 | 0.0421 | +30.2 | +90 | 0.0425 | +36.8 | +89 |
| 0.64 | .. | .. | .. | .. | .. | .. | .. | .. | .. | .. | .. | .. |
| 0.60 | 0.0488 | +16.1 | +82 | 0.0482 | +23.6 | +85 | 0.0477 | +30.5 | +88 | 0.0475 | +36.9 | +90 |
| 0.56 | 0.0529 | +16.4 | +78 | 0.0520 | +23.9 | +82 | 0.0513 | +30.8 | +86 | 0.0508 | +37.1 | +89 |
| 0.52 | 0.0577 | +16.7 | +74 | 0.0566 | +24.3 | +79 | 0.0554 | +31.2 | +84 | 0.0546 | +37.3 | +88 |
| 0.48 | 0.0632 | +17.0 | +70 | 0.0618 | +24.7 | +75 | 0.0603 | +31.5 | +81 | 0.0590 | +37.6 | +86 |
| 0.44 | 0.0698 | +17.4 | +66 | 0.0680 | +25.1 | +72 | 0.0660 | +32.0 | +78 | 0.0642 | +37.9 | +84 |
| 0.40 | 0.0775 | +17.9 | +62 | 0.0752 | +25.7 | +68 | 0.0727 | +32.5 | +75 | 0.0704 | +38.3 | +82 |
| 0.36 | 0.0866 | +18.3 | +58 | 0.0839 | +26.2 | +64 | 0.0808 | +33.0 | +72 | 0.0778 | +38.7 | +80 |
| 0.32 | 0.0976 | +18.9 | +54 | 0.0943 | +26.9 | +60 | 0.0905 | +33.6 | +69 | 0.0868 | +39.2 | +77 |
| 0.28 | 0.1111 | +19.5 | +50 | 0.1070 | +27.5 | +57 | 0.1023 | +34.3 | +65 | 0.0977 | +39.8 | +75 |
| 0.24 | 0.1276 | +20.1 | +46 | 0.1228 | +28.3 | +53 | 0.1170 | +35.0 | +62 | 0.1112 | +40.4 | +72 |
| 0.20 | 0.1485 | +20.9 | +42 | 0.1426 | +29.1 | +49 | 0.1355 | +35.8 | +59 | 0.1283 | +41.0 | +69 |
| 0.16 | 0.1752 | +21.7 | +39 | 0.1679 | +30.0 | +46 | 0.1592 | +36.6 | +56 | 0.1502 | +41.7 | +66 |
| 0.12 | 0.2097 | +22.6 | +35 | 0.2007 | +31.0 | +43 | 0.1898 | +37.5 | +53 | 0.1786 | +42.4 | +64 |
| 0.08 | 0.2525 | +23.7 | +32 | 0.2414 | +32.1 | +40 | 0.2279 | +38.5 | +50 | 0.2138 | +43.3 | +61 |
| 0.04 | 0.2909 | +24.8 | +29 | 0.2778 | +33.3 | +37 | 0.2618 | +39.5 | +47 | 0.2449 | +44.1 | +58 |
| 0 | 0.2816 | +25.4 | +27 | 0.2686 | +34.0 | +34 | 0.2525 | +40.2 | +44 | 0.2354 | +44.6 | +56 |

TABLE 5 (continued)

| μ | $\mu_o = 0.64; \varphi = 60°$ | | | $\mu_o = 0.64; \varphi = 70°$ | | | $\mu_o = 0.64; \varphi = 80°$ | | | $\mu_o = 0.64; \varphi = 90°$ | | |
|---|---|---|---|---|---|---|---|---|---|---|---|---|
| | $I_1+I_r$ | $\chi°$ | $\delta$ | $I_1+I_r$ | $\chi°$ | $\delta$ | $I_1+I_r$ | $\chi°$ | $\delta$ | $I_1+I_r$ | $\chi°$ | $\delta$ |
| 0 | 0.2194 | +42.2 | -68 | 0.2063 | +40.2 | -79 | 0.1978 | +39.0 | -87 | 0.1948 | +38.6 | -90 |
| 0.04 | 0.2365 | +41.2 | -65 | 0.2217 | +39.4 | -77 | 0.2116 | +38.5 | -85 | 0.2073 | +38.3 | -89 |
| 0.08 | 0.2136 | +40.3 | -63 | 0.1996 | +38.8 | -75 | 0.1895 | +38.2 | -84 | 0.1846 | +38.3 | -89 |
| 0.12 | 0.1847 | +39.5 | -60 | 0.1719 | +38.2 | -73 | 0.1626 | +37.8 | -83 | 0.1574 | +38.2 | -89 |
| 0.16 | 0.1608 | +38.6 | -58 | 0.1492 | +37.6 | -71 | 0.1405 | +37.4 | -81 | 0.1353 | +38.0 | -88 |
| 0.20 | 0.1421 | +37.6 | -56 | 0.1315 | +36.9 | -68 | 0.1233 | +37.0 | -80 | 0.1182 | +37.9 | -87 |
| 0.24 | 0.1273 | +36.5 | -53 | 0.1176 | +36.1 | -66 | 0.1099 | +36.5 | -78 | 0.1048 | +37.7 | -86 |
| 0.28 | 0.1155 | +35.4 | -51 | 0.1065 | +35.3 | -64 | 0.0992 | +36.0 | -76 | 0.0942 | +37.4 | -85 |
| 0.32 | 0.1059 | +34.2 | -49 | 0.0975 | +34.4 | -62 | 0.0906 | +35.4 | -74 | 0.0856 | +37.1 | -84 |
| 0.36 | 0.0979 | +32.8 | -47 | 0.0900 | +33.5 | -60 | 0.0835 | +34.8 | -72 | 0.0786 | +36.7 | -82 |
| 0.40 | 0.0912 | +31.4 | -45 | 0.0838 | +32.4 | -57 | 0.0775 | +34.1 | -70 | 0.0728 | +36.3 | -80 |
| 0.44 | 0.0854 | +29.9 | -43 | 0.0785 | +31.3 | -55 | 0.0725 | +33.3 | -67 | 0.0679 | +35.8 | -78 |
| 0.48 | 0.0804 | +28.3 | -41 | 0.0739 | +30.1 | -53 | 0.0682 | +32.4 | -65 | 0.0637 | +35.3 | -76 |
| 0.52 | 0.0760 | +26.5 | -39 | 0.0699 | +28.7 | -51 | 0.0645 | +31.4 | -63 | 0.0602 | +34.6 | -74 |
| 0.56 | 0.0721 | +24.5 | -37 | 0.0664 | +27.3 | -49 | 0.0613 | +30.4 | -61 | 0.0570 | +33.9 | -72 |
| 0.60 | 0.0686 | +22.5 | -35 | 0.0633 | +25.7 | -46 | 0.0584 | +29.2 | -58 | 0.0544 | +33.1 | -69 |
| 0.64 | .. | .. | .. | .. | .. | .. | .. | .. | .. | .. | .. | .. |
| 0.68 | 0.0625 | +17.5 | -32 | 0.0579 | +21.9 | -42 | 0.0536 | +26.4 | -53 | 0.0500 | +31.1 | -64 |
| 0.72 | 0.0598 | +14.7 | -31 | 0.0556 | +19.7 | -40 | 0.0516 | +24.7 | -51 | 0.0481 | +29.9 | -61 |
| 0.76 | 0.0573 | +11.5 | -30 | 0.0534 | +17.2 | -39 | 0.0498 | +22.8 | -48 | 0.0465 | +28.4 | -58 |
| 0.80 | 0.0548 | + 7.8 | -28 | 0.0514 | +14.3 | -37 | 0.0481 | +20.5 | -46 | 0.0451 | +26.7 | -55 |
| 0.84 | 0.0524 | + 3.6 | -28 | 0.0494 | +10.9 | -35 | 0.0465 | +17.9 | -44 | 0.0438 | +24.7 | -52 |
| 0.88 | 0.0501 | - 1.4 | -27 | 0.0475 | + 6.8 | -34 | 0.0450 | +14.6 | -41 | 0.0427 | +22.1 | -49 |
| 0.92 | 0.0476 | - 7.4 | -27 | 0.0456 | + 1.8 | -33 | 0.0436 | +10.4 | -39 | 0.0417 | +18.7 | -45 |
| 0.96 | 0.0449 | -15.0 | -29 | 0.0435 | - 5.0 | -33 | 0.0421 | + 4.6 | -37 | 0.0408 | +13.8 | -42 |
| 1.00 | 0.0400 | -30.3 | -38 | 0.0400 | -20.2 | -38 | 0.0400 | -10.1 | -38 | 0.0400 | 0 | -38 |
| 0.96 | 0.0373 | -39.9 | -55 | 0.0383 | -31.4 | -51 | 0.0395 | -22.7 | -46 | 0.0408 | +13.8 | -42 |
| 0.92 | 0.0370 | -42.5 | -63 | 0.0384 | -34.7 | -58 | 0.0399 | -26.8 | -52 | 0.0417 | +18.7 | -45 |
| 0.88 | 0.0373 | -44.1 | -69 | 0.0388 | -36.8 | -63 | 0.0406 | -29.5 | -56 | 0.0427 | +22.1 | -49 |
| 0.84 | 0.0379 | +44.8 | +74 | 0.0395 | -38.3 | -68 | 0.0415 | -31.5 | -60 | 0.0438 | +24.7 | -52 |
| 0.80 | 0.0388 | +44.1 | +79 | 0.0404 | -39.4 | -72 | 0.0425 | -33.0 | -64 | 0.0451 | +26.7 | -55 |
| 0.76 | 0.0400 | +43.6 | +82 | 0.0416 | -40.2 | -76 | 0.0438 | -34.2 | -68 | 0.0465 | +28.4 | -58 |
| 0.72 | 0.0415 | +43.2 | +85 | 0.0430 | -40.9 | -79 | 0.0453 | -35.2 | -71 | 0.0481 | +29.9 | -61 |
| 0.68 | 0.0433 | +42.9 | +87 | 0.0447 | -41.4 | -82 | 0.0470 | -36.1 | -74 | 0.0500 | +31.1 | -64 |
| 0.64 | .. | .. | .. | .. | .. | .. | .. | .. | .. | .. | .. | .. |
| 0.60 | 0.0479 | +42.6 | +89 | 0.0491 | -42.1 | -86 | 0.0512 | -37.4 | -79 | 0.0544 | +33.1 | -69 |
| 0.56 | 0.0509 | +42.7 | +90 | 0.0519 | -42.3 | -87 | 0.0539 | -37.9 | -81 | 0.0570 | +33.9 | -72 |
| 0.52 | 0.0544 | +42.7 | +90 | 0.0551 | -42.5 | -88 | 0.0570 | -38.3 | -83 | 0.0602 | +34.6 | -74 |
| 0.48 | 0.0584 | +42.8 | +89 | 0.0589 | -42.6 | -89 | 0.0606 | -38.6 | -85 | 0.0637 | +35.3 | -76 |
| 0.44 | 0.0632 | +43.0 | +89 | 0.0633 | -42.6 | -90 | 0.0648 | -38.9 | -86 | 0.0679 | +35.8 | -78 |
| 0.40 | 0.0689 | +43.2 | +88 | 0.0686 | -42.6 | -90 | 0.0698 | -39.1 | -87 | 0.0728 | +36.3 | -80 |
| 0.36 | 0.0757 | +43.5 | +86 | 0.0748 | -42.5 | -90 | 0.0757 | -39.3 | -88 | 0.0786 | +36.7 | -82 |
| 0.32 | 0.0839 | +43.8 | +85 | 0.0824 | -42.4 | -89 | 0.0829 | -39.4 | -89 | 0.0856 | +37.1 | -84 |
| 0.28 | 0.0939 | +44.2 | +83 | 0.0917 | -42.2 | -88 | 0.0917 | -39.5 | -89 | 0.0942 | +37.4 | -85 |
| 0.24 | 0.1063 | +44.6 | +81 | 0.1032 | -42.0 | -88 | 0.1026 | -39.5 | -90 | 0.1048 | +37.7 | -86 |
| 0.20 | 0.1221 | -44.9 | -79 | 0.1178 | -41.8 | -86 | 0.1164 | -39.4 | -90 | 0.1182 | +37.9 | -87 |
| 0.16 | 0.1423 | -44.4 | -77 | 0.1366 | -41.5 | -85 | 0.1341 | -39.4 | -89 | 0.1353 | +38.0 | -88 |
| 0.12 | 0.1684 | -43.8 | -75 | 0.1608 | -41.1 | -84 | 0.1569 | -39.3 | -89 | 0.1574 | +38.2 | -89 |
| 0.08 | 0.2008 | -43.2 | -72 | 0.1908 | -40.8 | -82 | 0.1851 | -39.1 | -88 | 0.1846 | +38.3 | -89 |
| 0.04 | 0.2293 | -42.6 | -70 | 0.2168 | -40.3 | -80 | 0.2091 | -39.0 | -88 | 0.2073 | +38.3 | -89 |
| 0 | 0.2194 | -42.2 | -68 | 0.2063 | -40.2 | -79 | 0.1978 | -39.0 | -87 | 0.1948 | +38.6 | -90 |

TABLE 5 (continued)

| μ | $\mu_o = 0.72; \varphi = 0°$ | | $\mu_o = 0.72; \varphi = 10°$ | | | $\mu_o = 0.72; \varphi = 20°$ | | | $\mu_o = 0.72; \varphi = 30°$ | | |
|---|---|---|---|---|---|---|---|---|---|---|---|
| | $I_1 + I_r$ | δ | $I_1 + I_r$ | $\chi°$ | δ | $I_1 + I_r$ | $\chi°$ | δ | $I_1 + I_r$ | $\chi°$ | δ |
| 0 | 0.2795 | +29 | 0.2771 | -10.6 | +31 | 0.2699 | -20.1 | +35 | 0.2590 | -27.7 | +42 |
| 0.04 | 0.3010 | +25 | 0.2983 | -11.5 | +27 | 0.2906 | -21.4 | +31 | 0.2787 | -29.1 | +38 |
| 0.08 | 0.2736 | +22 | 0.2711 | -12.2 | +24 | 0.2640 | -22.4 | +28 | 0.2530 | -30.2 | +35 |
| 0.12 | 0.2381 | +19 | 0.2360 | -12.9 | +21 | 0.2297 | -23.6 | +25 | 0.2200 | -31.5 | +32 |
| 0.16 | 0.2086 | +17 | 0.2067 | -13.8 | +18 | 0.2012 | -24.9 | +23 | 0.1926 | -32.8 | +30 |
| 0.20 | 0.1854 | +14 | 0.1837 | -14.9 | +16 | 0.1788 | -26.4 | +20 | 0.1711 | -34.3 | +27 |
| 0.24 | 0.1670 | +12 | 0.1654 | -16.3 | +14 | 0.1610 | -28.2 | +18 | 0.1541 | -36.0 | +25 |
| 0.28 | 0.1521 | +10 | 0.1507 | -17.9 | +12 | 0.1467 | -30.2 | +16 | 0.1404 | -37.8 | +22 |
| 0.32 | 0.1399 | + 8 | 0.1387 | -20.0 | +10 | 0.1350 | -32.6 | +14 | 0.1292 | -39.9 | +20 |
| 0.36 | 0.1297 | + 6 | 0.1285 | -22.6 | + 8 | 0.1252 | -35.3 | +12 | 0.1199 | -42.2 | +19 |
| 0.40 | 0.1210 | + 4 | 0.1199 | -26.2 | + 6 | 0.1168 | -38.5 | +11 | 0.1119 | -44.8 | +17 |
| 0.44 | 0.1184 | + 3 | 0.1125 | -31.0 | + 5 | 0.1096 | -42.3 | + 9 | 0.1051 | +42.3 | -15 |
| 0.48 | 0.1068 | + 2 | 0.1059 | -37.4 | + 4 | 0.1032 | +43.5 | - 8 | 0.0990 | +39.1 | -14 |
| 0.52 | 0.1009 | 0 | 0.1000 | +44.3 | - 3 | 0.0976 | +38.6 | - 7 | 0.0937 | +35.5 | -13 |
| 0.56 | 0.0955 | 0 | 0.0947 | +34.8 | - 3 | 0.0925 | +33.2 | - 7 | 0.0889 | +31.6 | -11 |
| 0.60 | 0.0906 | - 1 | 0.0899 | +25.5 | - 3 | 0.0878 | +27.4 | - 6 | 0.0845 | +27.3 | -10 |
| 0.64 | 0.0861 | - 2 | 0.0854 | +17.4 | - 3 | 0.0835 | +21.2 | - 6 | 0.0805 | +22.5 | -10 |
| 0.68 | 0.0818 | - 2 | 0.0812 | +10.5 | - 3 | 0.0795 | +14.7 | - 5 | 0.0767 | +17.2 | - 9 |
| 0.72 | . . | . . | . . | . . | . . | | | | . . | . . | . . |
| 0.76 | 0.0738 | - 2 | 0.0733 | - 3.2 | - 2 | 0.0719 | - 0.1 | - 5 | 0.0697 | + 4.7 | - 8 |
| 0.80 | 0.0699 | - 1 | 0.0695 | -14.9 | - 2 | 0.0683 | -10.2 | - 4 | 0.0663 | - 3.2 | - 7 |
| 0.84 | 0.0661 | 0 | 0.0657 | -35.5 | - 2 | 0.0647 | -22.7 | - 4 | 0.0630 | -12.5 | - 7 |
| 0.88 | 0.0621 | + 2 | 0.0618 | +33.0 | + 3 | 0.0609 | -37.1 | - 5 | 0.0595 | -23.3 | - 8 |
| 0.92 | 0.0579 | + 5 | 0.0576 | +21.5 | + 6 | 0.0570 | +39.9 | + 7 | 0.0559 | -35.0 | -10 |
| 0.96 | 0.0529 | +11 | 0.0528 | +15.3 | +11 | 0.0523 | +30.0 | +12 | 0.0516 | +43.7 | +13 |
| 1.00 | 0.0434 | +30 | 0.0434 | + 9.8 | +30 | 0.0434 | +19.7 | +30 | 0.0434 | +29.6 | +30 |
| 0.96 | 0.0373 | +57 | 0.0374 | + 7.9 | +57 | 0.0377 | +15.8 | +56 | 0.0381 | +23.7 | +54 |
| 0.92 | 0.0361 | +69 | 0.0362 | + 7.5 | +68 | 0.0365 | +15.0 | +67 | 0.0370 | +22.4 | +65 |
| 0.88 | 0.0358 | +77 | 0.0359 | + 7.3 | +77 | 0.0362 | +14.5 | +75 | 0.0368 | +21.7 | +73 |
| 0.84 | 0.0361 | +83 | 0.0362 | + 7.1 | +83 | 0.0365 | +14.2 | +81 | 0.0370 | +21.2 | +79 |
| 0.80 | 0.0370 | +87 | 0.0370 | + 7.1 | +87 | 0.0373 | +14.1 | +86 | 0.0378 | +20.9 | +84 |
| 0.76 | 0.0383 | +89 | 0.0383 | + 7.0 | +89 | 0.0385 | +14.0 | +89 | 0.0389 | +20.8 | +87 |
| 0.72 | . . | . . | . . | . . | . . | | | | . . | . . | . . |
| 0.68 | 0.0422 | +90 | 0.0422 | + 7.1 | +90 | 0.0423 | +14.0 | +90 | 0.0424 | +20.7 | +90 |
| 0.64 | 0.0448 | +89 | 0.0448 | + 7.1 | +89 | 0.0448 | +14.1 | +89 | 0.0448 | +20.7 | +90 |
| 0.60 | 0.0480 | +86 | 0.0480 | + 7.2 | +87 | 0.0478 | +14.2 | +88 | 0.0476 | +20.8 | +89 |
| 0.56 | 0.0518 | +84 | 0.0517 | + 7.3 | +84 | 0.0514 | +14.3 | +86 | 0.0510 | +21.0 | +88 |
| 0.52 | 0.0562 | +80 | 0.0561 | + 7.4 | +81 | 0.0556 | +14.5 | +83 | 0.0551 | +21.2 | +86 |
| 0.48 | 0.0615 | +76 | 0.0613 | + 7.5 | +77 | 0.0607 | +14.7 | +80 | 0.0598 | +21.4 | +83 |
| 0.44 | 0.0677 | +72 | 0.0674 | + 7.6 | +73 | 0.0666 | +14.9 | +76 | 0.0655 | +21.7 | +80 |
| 0.40 | 0.0751 | +68 | 0.0747 | + 7.8 | +69 | 0.0737 | +15.2 | +72 | 0.0722 | +22.0 | +77 |
| 0.36 | 0.0839 | +64 | 0.0834 | + 7.9 | +65 | 0.0821 | +15.5 | +68 | 0.0802 | +22.4 | +73 |
| 0.32 | 0.0946 | +60 | 0.0940 | + 8.1 | +61 | 0.0924 | +15.8 | +64 | 0.0900 | +22.8 | +70 |
| 0.28 | 0.1076 | +55 | 0.1069 | + 8.4 | +57 | 0.1049 | +16.2 | +60 | 0.1019 | +23.3 | +66 |
| 0.24 | 0.1238 | +51 | 0.1229 | + 8.6 | +52 | 0.1204 | +16.7 | +56 | 0.1167 | +23.8 | +62 |
| 0.20 | 0.1442 | +47 | 0.1431 | + 8.9 | +48 | 0.1401 | +17.1 | +52 | 0.1355 | +24.3 | +59 |
| 0.16 | 0.1705 | +43 | 0.1692 | + 9.2 | +44 | 0.1654 | +17.7 | +49 | 0.1596 | +25.0 | +55 |
| 0.12 | 0.2045 | +39 | 0.2029 | + 9.6 | +41 | 0.1981 | +18.3 | +45 | 0.1910 | +25.6 | +51 |
| 0.08 | 0.2471 | +35 | 0.2451 | + 9.9 | +37 | 0.2391 | +18.9 | +41 | 0.2301 | +26.4 | +48 |
| 0.04 | 0.2861 | +32 | 0.2836 | +10.4 | +33 | 0.2766 | +19.6 | +38 | 0.2658 | +27.2 | +45 |
| 0 | 0.2795 | +29 | 0.2771 | +10.6 | +31 | 0.2699 | +20.1 | +35 | 0.2590 | +27.7 | +42 |

TABLE 5 (continued)

| μ | $\mu_o = 0.72; \varphi = 40°$ | | | $\mu_o = 0.72; \varphi = 50°$ | | | $\mu_o = 0.72; \varphi = 60°$ | | | $\mu_o = 0.72; \varphi = 70°$ | | |
|---|---|---|---|---|---|---|---|---|---|---|---|---|
| | $I_1{+}I_r$ | $\chi°$ | $\delta$ | $I_1{+}I_r$ | $\chi°$ | $\delta$ | $I_1{+}I_r$ | $\chi°$ | $\delta$ | $I_1{+}I_r$ | $\chi°$ | $\delta$ |
| 0 | 0.2456 | -33.6 | +51 | 0.2313 | -38.0 | +61 | 0.2179 | -41.2 | +72 | 0.2069 | -43.4 | +81 |
| 0.04 | 0.2641 | -34.9 | +47 | 0.2484 | -39.2 | +57 | 0.2335 | -42.1 | +68 | 0.2210 | -44.0 | +78 |
| 0.08 | 0.2394 | -36.0 | +44 | 0.2248 | -40.0 | +54 | 0.2106 | -42.8 | +66 | 0.1986 | -44.5 | +76 |
| 0.12 | 0.2080 | -37.1 | +41 | 0.1949 | -41.0 | +52 | 0.1822 | -43.5 | +63 | 0.1711 | +45.0 | -74 |
| 0.16 | 0.1820 | -38.3 | +38 | 0.1708 | -42.0 | +49 | 0.1588 | -44.3 | +60 | 0.1487 | +44.4 | -72 |
| 0.20 | 0.1616 | -39.6 | +36 | 0.1510 | -43.1 | +46 | 0.1406 | +44.8 | -57 | 0.1312 | +43.8 | -69 |
| 0.24 | 0.1455 | -41.0 | +33 | 0.1359 | -44.2 | +43 | 0.1263 | +43.9 | -55 | 0.1176 | +43.1 | -66 |
| 0.28 | 0.1325 | -42.6 | +31 | 0.1237 | +44.5 | -41 | 0.1149 | +42.9 | -52 | 0.1067 | +42.3 | -64 |
| 0.32 | 0.1220 | -44.4 | +29 | 0.1139 | +43.1 | -38 | 0.1056 | +41.8 | -49 | 0.0980 | +41.5 | -61 |
| 0.36 | 0.1132 | +43.8 | -26 | 0.1056 | +41.5 | -36 | 0.0980 | +40.6 | -47 | 0.0907 | +40.6 | -59 |
| 0.40 | 0.1057 | +41.7 | -24 | 0.0987 | +39.9 | -34 | 0.0915 | +39.3 | -44 | 0.0847 | +39.6 | -56 |
| 0.44 | 0.0993 | +39.4 | -23 | 0.0928 | +38.0 | -32 | 0.0860 | +37.9 | -42 | 0.0796 | +38.5 | -53 |
| 0.48 | 0.0937 | +36.9 | -21 | 0.0876 | +36.1 | -29 | 0.0813 | +36.3 | -40 | 0.0752 | +37.4 | -51 |
| 0.52 | 0.0887 | +34.1 | -19 | 0.0831 | +33.9 | -27 | 0.0772 | +34.6 | -37 | 0.0714 | +36.0 | -48 |
| 0.56 | 0.0843 | +31.0 | -18 | 0.0790 | +31.5 | -26 | 0.0735 | +32.7 | -35 | 0.0681 | +34.6 | -45 |
| 0.60 | 0.0803 | +27.7 | -16 | 0.0754 | +28.8 | -24 | 0.0703 | +30.6 | -33 | 0.0652 | +33.0 | -43 |
| 0.64 | 0.0766 | +23.9 | -15 | 0.0721 | +25.8 | -22 | 0.0673 | +28.3 | -31 | 0.0625 | +31.1 | -40 |
| 0.68 | 0.0732 | +19.6 | -14 | 0.0690 | +22.5 | -21 | 0.0646 | +25.6 | -29 | 0.0602 | +29.1 | -38 |
| 0.72 | .. | .. | | .. | .. | | .. | .. | | .. | .. | |
| 0.76 | 0.0668 | + 9.7 | -12 | 0.0634 | +14.5 | -18 | 0.0597 | +19.3 | -25 | 0.0560 | +24.1 | -33 |
| 0.80 | 0.0637 | + 3.4 | -12 | 0.0607 | + 9.5 | -17 | 0.0575 | +15.3 | -24 | 0.0541 | +21.0 | -31 |
| 0.84 | 0.0607 | - 3.9 | -12 | 0.0581 | + 3.8 | -17 | 0.0553 | +10.7 | -22 | 0.0523 | +17.3 | -29 |
| 0.88 | 0.0577 | -12.4 | -12 | 0.0555 | - 3.1 | -16 | 0.0531 | + 5.1 | -21 | 0.0506 | +12.7 | -27 |
| 0.92 | 0.0544 | -22.4 | -13 | 0.0527 | -11.6 | -16 | 0.0508 | - 2.0 | -21 | 0.0488 | + 6.8 | -26 |
| 0.96 | 0.0507 | -33.6 | -16 | 0.0495 | -22.0 | -18 | 0.0483 | -11.3 | -21 | 0.0469 | - 1.3 | -25 |
| 1.00 | 0.0434 | +39.5 | +29 | 0.0434 | -40.5 | -29 | 0.0434 | -30.4 | -29 | 0.0434 | -20.3 | -29 |
| 0.96 | 0.0387 | +31.7 | +51 | 0.0395 | +39.7 | +48 | 0.0404 | -42.2 | -45 | 0.0415 | -34.0 | -41 |
| 0.92 | 0.0377 | +29.8 | +62 | 0.0387 | +37.2 | +58 | 0.0399 | +44.6 | +53 | 0.0414 | -38.0 | -48 |
| 0.88 | 0.0375 | +28.7 | +70 | 0.0386 | +35.7 | +65 | 0.0400 | +42.6 | +60 | 0.0416 | -40.5 | -54 |
| 0.84 | 0.0378 | +28.1 | +76 | 0.0389 | +34.8 | +71 | 0.0403 | +41.3 | +66 | 0.0421 | -42.3 | -59 |
| 0.80 | 0.0385 | +27.6 | +81 | 0.0395 | +34.1 | +77 | 0.0410 | +40.3 | +71 | 0.0428 | -43.6 | -64 |
| 0.76 | 0.0396 | +27.3 | +85 | 0.0405 | +33.6 | +81 | 0.0420 | +39.6 | +75 | 0.0438 | -44.7 | -68 |
| 0.72 | .. | .. | | .. | .. | | .. | .. | | .. | .. | |
| 0.68 | 0.0428 | +27.1 | +89 | 0.0436 | +33.1 | +87 | 0.0448 | +38.6 | +82 | 0.0466 | +43.8 | +76 |
| 0.64 | 0.0450 | +27.0 | +90 | 0.0456 | +32.9 | +88 | 0.0467 | +38.3 | +85 | 0.0484 | +43.3 | +79 |
| 0.60 | 0.0477 | +27.1 | +90 | 0.0480 | +32.8 | +89 | 0.0490 | +38.1 | +87 | 0.0506 | +42.8 | +82 |
| 0.56 | 0.0508 | +27.2 | +89 | 0.0509 | +32.8 | +90 | 0.0517 | +37.9 | +88 | 0.0531 | +42.5 | +84 |
| 0.52 | 0.0546 | +27.3 | +88 | 0.0544 | +32.9 | +90 | 0.0549 | +37.9 | +89 | 0.0562 | +42.2 | +86 |
| 0.48 | 0.0590 | +27.6 | +87 | 0.0585 | +33.0 | +89 | 0.0587 | +37.8 | +90 | 0.0597 | +42.0 | +87 |
| 0.44 | 0.0643 | +27.8 | +85 | 0.0634 | +33.2 | +88 | 0.0632 | +37.9 | +90 | 0.0640 | +41.9 | +89 |
| 0.40 | 0.0705 | +28.1 | +82 | 0.0692 | +33.4 | +87 | 0.0686 | +38.0 | +90 | 0.0690 | +41.8 | +89 |
| 0.36 | 0.0781 | +28.5 | +79 | 0.0762 | +33.7 | +85 | 0.0751 | +38.1 | +89 | 0.0751 | +41.8 | +90 |
| 0.32 | 0.0872 | +28.9 | +76 | 0.0847 | +34.0 | +83 | 0.0829 | +38.3 | +88 | 0.0824 | +41.8 | +90 |
| 0.28 | 0.0984 | +29.3 | +73 | 0.0951 | +34.4 | +80 | 0.0926 | +38.6 | +86 | 0.0915 | +41.9 | +90 |
| 0.24 | 0.1124 | +29.8 | +70 | 0.1081 | +34.8 | +78 | 0.1047 | +38.8 | +85 | 0.1028 | +42.0 | +89 |
| 0.20 | 0.1300 | +30.4 | +67 | 0.1246 | +35.3 | +75 | 0.1200 | +39.2 | +83 | 0.1172 | +42.2 | +88 |
| 0.16 | 0.1528 | +31.0 | +63 | 0.1458 | +35.8 | +72 | 0.1398 | +39.5 | +81 | 0.1356 | +42.4 | +87 |
| 0.12 | 0.1823 | +31.6 | +60 | 0.1734 | +36.4 | +69 | 0.1654 | +40.0 | +78 | 0.1596 | +42.6 | +86 |
| 0.08 | 0.2192 | +32.4 | +57 | 0.2078 | +37.0 | +66 | 0.1974 | +40.4 | +76 | 0.1895 | +42.9 | +84 |
| 0.04 | 0.2527 | +33.1 | +53 | 0.2388 | +37.6 | +63 | 0.2260 | +40.9 | +74 | 0.2159 | +43.2 | +82 |
| 0 | 0.2456 | +33.6 | +51 | 0.2313 | +38.0 | +61 | 0.2179 | +41.2 | +72 | 0.2069 | +43.4 | +81 |

TABLE 5 (continued)

| μ | $\mu_0 = 0.72; \varphi = 80°$ | | | $\mu_0 = 0.72; \varphi = 90°$ | | | $\mu_0 = 0.80; \varphi = 0°$ | | $\mu_0 = 0.80; \varphi = 10°$ | | |
|---|---|---|---|---|---|---|---|---|---|---|---|
| | $I_1+I_r$ | $\chi°$ | $\delta$ | $I_1+I_r$ | $\chi°$ | $\delta$ | $I_1+I_r$ | $\delta$ | $I_1+I_r$ | $\chi°$ | $\delta$ |
| 0 | 0.1998 | -44.6 | +88 | 0.1973 | +45.0 | -90 | 0.2614 | +41 | 0.2595 | - 8.0 | +43 |
| 0.04 | 0.2123 | +45.0 | -86 | 0.2085 | +44.8 | -90 | 0.2804 | +37 | 0.2783 | - 8.4 | +38 |
| 0.08 | 0.1898 | +44.7 | -85 | 0.1853 | +44.7 | -89 | 0.2548 | +33 | 0.2529 | - 8.8 | +35 |
| 0.12 | 0.1628 | +44.4 | -83 | 0.1580 | +44.6 | -89 | 0.2221 | +30 | 0.2204 | - 9.2 | +31 |
| 0.16 | 0.1408 | +44.0 | -81 | 0.1359 | +44.4 | -88 | 0.1950 | +27 | 0.1935 | - 9.6 | +28 |
| 0.20 | 0.1238 | +43.6 | -79 | 0.1188 | +44.3 | -87 | 0.1738 | +24 | 0.1724 | -10.1 | +25 |
| 0.24 | 0.1105 | +43.2 | -77 | 0.1055 | +44.0 | -85 | 0.1570 | +21 | 0.1558 | -10.8 | +22 |
| 0.28 | 0.1000 | +42.7 | -75 | 0.0950 | +43.8 | -84 | 0.1436 | +18 | 0.1424 | -11.5 | +19 |
| 0.32 | 0.0915 | +42.1 | -73 | 0.0866 | +43.4 | -82 | 0.1326 | +15 | 0.1315 | -12.3 | +17 |
| 0.36 | 0.0845 | +41.5 | -70 | 0.0797 | +43.0 | -80 | 0.1234 | +13 | 0.1224 | -13.4 | +14 |
| 0.40 | 0.0788 | +40.8 | -68 | 0.0740 | +42.6 | -78 | 0.1156 | +11 | 0.1147 | -14.7 | +12 |
| 0.44 | 0.0739 | +40.0 | -65 | 0.0693 | +42.1 | -75 | 0.1089 | + 9 | 0.1081 | -16.3 | +10 |
| 0.48 | 0.0698 | +39.1 | -62 | 0.0653 | +41.5 | -73 | 0.1030 | + 7 | 0.1022 | -18.5 | + 8 |
| 0.52 | 0.0662 | +38.1 | -59 | 0.0618 | +40.8 | -70 | 0.0978 | + 5 | 0.0971 | -21.4 | + 6 |
| 0.56 | 0.0631 | +37.0 | -57 | 0.0589 | +40.0 | -67 | 0.0931 | + 3 | 0.0925 | -25.5 | + 5 |
| 0.60 | 0.0605 | +35.8 | -54 | 0.0564 | +39.1 | -64 | 0.0889 | + 2 | 0.0883 | -31.5 | + 4 |
| 0.64 | 0.0581 | +34.4 | -51 | 0.0542 | +38.1 | -61 | 0.0849 | + 1 | 0.0844 | -40.2 | + 3 |
| 0.68 | 0.0560 | +32.8 | -48 | 0.0523 | +36.9 | -58 | 0.0812 | 0 | 0.0807 | +38.3 | - 2 |
| 0.72 | .. | .. | .. | .. | .. | .. | 0.0777 | - 1 | 0.0772 | +25.8 | - 2 |
| 0.76 | 0.0524 | +29.0 | -42 | 0.0491 | +34.0 | -51 | 0.0743 | - 1 | 0.0739 | +15.0 | - 2 |
| 0.80 | 0.0508 | +26.5 | -39 | 0.0478 | +32.1 | -48 | .. | .. | .. | .. | .. |
| 0.84 | 0.0494 | +23.6 | -37 | 0.0467 | +29.8 | -44 | 0.0676 | - 1 | 0.0673 | - 7.4 | - 1 |
| 0.88 | 0.0481 | +19.9 | -34 | 0.0457 | +26.9 | -41 | 0.0642 | 0 | 0.0639 | -32.8 | - 1 |
| 0.92 | 0.0468 | +15.0 | -31 | 0.0449 | +22.9 | -37 | 0.0605 | + 2 | 0.0603 | +29.0 | + 2 |
| 0.96 | 0.0455 | + 8.1 | -29 | 0.0441 | +17.0 | -33 | 0.0561 | + 6 | 0.0559 | +17.3 | + 6 |
| 1.00 | 0.0434 | -10.2 | -29 | 0.0434 | 0 | -29 | 0.0472 | +21 | 0.0472 | + 9.8 | +21 |
| 0.96 | 0.0428 | -25.7 | -37 | 0.0441 | +17.0 | -33 | 0.0409 | +45 | 0.0410 | + 7.4 | +45 |
| 0.92 | 0.0430 | -30.5 | -42 | 0.0449 | +22.9 | -37 | 0.0393 | +57 | 0.0394 | + 6.9 | +56 |
| 0.88 | 0.0435 | -33.7 | -47 | 0.0457 | +26.9 | -41 | 0.0387 | +66 | 0.0388 | + 6.6 | +66 |
| 0.84 | 0.0442 | -36.0 | -52 | 0.0467 | +29.8 | -44 | 0.0386 | +74 | 0.0387 | + 6.4 | +73 |
| 0.80 | 0.0451 | -37.8 | -56 | 0.0478 | +32.1 | -48 | .. | .. | .. | .. | .. |
| 0.76 | 0.0462 | -39.3 | -60 | 0.0491 | +34.0 | -51 | 0.0398 | +84 | 0.0399 | + 6.2 | +84 |
| 0.72 | .. | .. | .. | .. | .. | .. | 0.0411 | +88 | 0.0412 | + 6.1 | +87 |
| 0.68 | 0.0491 | -41.4 | -68 | 0.0523 | +36.9 | -58 | 0.0428 | +90 | 0.0428 | + 6.1 | +89 |
| 0.64 | 0.0509 | -42.2 | -71 | 0.0542 | +38.1 | -61 | 0.0449 | +91 | 0.0449 | + 6.1 | +91 |
| 0.60 | 0.0531 | -42.9 | -74 | 0.0564 | +39.1 | -64 | 0.0475 | +91 | 0.0475 | + 6.1 | +91 |
| 0.56 | 0.0555 | -43.5 | -77 | 0.0589 | +40.0 | -67 | 0.0507 | +90 | 0.0507 | + 6.1 | +90 |
| 0.52 | 0.0585 | -44.0 | -79 | 0.0618 | +40.8 | -70 | 0.0545 | +88 | 0.0544 | + 6.2 | +89 |
| 0.48 | 0.0619 | -44.4 | -82 | 0.0653 | +41.5 | -73 | 0.0590 | +86 | 0.0589 | + 6.2 | +87 |
| 0.44 | 0.0660 | -44.8 | -84 | 0.0693 | +42.1 | -75 | 0.0645 | +83 | 0.0643 | + 6.3 | +84 |
| 0.40 | 0.0708 | +44.9 | +85 | 0.0740 | +42.6 | -78 | 0.0710 | +80 | 0.0708 | + 6.4 | +81 |
| 0.36 | 0.0766 | +44.7 | +87 | 0.0797 | +43.0 | -80 | 0.0789 | +77 | 0.0786 | + 6.5 | +77 |
| 0.32 | 0.0836 | +44.5 | +88 | 0.0866 | +43.4 | -82 | 0.0885 | +73 | 0.0881 | + 6.6 | +74 |
| 0.28 | 0.0922 | +44.4 | +89 | 0.0950 | +43.8 | -84 | 0.1003 | +69 | 0.0999 | + 6.7 | +70 |
| 0.24 | 0.1030 | +44.4 | +89 | 0.1055 | +44.0 | -85 | 0.1151 | +65 | 0.1145 | + 6.8 | +66 |
| 0.20 | 0.1166 | +44.3 | +90 | 0.1188 | +44.3 | -87 | 0.1338 | +61 | 0.1330 | + 7.0 | +62 |
| 0.16 | 0.1342 | +44.4 | +90 | 0.1359 | +44.4 | -88 | 0.1580 | +56 | 0.1570 | + 7.2 | +58 |
| 0.12 | 0.1570 | +44.4 | +89 | 0.1580 | +44.6 | -89 | 0.1895 | +52 | 0.1883 | + 7.4 | +53 |
| 0.08 | 0.1852 | +44.5 | +89 | 0.1853 | +44.7 | -89 | 0.2291 | +48 | 0.2276 | + 7.6 | +49 |
| 0.04 | 0.2097 | +44.6 | +88 | 0.2085 | +44.8 | -90 | 0.2658 | +44 | 0.2640 | + 7.8 | +46 |
| 0 | 0.1998 | +44.6 | +88 | 0.1973 | +45.0 | -90 | 0.2614 | +41 | 0.2595 | + 8.0 | +43 |

TABLE 5 (continued)

| μ | $\mu_o = 0.80; \varphi = 20°$ | | | $\mu_o = 0.80; \varphi = 30°$ | | | $\mu_o = 0.80; \varphi = 40°$ | | | $\mu_o = 0.80; \varphi = 50°$ | | |
|---|---|---|---|---|---|---|---|---|---|---|---|---|
| | $I_1+I_r$ | $\chi°$ | δ | $I_1+I_r$ | $\chi°$ | δ | $I_1+I_r$ | $\chi°$ | δ | $I_1+I_r$ | $\chi°$ | δ |
| 0 | 0.2541 | -15.4 | +46 | 0.2457 | -21.7 | +52 | 0.2355 | -26.9 | +59 | 0.2246 | -31.0 | +68 |
| 0.04 | 0.2724 | -16.1 | +42 | 0.2633 | -22.6 | +48 | 0.2520 | -27.8 | +55 | 0.2399 | -31.8 | +64 |
| 0.08 | 0.2474 | -16.7 | +38 | 0.2389 | -23.3 | +44 | 0.2283 | -28.5 | +52 | 0.2168 | -32.4 | +60 |
| 0.12 | 0.2155 | -17.4 | +35 | 0.2079 | -24.1 | +41 | 0.1985 | -29.3 | +48 | 0.1882 | -33.1 | +57 |
| 0.16 | 0.1892 | -18.1 | +32 | 0.1824 | -25.0 | +37 | 0.1739 | -30.1 | +45 | 0.1646 | -33.9 | +54 |
| 0.20 | 0.1685 | -19.0 | +28 | 0.1624 | -25.9 | +34 | 0.1547 | -31.1 | +42 | 0.1462 | -34.7 | +51 |
| 0.24 | 0.1522 | -20.0 | +26 | 0.1467 | -27.0 | +31 | 0.1397 | -32.1 | +39 | 0.1319 | -35.6 | +48 |
| 0.28 | 0.1392 | -21.1 | +23 | 0.1341 | -28.2 | +28 | 0.1276 | -33.3 | +36 | 0.1204 | -36.6 | +44 |
| 0.32 | 0.1285 | -22.4 | +20 | 0.1238 | -29.6 | +26 | 0.1178 | -34.6 | +33 | 0.1111 | -37.7 | +41 |
| 0.36 | 0.1196 | -23.9 | +18 | 0.1153 | -31.2 | +23 | 0.1097 | -36.0 | +30 | 0.1034 | -39.0 | +39 |
| 0.40 | 0.1121 | -25.7 | +15 | 0.1081 | -33.0 | +21 | 0.1029 | -37.6 | +27 | 0.0970 | -40.3 | +36 |
| 0.44 | 0.1057 | -27.9 | +13 | 0.1019 | -35.1 | +18 | 0.0970 | -39.4 | +25 | 0.0915 | -41.8 | +33 |
| 0.48 | 0.1000 | -30.5 | +11 | 0.0965 | -37.4 | +16 | 0.0920 | -41.4 | +23 | 0.0868 | -43.5 | +30 |
| 0.52 | 0.0950 | -33.6 | +10 | 0.0917 | -40.2 | +14 | 0.0875 | -43.7 | +20 | 0.0827 | +44.6 | -28 |
| 0.56 | 0.0905 | -37.5 | + 8 | 0.0875 | -43.3 | +13 | 0.0836 | +43.7 | -18 | 0.0790 | +42.4 | -26 |
| 0.60 | 0.0865 | -42.2 | + 7 | 0.0836 | +43.0 | -11 | 0.0800 | +40.7 | -17 | 0.0757 | +40.0 | -23 |
| 0.64 | 0.0827 | +42.1 | - 6 | 0.0801 | +38.7 | -10 | 0.0767 | +37.3 | -15 | 0.0728 | +37.3 | -21 |
| 0.68 | 0.0792 | +35.4 | - 5 | 0.0768 | +33.7 | - 9 | 0.0737 | +33.4 | -13 | 0.0700 | +34.1 | -19 |
| 0.72 | 0.0759 | +27.7 | - 4 | 0.0737 | +27.9 | - 8 | 0.0708 | +28.8 | -12 | 0.0675 | +30.5 | -17 |
| 0.76 | 0.0727 | +19.1 | - 4 | 0.0707 | +21.2 | - 7 | 0.0681 | +23.5 | -11 | 0.0651 | +26.2 | -16 |
| 0.80 | .. | .. | .. | .. | .. | .. | .. | .. | .. | .. | .. | .. |
| 0.84 | 0.0664 | - 2.9 | - 3 | 0.0649 | + 3.4 | - 5 | 0.0629 | + 9.4 | - 9 | 0.0605 | +15.0 | -13 |
| 0.88 | 0.0632 | -19.7 | - 3 | 0.0619 | - 9.2 | - 5 | 0.0602 | - 0.4 | - 8 | 0.0582 | + 7.2 | -12 |
| 0.92 | 0.0596 | -40.0 | - 4 | 0.0587 | -24.8 | - 6 | 0.0573 | -12.9 | - 8 | 0.0558 | - 2.9 | -11 |
| 0.96 | 0.0555 | +33.7 | + 7 | 0.0549 | -41.4 | - 8 | 0.0540 | -28.1 | -10 | 0.0529 | -16.3 | -12 |
| 1.00 | 0.0472 | +19.5 | +21 | 0.0472 | +29.4 | +21 | 0.0472 | +39.3 | +21 | 0.0472 | -40.7 | -20 |
| 0.96 | 0.0413 | +14.8 | +44 | 0.0417 | +22.2 | +42 | 0.0424 | +29.6 | +40 | 0.0432 | +37.1 | +37 |
| 0.92 | 0.0398 | +13.7 | +55 | 0.0403 | +20.5 | +53 | 0.0411 | +27.3 | +50 | 0.0422 | +34.1 | +47 |
| 0.88 | 0.0392 | +13.1 | +64 | 0.0398 | +19.6 | +62 | 0.0407 | +26.0 | +58 | 0.0418 | +32.3 | +54 |
| 0.84 | 0.0391 | +12.7 | +72 | 0.0397 | +18.9 | +69 | 0.0406 | +25.1 | +66 | 0.0418 | +31.1 | +61 |
| 0.80 | .. | .. | .. | .. | .. | .. | .. | .. | .. | .. | .. | .. |
| 0.76 | 0.0403 | +12.3 | +83 | 0.0408 | +18.2 | +80 | 0.0417 | +24.0 | +77 | 0.0429 | +29.6 | +72 |
| 0.72 | 0.0415 | +12.2 | +86 | 0.0420 | +18.0 | +84 | 0.0428 | +23.7 | +81 | 0.0440 | +29.1 | +77 |
| 0.68 | 0.0431 | +12.1 | +89 | 0.0435 | +17.9 | +87 | 0.0442 | +23.5 | +85 | 0.0453 | +28.7 | +81 |
| 0.64 | 0.0451 | +12.1 | +90 | 0.0454 | +17.8 | +89 | 0.0460 | +23.3 | +87 | 0.0470 | +28.4 | +84 |
| 0.60 | 0.0476 | +12.1 | +91 | 0.0478 | +17.8 | +90 | 0.0483 | +23.2 | +89 | 0.0492 | +28.2 | +87 |
| 0.56 | 0.0507 | +12.1 | +90 | 0.0507 | +17.8 | +91 | 0.0510 | +23.2 | +90 | 0.0517 | +28.1 | +88 |
| 0.52 | 0.0543 | +12.2 | +89 | 0.0542 | +17.9 | +90 | 0.0543 | +23.2 | +91 | 0.0548 | +28.0 | +90 |
| 0.48 | 0.0587 | +12.3 | +88 | 0.0584 | +18.0 | +89 | 0.0583 | +23.2 | +90 | 0.0585 | +28.0 | +90 |
| 0.44 | 0.0640 | +12.4 | +85 | 0.0634 | +18.1 | +88 | 0.0630 | +23.3 | +90 | 0.0630 | +28.0 | +91 |
| 0.40 | 0.0702 | +12.5 | +83 | 0.0695 | +18.3 | +85 | 0.0687 | +23.5 | +88 | 0.0684 | +28.1 | +90 |
| 0.36 | 0.0779 | +12.7 | +80 | 0.0768 | +18.5 | +83 | 0.0757 | +23.7 | +86 | 0.0749 | +28.2 | +89 |
| 0.32 | 0.0871 | +12.9 | +76 | 0.0857 | +18.7 | +80 | 0.0841 | +23.9 | +84 | 0.0828 | +28.4 | +88 |
| 0.28 | 0.0985 | +13.1 | +73 | 0.0966 | +19.0 | +77 | 0.0945 | +24.1 | +82 | 0.0926 | +28.6 | +86 |
| 0.24 | 0.1128 | +13.3 | +69 | 0.1103 | +19.3 | +73 | 0.1075 | +24.4 | +79 | 0.1049 | +28.8 | +84 |
| 0.20 | 0.1309 | +13.6 | +65 | 0.1278 | +19.6 | +70 | 0.1241 | +24.8 | +76 | 0.1205 | +29.1 | +82 |
| 0.16 | 0.1543 | +13.9 | +61 | 0.1503 | +20.0 | +66 | 0.1455 | +25.2 | +72 | 0.1408 | +29.5 | +79 |
| 0.12 | 0.1848 | +14.3 | +57 | 0.1797 | +20.4 | +62 | 0.1735 | +25.6 | +69 | 0.1672 | +29.8 | +76 |
| 0.08 | 0.2232 | +14.7 | +53 | 0.2166 | +20.9 | +58 | 0.2086 | +26.1 | +65 | 0.2003 | +30.3 | +73 |
| 0.04 | 0.2587 | +15.1 | +49 | 0.2507 | +21.4 | +55 | 0.2409 | +26.6 | +62 | 0.2306 | +30.7 | +70 |
| 0 | 0.2541 | +15.4 | +46 | 0.2457 | +21.7 | +52 | 0.2355 | +26.9 | +59 | 0.2246 | +31.0 | +68 |

TABLE 5 (continued)

| μ | $\mu_0 = 0.80; \varphi = 60°$ | | | $\mu_0 = 0.80; \varphi = 70°$ | | | $\mu_0 = 0.80; \varphi = 80°$ | | | $\mu_0 = 0.80; \varphi = 90°$ | | |
|---|---|---|---|---|---|---|---|---|---|---|---|---|
| | $I_1+I_r$ | $\chi°$ | $\delta$ | $I_1+I_r$ | $\chi°$ | $\delta$ | $I_1+I_r$ | $\chi°$ | $\delta$ | $I_1+I_r$ | $\chi°$ | $\delta$ |
| 0 | 0.2144 | -34.0 | +76 | 0.2060 | -36.1 | +84 | 0.2006 | -37.3 | +89 | 0.1987 | -37.7 | +91 |
| 0.04 | 0.2283 | -34.7 | +73 | 0.2186 | -36.6 | +81 | 0.2118 | -37.7 | +87 | 0.2086 | -37.9 | +90 |
| 0.08 | 0.2057 | -35.2 | +70 | 0.1961 | -37.0 | +79 | 0.1890 | -37.9 | +86 | 0.1851 | -38.0 | +90 |
| 0.12 | 0.1780 | -35.8 | +67 | 0.1690 | -37.4 | +76 | 0.1621 | -38.2 | +84 | 0.1579 | -38.1 | +89 |
| 0.16 | 0.1553 | -36.4 | +64 | 0.1470 | -37.9 | +73 | 0.1404 | -38.5 | +82 | 0.1359 | -38.2 | +88 |
| 0.20 | 0.1377 | -37.1 | +61 | 0.1299 | -38.4 | +71 | 0.1235 | -38.8 | +80 | 0.1190 | -38.4 | +86 |
| 0.24 | 0.1239 | -37.8 | +57 | 0.1166 | -39.0 | +68 | 0.1105 | -39.2 | +77 | 0.1059 | -38.6 | +85 |
| 0.28 | 0.1130 | -38.7 | +54 | 0.1061 | -39.6 | +65 | 0.1002 | -39.7 | +74 | 0.0956 | -38.9 | +83 |
| 0.32 | 0.1042 | -39.6 | +51 | 0.0976 | -40.3 | +62 | 0.0919 | -40.2 | +72 | 0.0874 | -39.3 | +80 |
| 0.36 | 0.0969 | -40.6 | +48 | 0.0907 | -41.1 | +59 | 0.0852 | -40.8 | +69 | 0.0807 | -39.7 | +78 |
| 0.40 | 0.0909 | -41.7 | +45 | 0.0849 | -42.0 | +55 | 0.0796 | -41.4 | +66 | 0.0752 | -40.1 | +75 |
| 0.44 | 0.0857 | -42.9 | +42 | 0.0801 | -43.0 | +52 | 0.0749 | -42.2 | +63 | 0.0706 | -40.6 | +72 |
| 0.48 | 0.0813 | -44.3 | +39 | 0.0760 | -44.1 | +49 | 0.0710 | -43.0 | +59 | 0.0668 | -41.2 | +69 |
| 0.52 | 0.0775 | +44.1 | -36 | 0.0724 | +44.7 | -46 | 0.0677 | +44.1 | -56 | 0.0635 | -41.9 | +66 |
| 0.56 | 0.0742 | +42.4 | -34 | 0.0693 | +43.3 | -43 | 0.0648 | +45.0 | -53 | 0.0608 | -42.7 | +63 |
| 0.60 | 0.0712 | +40.5 | -31 | 0.0666 | +41.8 | -40 | 0.0623 | +43.8 | -49 | 0.0584 | -43.6 | +59 |
| 0.64 | 0.0685 | +38.3 | -29 | 0.0642 | +40.0 | -37 | 0.0601 | +42.4 | -46 | 0.0564 | -44.7 | +55 |
| 0.68 | 0.0661 | +35.7 | -26 | 0.0621 | +38.0 | -34 | 0.0582 | +40.8 | -43 | 0.0546 | +44.1 | -52 |
| 0.72 | 0.0639 | +32.8 | -24 | 0.0601 | +35.7 | -31 | 0.0565 | +39.0 | -40 | 0.0531 | +42.7 | -48 |
| 0.76 | 0.0618 | +29.4 | -22 | 0.0583 | +32.9 | -29 | 0.0550 | +36.8 | -36 | 0.0518 | +41.0 | -44 |
| 0.80 | .. | .. | .. | .. | .. | .. | .. | .. | .. | .. | .. | .. |
| 0.84 | 0.0579 | +20.4 | -18 | 0.0551 | +25.8 | -23 | 0.0524 | +31.1 | -30 | 0.0498 | +36.4 | -36 |
| 0.88 | 0.0560 | +14.1 | -16 | 0.0537 | +20.6 | -21 | 0.0513 | +26.9 | -27 | 0.0490 | +33.1 | -32 |
| 0.92 | 0.0540 | + 5.9 | -15 | 0.0521 | +13.9 | -19 | 0.0502 | +21.4 | -24 | 0.0483 | +28.6 | -28 |
| 0.96 | 0.0517 | - 5.7 | -15 | 0.0504 | + 4.0 | -18 | 0.0491 | +13.1 | -21 | 0.0477 | +21.6 | -24 |
| 1.00 | 0.0472 | -30.6 | -20 | 0.0472 | -20.5 | -20 | 0.0472 | -10.3 | -20 | 0.0472 | 0 | -20 |
| 0.96 | 0.0441 | +44.7 | +84 | 0.0452 | -37.6 | -31 | 0.0464 | -29.7 | -28 | 0.0477 | +21.6 | -24 |
| 0.92 | 0.0434 | +40.9 | +42 | 0.0449 | -42.4 | -38 | 0.0465 | -35.5 | -33 | 0.0483 | +28.6 | -28 |
| 0.88 | 0.0432 | +38.6 | +49 | 0.0449 | -44.7 | +44 | 0.0469 | -39.2 | -38 | 0.0490 | +33.1 | -32 |
| 0.84 | 0.0434 | +36.9 | +56 | 0.0452 | +42.6 | +50 | 0.0474 | -41.8 | -43 | 0.0498 | +36.4 | -36 |
| 0.80 | .. | .. | .. | .. | .. | .. | .. | .. | .. | .. | .. | .. |
| 0.76 | 0.0445 | +34.9 | +67 | 0.0465 | +39.9 | +60 | 0.0490 | +44.6 | +52 | 0.0518 | +41.0 | -44 |
| 0.72 | 0.0455 | +34.1 | +71 | 0.0476 | +38.9 | +64 | 0.0501 | +43.3 | +56 | 0.0531 | +42.7 | -48 |
| 0.68 | 0.0469 | +33.6 | +75 | 0.0489 | +38.1 | +69 | 0.0515 | +42.2 | +61 | 0.0546 | +44.1 | -52 |
| 0.64 | 0.0485 | +33.1 | +79 | 0.0505 | +37.4 | +72 | 0.0532 | +41.3 | +64 | 0.0564 | -44.7 | +55 |
| 0.60 | 0.0505 | +32.8 | +82 | 0.0525 | +36.9 | +76 | 0.0551 | +40.5 | +68 | 0.0584 | -43.6 | +59 |
| 0.56 | 0.0529 | +32.5 | +85 | 0.0548 | +36.4 | +79 | 0.0574 | +39.8 | +72 | 0.0608 | -42.7 | +63 |
| 0.52 | 0.0558 | +32.3 | +87 | 0.0576 | +36.1 | +82 | 0.0601 | +39.3 | +75 | 0.0635 | -41.9 | +66 |
| 0.48 | 0.0593 | +32.2 | +89 | 0.0609 | +35.8 | +84 | 0.0634 | +38.8 | +78 | 0.0668 | -40.6 | +69 |
| 0.44 | 0.0635 | +32.1 | +90 | 0.0649 | +35.6 | +86 | 0.0672 | +38.4 | +80 | 0.0706 | -40.6 | +72 |
| 0.40 | 0.0686 | +32.1 | +90 | 0.0697 | +35.4 | +88 | 0.0719 | +38.1 | +83 | 0.0752 | -40.1 | +75 |
| 0.36 | 0.0747 | +32.1 | +90 | 0.0755 | +35.3 | +89 | 0.0774 | +37.8 | +85 | 0.0807 | -39.7 | +78 |
| 0.32 | 0.0822 | +32.2 | +90 | 0.0826 | +35.2 | +90 | 0.0843 | +37.6 | +87 | 0.0874 | -39.3 | +80 |
| 0.28 | 0.0914 | +32.3 | +89 | 0.0913 | +35.2 | +90 | 0.0927 | +37.4 | +88 | 0.0956 | -38.9 | +83 |
| 0.24 | 0.1030 | +32.4 | +88 | 0.1023 | +35.2 | +90 | 0.1032 | +37.3 | +89 | 0.1059 | -38.6 | +85 |
| 0.20 | 0.1177 | +32.6 | +87 | 0.1162 | +35.3 | +90 | 0.1166 | +37.2 | +90 | 0.1190 | -38.4 | +86 |
| 0.16 | 0.1368 | +32.9 | +85 | 0.1343 | +35.4 | +89 | 0.1339 | +37.2 | +90 | 0.1359 | -38.2 | +88 |
| 0.12 | 0.1617 | +33.1 | +83 | 0.1579 | +35.6 | +88 | 0.1565 | +37.2 | +90 | 0.1579 | -38.1 | +89 |
| 0.08 | 0.1929 | +33.5 | +81 | 0.1873 | +35.8 | +87 | 0.1846 | +37.3 | +90 | 0.1851 | -38.0 | +90 |
| 0.04 | 0.2211 | +33.8 | +78 | 0.2136 | +36.0 | +85 | 0.2092 | +37.4 | +89 | 0.2086 | -37.9 | +90 |
| 0 | 0.2144 | +34.0 | +76 | 0.2060 | +36.1 | +84 | 0.2006 | +37.3 | +89 | 0.1987 | -37.7 | +91 |

TABLE 5 (continued)

| μ | μ₀= 0.94; φ = 0° | | μ₀ = 0.94; φ = 10° | | | μ₀ = 0.94; φ = 20° | | | μ₀ = 0.94; φ = 30° | | |
|---|---|---|---|---|---|---|---|---|---|---|---|
| | $I_1+I_r$ | δ | $I_1+I_r$ | χ° | δ | $I_1+I_r$ | χ° | δ | $I_1+I_r$ | χ° | δ |
| 0 | 0.2198 | +73 | 0.2192 | − 3.7 | +74 | 0.2174 | − 7.3 | +75 | 0.2146 | −10.6 | +78 |
| 0.04 | 0.2330 | +69 | 0.2322 | − 3.8 | +69 | 0.2302 | − 7.4 | +71 | 0.2270 | −10.7 | +74 |
| 0.08 | 0.2106 | +65 | 0.2099 | − 3.9 | +65 | 0.2078 | − 7.6 | +67 | 0.2047 | −10.9 | +70 |
| 0.12 | 0.1832 | +61 | 0.1825 | − 3.9 | +61 | 0.1807 | − 7.7 | +63 | 0.1777 | −11.1 | +66 |
| 0.16 | 0.1608 | +57 | 0.1602 | − 4.0 | +57 | 0.1585 | − 7.9 | +59 | 0.1558 | −11.3 | +62 |
| 0.20 | 0.1435 | +53 | 0.1430 | − 4.1 | +53 | 0.1414 | − 8.1 | +55 | 0.1388 | −11.6 | +58 |
| 0.24 | 0.1300 | +49 | 0.1295 | − 4.3 | +49 | 0.1280 | − 8.3 | +51 | 0.1257 | −11.9 | +54 |
| 0.28 | 0.1194 | +45 | 0.1189 | − 4.4 | +45 | 0.1175 | − 8.5 | +47 | 0.1153 | −12.2 | +50 |
| 0.32 | 0.1108 | +41 | 0.1104 | − 4.5 | +41 | 0.1091 | − 8.8 | +43 | 0.1069 | −12.6 | +46 |
| 0.36 | 0.1038 | +37 | 0.1034 | − 4.7 | +37 | 0.1021 | − 9.1 | +39 | 0.1001 | −13.0 | +42 |
| 0.40 | 0.0980 | +33 | 0.0976 | − 4.9 | +34 | 0.0964 | − 9.5 | +35 | 0.0945 | −13.5 | +38 |
| 0.44 | 0.0931 | +29 | 0.0927 | − 5.1 | +30 | 0.0916 | − 9.9 | +32 | 0.0898 | −14.0 | +35 |
| 0.48 | 0.0889 | +26 | 0.0886 | − 5.4 | +26 | 0.0875 | −10.4 | +28 | 0.0858 | −14.7 | +31 |
| 0.52 | 0.0853 | +22 | 0.0850 | − 5.7 | +23 | 0.0840 | −11.0 | +25 | 0.0823 | −15.4 | +27 |
| 0.56 | 0.0822 | +19 | 0.0819 | − 6.1 | +20 | 0.0809 | −11.7 | +21 | 0.0793 | −16.3 | +24 |
| 0.60 | 0.0794 | +16 | 0.0791 | − 6.6 | +17 | 0.0782 | −12.5 | +18 | 0.0767 | −17.3 | +21 |
| 0.64 | 0.0769 | +13 | 0.0766 | − 7.2 | +14 | 0.0757 | −13.5 | +15 | 0.0744 | −18.6 | +18 |
| 0.68 | 0.0746 | +11 | 0.0743 | − 7.9 | +11 | 0.0735 | −14.9 | +13 | 0.0723 | −20.2 | +15 |
| 0.72 | 0.0725 | + 8 | 0.0723 | − 9.0 | + 9 | 0.0715 | −16.6 | +10 | 0.0703 | −22.3 | +12 |
| 0.76 | 0.0705 | + 6 | 0.0703 | −10.4 | + 6 | 0.0696 | −19.0 | + 8 | 0.0685 | −25.0 | +10 |
| 0.80 | 0.0686 | + 4 | 0.0684 | −12.7 | + 4 | 0.0678 | −22.4 | + 6 | 0.0668 | −28.7 | + 7 |
| 0.84 | 0.0667 | + 2 | 0.0666 | −16.6 | + 3 | 0.0660 | −27.7 | + 4 | 0.0652 | −34.1 | + 5 |
| 0.88 | 0.0648 | + 1 | 0.0647 | −24.7 | + 1 | 0.0642 | −37.0 | + 2 | 0.0635 | −42.6 | + 3 |
| 0.92 | 0.0627 | 0 | 0.0626 | −43.2 | 0 | 0.0622 | +36.0 | − 1 | 0.0616 | +33.1 | − 2 |
| 0.96 | 0.0601 | 0 | 0.0600 | +22.8 | 0 | 0.0598 | −24.1 | 0 | 0.0594 | − 2.3 | − 1 |
| 1.00 | 0.0546 | + 6 | 0.0546 | + 9.1 | + 6 | 0.0546 | +18.3 | + 6 | 0.0546 | +27.6 | + 6 |
| 0.96 | 0.0498 | +21 | 0.0499 | + 5.5 | +21 | 0.0501 | +11.0 | +20 | 0.0505 | +16.6 | +20 |
| 0.92 | 0.0484 | +30 | 0.0485 | + 4.9 | +29 | 0.0488 | + 9.7 | +29 | 0.0492 | +14.5 | +28 |
| 0.88 | 0.0475 | +38 | 0.0476 | + 4.5 | +37 | 0.0479 | + 8.9 | +36 | 0.0485 | +13.3 | +35 |
| 0.84 | 0.0471 | +45 | 0.0472 | + 4.2 | +45 | 0.0475 | + 8.4 | +44 | 0.0481 | +12.5 | +42 |
| 0.80 | 0.0470 | +52 | 0.0471 | + 4.0 | +52 | 0.0475 | + 8.0 | +50 | 0.0481 | +11.9 | +49 |
| 0.76 | 0.0472 | +58 | 0.0473 | + 3.9 | +58 | 0.0477 | + 7.7 | +57 | 0.0483 | +11.5 | +55 |
| 0.72 | 0.0477 | +65 | 0.0478 | + 3.8 | +64 | 0.0482 | + 7.5 | +63 | 0.0488 | +11.1 | +61 |
| 0.68 | 0.0486 | +70 | 0.0487 | + 3.7 | +70 | 0.0491 | + 7.3 | +68 | 0.0497 | +10.9 | +66 |
| 0.64 | 0.0498 | +75 | 0.0499 | + 3.6 | +75 | 0.0503 | + 7.2 | +74 | 0.0509 | +10.6 | +72 |
| 0.60 | 0.0514 | +80 | 0.0515 | + 3.6 | +79 | 0.0518 | + 7.1 | +78 | 0.0525 | +10.5 | +76 |
| 0.56 | 0.0534 | +84 | 0.0535 | + 3.5 | +83 | 0.0539 | + 7.0 | +82 | 0.0544 | +10.3 | +80 |
| 0.52 | 0.0560 | +87 | 0.0561 | + 3.5 | +86 | 0.0564 | + 6.9 | +86 | 0.0569 | +10.2 | +84 |
| 0.48 | 0.0591 | +89 | 0.0592 | + 3.5 | +89 | 0.0595 | + 6.9 | +88 | 0.0600 | +10.1 | +87 |
| 0.44 | 0.0630 | +91 | 0.0631 | + 3.5 | +91 | 0.0633 | + 6.8 | +90 | 0.0637 | +10.1 | +89 |
| 0.40 | 0.0678 | +92 | 0.0679 | + 3.5 | +92 | 0.0680 | + 6.8 | +92 | 0.0684 | +10.0 | +91 |
| 0.36 | 0.0737 | +93 | 0.0737 | + 3.5 | +93 | 0.0738 | + 6.8 | +92 | 0.0740 | +10.0 | +92 |
| 0.32 | 0.0810 | +92 | 0.0810 | + 3.5 | +92 | 0.0810 | + 6.8 | +93 | 0.0811 | +10.0 | +93 |
| 0.28 | 0.0901 | +92 | 0.0900 | + 3.5 | +92 | 0.0900 | + 6.8 | +92 | 0.0899 | +10.0 | +92 |
| 0.24 | 0.1016 | +90 | 0.1015 | + 3.5 | +90 | 0.1013 | + 6.9 | +91 | 0.1010 | +10.0 | +92 |
| 0.20 | 0.1164 | +88 | 0.1162 | + 3.5 | +89 | 0.1159 | + 6.9 | +89 | 0.1153 | +10.1 | +90 |
| 0.16 | 0.1357 | +86 | 0.1355 | + 3.5 | +86 | 0.1349 | + 7.0 | +87 | 0.1340 | +10.2 | +88 |
| 0.12 | 0.1610 | +83 | 0.1607 | + 3.6 | +83 | 0.1598 | + 7.0 | +84 | 0.1585 | +10.2 | +86 |
| 0.08 | 0.1931 | +80 | 0.1927 | + 3.6 | +80 | 0.1914 | + 7.1 | +82 | 0.1896 | +10.3 | +84 |
| 0.04 | 0.2230 | +76 | 0.2225 | + 3.7 | +77 | 0.2208 | + 7.2 | +78 | 0.2184 | +10.5 | +80 |
| 0 | 0.2198 | +73 | 0.2192 | + 3.7 | +74 | 0.2174 | + 7.3 | +75 | 0.2146 | +10.6 | +78 |

TABLE 5 (continued)

| μ | $\mu_0 = 0.94; \varphi = 40°$ | | | $\mu_0 = 0.94; \varphi = 50°$ | | | $\mu_0 = 0.94; \varphi = 60°$ | | | $\mu_0 = 0.94; \varphi = 70°$ | | |
|---|---|---|---|---|---|---|---|---|---|---|---|---|
| | $I_1+I_r$ | $\chi°$ | $\delta$ | $I_1+I_r$ | $\chi°$ | $\delta$ | $I_1+I_r$ | $\chi°$ | $\delta$ | $I_1+I_r$ | $\chi°$ | $\delta$ |
| 0 | 0.2112 | -13.4 | +81 | 0.2076 | -15.9 | +84 | 0.2042 | -17.8 | +87 | 0.2014 | -19.2 | +90 |
| 0.04 | 0.2230 | -13.7 | +77 | 0.2187 | -16.1 | +81 | 0.2145 | -18.0 | +85 | 0.2109 | -19.4 | +88 |
| 0.08 | 0.2008 | -13.9 | +74 | 0.1965 | -16.3 | +78 | 0.1922 | -18.2 | +82 | 0.1883 | -19.5 | +86 |
| 0.12 | 0.1741 | -14.1 | +70 | 0.1700 | -16.5 | +74 | 0.1658 | -18.4 | +79 | 0.1620 | -19.7 | +83 |
| 0.16 | 0.1524 | -14.3 | +66 | 0.1485 | -16.8 | +71 | 0.1445 | -18.6 | +76 | 0.1408 | -19.9 | +80 |
| 0.20 | 0.1356 | -14.6 | +62 | 0.1320 | -17.1 | +67 | 0.1282 | -18.9 | +72 | 0.1245 | -20.1 | +77 |
| 0.24 | 0.1226 | -15.0 | +58 | 0.1192 | -17.4 | +63 | 0.1155 | -19.2 | +68 | 0.1120 | -20.4 | +74 |
| 0.28 | 0.1124 | -15.3 | +54 | 0.1091 | -17.8 | +59 | 0.1056 | -19.6 | +65 | 0.1021 | -20.7 | +70 |
| 0.32 | 0.1042 | -15.7 | +50 | 0.1011 | -18.2 | +55 | 0.0977 | -20.0 | +61 | 0.0943 | -21.0 | +67 |
| 0.36 | 0.0975 | -16.2 | +46 | 0.0945 | -18.7 | +51 | 0.0913 | -20.4 | +57 | 0.0880 | -21.5 | +63 |
| 0.40 | 0.0920 | -16.7 | +42 | 0.0891 | -19.2 | +47 | 0.0860 | -20.9 | +53 | 0.0828 | -21.9 | +59 |
| 0.44 | 0.0874 | -17.4 | +38 | 0.0847 | -19.8 | +43 | 0.0817 | -21.5 | +49 | 0.0786 | -22.5 | +55 |
| 0.48 | 0.0835 | -18.1 | +35 | 0.0809 | -20.6 | +39 | 0.0780 | -22.2 | +45 | 0.0750 | -23.1 | +50 |
| 0.52 | 0.0802 | -18.9 | +31 | 0.0777 | -21.4 | +35 | 0.0749 | -23.0 | +41 | 0.0720 | -23.8 | +46 |
| 0.56 | 0.0773 | -19.8 | +27 | 0.0749 | -22.4 | +32 | 0.0722 | -23.9 | +37 | 0.0695 | -24.6 | +42 |
| 0.60 | 0.0748 | -21.0 | +24 | 0.0725 | -23.5 | +28 | 0.0700 | -25.0 | +33 | 0.0673 | -25.6 | +38 |
| 0.64 | 0.0726 | -22.4 | +21 | 0.0704 | -24.8 | +25 | 0.0680 | -26.2 | +29 | 0.0655 | -26.7 | +34 |
| 0.68 | 0.0706 | -24.0 | +18 | 0.0685 | -26.5 | +21 | 0.0663 | -27.7 | +26 | 0.0639 | -28.0 | +30 |
| 0.72 | 0.0688 | -26.2 | +15 | 0.0669 | -28.5 | +18 | 0.0648 | -29.5 | +22 | 0.0625 | -29.7 | +26 |
| 0.76 | 0.0671 | -28.8 | +12 | 0.0654 | -31.0 | +15 | 0.0634 | -31.8 | +19 | 0.0613 | -31.6 | +23 |
| 0.80 | 0.0655 | -32.4 | +10 | 0.0639 | -34.2 | +12 | 0.0622 | -34.7 | +16 | 0.0603 | -34.2 | +19 |
| 0.84 | 0.0640 | -37.4 | + 7 | 0.0626 | -38.7 | +10 | 0.0610 | -38.6 | +13 | 0.0593 | -37.5 | +16 |
| 0.88 | 0.0625 | -44.8 | + 5 | 0.0613 | +45.0 | - 7 | 0.0599 | -44.0 | +10 | 0.0584 | -42.1 | +12 |
| 0.92 | 0.0608 | +33.2 | - 3 | 0.0599 | +35.0 | - 5 | 0.0587 | +37.7 | - 7 | 0.0575 | +40.9 | - 9 |
| 0.96 | 0.0588 | + 8.3 | - 2 | 0.0581 | +16.0 | - 3 | 0.0574 | +22.6 | - 5 | 0.0565 | +28.7 | - 6 |
| 1.00 | 0.0546 | +37.2 | + 6 | 0.0546 | -42.9 | - 6 | 0.0546 | -32.6 | - 6 | 0.0546 | -22.0 | - 5 |
| 0.96 | 0.0509 | +22.1 | +18 | 0.0515 | +27.6 | +17 | 0.0522 | +33.1 | +15 | 0.0530 | +38.6 | +14 |
| 0.92 | 0.0499 | +19.3 | +26 | 0.0507 | +24.0 | +24 | 0.0516 | +28.6 | +22 | 0.0527 | +33.0 | +19 |
| 0.88 | 0.0492 | +17.6 | +33 | 0.0501 | +21.8 | +31 | 0.0512 | +25.9 | +28 | 0.0525 | +29.7 | +25 |
| 0.84 | 0.0489 | +16.5 | +40 | 0.0499 | +20.4 | +37 | 0.0512 | +24.0 | +34 | 0.0526 | +27.5 | +30 |
| 0.80 | 0.0489 | +15.7 | +46 | 0.0500 | +19.3 | +43 | 0.0513 | +22.7 | +39 | 0.0529 | +25.8 | +36 |
| 0.76 | 0.0492 | +15.1 | +52 | 0.0503 | +18.5 | +49 | 0.0517 | +21.7 | +45 | 0.0533 | +24.5 | +41 |
| 0.72 | 0.0498 | +14.6 | +58 | 0.0509 | +17.9 | +55 | 0.0524 | +20.8 | +50 | 0.0540 | +23.5 | +46 |
| 0.68 | 0.0506 | +14.2 | +64 | 0.0518 | +17.3 | +60 | 0.0533 | +20.2 | +56 | 0.0550 | +22.7 | +51 |
| 0.64 | 0.0518 | +13.9 | +69 | 0.0530 | +16.9 | +65 | 0.0545 | +19.6 | +61 | 0.0562 | +22.0 | +56 |
| 0.60 | 0.0533 | +13.6 | +74 | 0.0545 | +16.6 | +70 | 0.0560 | +19.2 | +66 | 0.0578 | +21.4 | +61 |
| 0.56 | 0.0553 | +13.4 | +76 | 0.0564 | +16.3 | +74 | 0.0579 | +18.8 | +70 | 0.0597 | +21.0 | +65 |
| 0.52 | 0.0577 | +13.3 | +82 | 0.0588 | +16.1 | +78 | 0.0602 | +18.5 | +74 | 0.0620 | +20.6 | +70 |
| 0.48 | 0.0607 | +13.1 | +85 | 0.0617 | +15.9 | +82 | 0.0631 | +18.2 | +78 | 0.0648 | +20.2 | +74 |
| 0.44 | 0.0644 | +13.0 | +88 | 0.0653 | +15.7 | +85 | 0.0666 | +18.0 | +82 | 0.0683 | +20.0 | +78 |
| 0.40 | 0.0689 | +13.0 | +90 | 0.0697 | +15.6 | +88 | 0.0709 | +17.9 | +85 | 0.0725 | +19.7 | +81 |
| 0.36 | 0.0745 | +12.9 | +91 | 0.0751 | +15.5 | +90 | 0.0762 | +17.7 | +87 | 0.0777 | +19.5 | +84 |
| 0.32 | 0.0813 | +12.9 | +92 | 0.0819 | +15.5 | +91 | 0.0828 | +17.6 | +89 | 0.0841 | +19.4 | +87 |
| 0.28 | 0.0899 | +12.9 | +93 | 0.0903 | +15.4 | +92 | 0.0909 | +17.6 | +91 | 0.0921 | +19.2 | +89 |
| 0.24 | 0.1008 | +12.9 | +92 | 0.1009 | +15.4 | +92 | 0.1013 | +17.5 | +92 | 0.1022 | +19.2 | +90 |
| 0.20 | 0.1148 | +13.0 | +91 | 0.1145 | +15.5 | +92 | 0.1146 | +17.5 | +92 | 0.1152 | +19.1 | +92 |
| 0.16 | 0.1331 | +13.0 | +90 | 0.1323 | +15.5 | +91 | 0.1320 | +17.5 | +92 | 0.1322 | +19.1 | +92 |
| 0.12 | 0.1571 | +13.1 | +88 | 0.1557 | +15.6 | +90 | 0.1547 | +17.6 | +92 | 0.1544 | +19.1 | +92 |
| 0.08 | 0.1874 | +13.2 | +86 | 0.1852 | +15.7 | +88 | 0.1834 | +17.7 | +90 | 0.1823 | +19.1 | +92 |
| 0.04 | 0.2154 | +13.3 | +83 | 0.2123 | +15.8 | +86 | 0.2095 | +17.7 | +89 | 0.2075 | +19.2 | +91 |
| 0 | 0.2112 | +13.4 | +81 | 0.2076 | +15.9 | +84 | 0.2042 | +17.8 | +87 | 0.2014 | +19.2 | +90 |

TABLE 5 (continued)

| $\mu$ | $\mu_o = 0.94; \varphi = 80°$ | | | $\mu_o = 0.94; \varphi = 90°$ | | | $\mu_o = 1.00$ | |
|---|---|---|---|---|---|---|---|---|
| | $I_1 + I_r$ | $\chi$ | $\delta$ | $I_1 + I_r$ | $\chi$ | $\delta$ | $I_1 + I_r$ | $\delta$ |
| 0 | 0.1996 | −20.0 | +92 | 0.1990 | −20.3 | +93 | 0.1984 | +94 |
| 0.04 | 0.2082 | −20.2 | +90 | 0.2067 | −20.4 | +92 | 0.2052 | +93 |
| 0.08 | 0.1852 | −20.2 | +89 | 0.1831 | −20.4 | +91 | 0.1817 | +93 |
| 0.12 | 0.1587 | −20.4 | +87 | 0.1563 | −20.5 | +90 | 0.1552 | +91 |
| 0.16 | 0.1375 | −20.5 | +85 | 0.1349 | −20.6 | +88 | 0.1341 | +89 |
| 0.20 | 0.1212 | −20.7 | +82 | 0.1185 | −20.7 | +86 | 0.1181 | +87 |
| 0.24 | 0.1087 | −20.9 | +79 | 0.1059 | −20.9 | +84 | 0.1057 | +84 |
| 0.28 | 0.0989 | −21.2 | +76 | 0.0961 | −21.1 | +81 | 0.0961 | +81 |
| 0.32 | 0.0911 | −21.5 | +72 | 0.0883 | −21.3 | +78 | 0.0886 | +77 |
| 0.36 | 0.0849 | −21.8 | +69 | 0.0821 | −21.6 | +74 | 0.0826 | +73 |
| 0.40 | 0.0798 | −22.2 | +65 | 0.0770 | −22.0 | +71 | 0.0777 | +69 |
| 0.44 | 0.0756 | −22.7 | +61 | 0.0728 | −22.3 | +67 | 0.0737 | +65 |
| 0.48 | 0.0721 | −23.2 | +57 | 0.0694 | −22.8 | +63 | 0.0705 | +60 |
| 0.52 | 0.0692 | −23.9 | +52 | 0.0665 | −23.3 | +58 | 0.0678 | +55 |
| 0.56 | 0.0668 | −24.6 | +48 | 0.0641 | −23.9 | +54 | 0.0657 | +50 |
| 0.60 | 0.0647 | −25.4 | +44 | 0.0622 | −24.6 | +50 | 0.0639 | +45 |
| 0.64 | 0.0630 | −26.4 | +40 | 0.0605 | −25.5 | +45 | 0.0624 | +41 |
| 0.68 | 0.0615 | −27.6 | +35 | 0.0592 | −26.5 | +41 | 0.0613 | +36 |
| 0.72 | 0.0603 | −29.0 | +31 | 0.0580 | −27.7 | +36 | 0.0603 | +31 |
| 0.76 | 0.0592 | −30.7 | +27 | 0.0571 | −29.1 | +32 | 0.0596 | +26 |
| 0.80 | 0.0583 | −32.9 | +23 | 0.0564 | −31.0 | +27 | 0.0590 | +22 |
| 0.84 | 0.0576 | −35.7 | +19 | 0.0558 | −33.4 | +23 | 0.0586 | +17 |
| 0.88 | 0.0569 | −39.6 | +15 | 0.0554 | −36.7 | +19 | 0.0584 | +13 |
| 0.92 | 0.0563 | +44.6 | −12 | 0.0550 | −41.5 | +14 | 0.0582 | + 9 |
| 0.96 | 0.0556 | +34.5 | − 8 | 0.0547 | +40.2 | −10 | 0.0561 | + 4 |
| 1.00 | 0.0546 | −11.1 | − 5 | 0.0546 | 0 | − 5 | 0 | 0 |
| 0.96 | 0.0538 | +44.2 | +12 | 0.0547 | +40.2 | −10 | 0.0581 | + 4 |
| 0.92 | 0.0538 | +37.4 | +17 | 0.0550 | −41.5 | +14 | 0.0582 | + 9 |
| 0.88 | 0.0539 | +33.3 | +22 | 0.0554 | −36.7 | +19 | 0.0584 | +13 |
| 0.84 | 0.0541 | +30.6 | +27 | 0.0558 | −33.4 | +23 | 0.0586 | +17 |
| 0.80 | 0.0546 | +28.6 | +31 | 0.0564 | −31.0 | +27 | 0.0590 | +22 |
| 0.76 | 0.0552 | +27.0 | +36 | 0.0571 | −29.1 | +32 | 0.0596 | +26 |
| 0.72 | 0.0560 | +25.8 | +41 | 0.0580 | −27.7 | +36 | 0.0603 | +31 |
| 0.68 | 0.0570 | +24.8 | +46 | 0.0592 | −26.5 | +41 | 0.0613 | +36 |
| 0.64 | 0.0583 | +24.0 | +51 | 0.0605 | −25.5 | +45 | 0.0624 | +41 |
| 0.60 | 0.0598 | +23.3 | +55 | 0.0622 | −24.6 | +50 | 0.0639 | +45 |
| 0.56 | 0.0618 | +22.7 | +60 | 0.0641 | −23.9 | +54 | 0.0657 | +50 |
| 0.52 | 0.0641 | +22.2 | +64 | 0.0665 | −23.3 | +58 | 0.0678 | +55 |
| 0.48 | 0.0669 | +21.8 | +68 | 0.0694 | −22.8 | +63 | 0.0705 | +60 |
| 0.44 | 0.0704 | +21.4 | +72 | 0.0728 | −22.3 | +67 | 0.0737 | +65 |
| 0.40 | 0.0745 | +21.1 | +76 | 0.0770 | −22.0 | +71 | 0.0777 | +69 |
| 0.36 | 0.0796 | +20.8 | +80 | 0.0821 | −21.6 | +74 | 0.0826 | +73 |
| 0.32 | 0.0860 | +20.6 | +83 | 0.0883 | −21.3 | +78 | 0.0886 | +77 |
| 0.28 | 0.0938 | +20.4 | +85 | 0.0961 | −21.1 | +81 | 0.0961 | +81 |
| 0.24 | 0.1038 | +20.3 | +88 | 0.1059 | −20.9 | +84 | 0.1057 | +84 |
| 0.20 | 0.1165 | +20.2 | +90 | 0.1185 | −20.7 | +86 | 0.1181 | +87 |
| 0.16 | 0.1331 | +20.1 | +91 | 0.1349 | −20.6 | +88 | 0.1341 | +89 |
| 0.12 | 0.1549 | +20.1 | +92 | 0.1563 | −20.5 | +90 | 0.1552 | +91 |
| 0.08 | 0.1822 | +20.0 | +92 | 0.1831 | −20.4 | +91 | 0.1817 | +93 |
| 0.04 | 0.2065 | +20.1 | +92 | 0.2067 | −20.4 | +92 | 0.2052 | +93 |
| 0 | 0.1996 | +20.0 | +92 | 0.1990 | −20.3 | +93 | 0.1984 | +94 |

TABLE 6

THE ILLUMINATION AND POLARIZATION OF THE SKY FOR VARIOUS ZENITH DISTANCES

OF THE SUN AND FOR AN ATMOSPHERE OF OPTICAL THICKNESS $\tau_1 = 0.10$

| μ | $\mu_0 = 0; \varphi = 0°$ | | $\mu_0 = 0; \varphi = 90°$ | | $\mu_0 = 0.16; \varphi = 0°$ | | $\mu_0 = 0.16; \varphi = 90°$ | | | $\mu_0 = 0.50; \varphi = 0°$ | |
|---|---|---|---|---|---|---|---|---|---|---|---|
| | $I_1+I_r$ | $\delta$ | $I_1+I_r$ | $\delta$ | $I_1+I_r$ | $\delta$ | $I_1+I_r$ | $\chi°$ | $\delta$ | $I_1+I_r$ | $\delta$ |
| 0 | .. | .. | .. | .. | 0.2405 | - 3 | 0.1322 | +8.7 | -92 | 0.3156 | +10 |
| 0.04 | 1.2394 | -10 | 0.7843 | -85 | 0.2715 | - 4 | 0.1487 | +8.7 | -93 | 0.3189 | + 7 |
| 0.08 | 1.7316 | - 6 | 0.9652 | -91 | 0.2229 | - 5 | 0.1213 | +8.7 | -93 | 0.2564 | + 6 |
| 0.12 | 1.6675 | - 4 | 0.9184 | -92 | 0.1803 | - 5 | 0.0978 | +8.6 | -93 | 0.2076 | + 4 |
| 0.16 | 1.5082 | - 4 | 0.8264 | -93 | .. | .. | .. | .. | .. | 0.1736 | + 2 |
| 0.20 | 1.3389 | - 3 | 0.7379 | -93 | 0.1273 | - 5 | 0.0689 | +8.6 | -93 | 0.1492 | + 1 |
| 0.24 | 1.1941 | - 2 | 0.6617 | -93 | 0.1104 | - 4 | 0.0598 | +8.5 | -93 | 0.1308 | 0 |
| 0.28 | 1.0699 | 0 | 0.5974 | -93 | 0.0971 | - 4 | 0.0528 | +8.4 | -93 | 0.1165 | - 1 |
| 0.32 | 0.9685 | + 1 | 0.5432 | -93 | 0.0864 | - 3 | 0.0472 | +8.3 | -93 | 0.1050 | - 2 |
| 0.36 | 0.8718 | + 2 | 0.4972 | -93 | 0.0775 | - 2 | 0.0426 | +8.2 | -93 | 0.0956 | - 2 |
| 0.40 | 0.7920 | + 4 | 0.4578 | -94 | 0.0701 | - 1 | 0.0389 | +8.1 | -93 | 0.0876 | - 3 |
| 0.44 | 0.7219 | + 6 | 0.4238 | -94 | 0.0637 | 0 | 0.0357 | +8.0 | -92 | 0.0808 | - 3 |
| 0.48 | 0.6597 | + 8 | 0.3941 | -94 | 0.0581 | + 1 | 0.0329 | +7.8 | -92 | 0.0748 | - 3 |
| 0.52 | 0.6040 | +11 | 0.3680 | -94 | 0.0532 | + 3 | 0.0306 | +7.6 | -92 | 0.0696 | - 3 |
| 0.56 | 0.5538 | +14 | 0.3449 | -94 | 0.0488 | + 5 | 0.0285 | +7.4 | -92 | 0.0649 | - 3 |
| 0.60 | 0.5082 | +17 | 0.3243 | -94 | 0.0448 | + 7 | 0.0267 | +7.2 | -92 | 0.0606 | - 2 |
| 0.64 | 0.4664 | +21 | 0.3057 | -94 | 0.0412 | +10 | 0.0251 | +7.0 | -91 | 0.0566 | - 1 |
| 0.68 | 0.4279 | +25 | 0.2890 | -94 | 0.0379 | +13 | 0.0236 | +6.7 | -91 | 0.0529 | 0 |
| 0.72 | 0.3923 | +30 | 0.2738 | -94 | 0.0348 | +17 | 0.0223 | +6.4 | -91 | 0.0494 | + 1 |
| 0.76 | 0.3591 | +35 | 0.2600 | -94 | 0.0319 | +22 | 0.0212 | +6.0 | -91 | 0.0461 | + 3 |
| 0.80 | 0.3280 | +42 | 0.2472 | -94 | 0.0291 | +27 | 0.0201 | +5.6 | -90 | 0.0429 | + 6 |
| 0.84 | 0.2988 | +49 | 0.2356 | -94 | 0.0265 | +33 | 0.0191 | +5.1 | -90 | 0.0397 | + 9 |
| 0.88 | 0.2713 | +57 | 0.2248 | -94 | 0.0240 | +41 | 0.0182 | +4.5 | -90 | 0.0366 | +13 |
| 0.92 | 0.2452 | +67 | 0.2148 | -94 | 0.0216 | +50 | 0.0174 | +3.7 | -89 | 0.0333 | +19 |
| 0.96 | 0.2204 | +79 | 0.2054 | -93 | 0.0191 | +63 | 0.0166 | +2.7 | -89 | 0.0297 | +29 |
| 1.00 | 0.1968 | +93 | 0.1968 | -93 | 0.0159 | +88 | 0.0159 | 0 | -88 | 0.0235 | +56 |
| 0.96 | 0.2204 | +79 | 0.2054 | -93 | 0.0164 | +89 | 0.0166 | +2.7 | -89 | 0.0208 | +84 |
| 0.92 | 0.2452 | +67 | 0.2148 | -94 | 0.0179 | +81 | 0.0174 | +3.7 | -89 | 0.0209 | +90 |
| 0.88 | 0.2713 | +57 | 0.2248 | -94 | 0.0195 | +73 | 0.0182 | +4.5 | -90 | 0.0216 | +92 |
| 0.84 | 0.2988 | +49 | 0.2356 | -94 | 0.0214 | +65 | 0.0191 | +5.1 | -90 | 0.0226 | +91 |
| 0.80 | 0.3280 | +42 | 0.2472 | -94 | 0.0235 | +57 | 0.0201 | +5.6 | -90 | 0.0240 | +89 |
| 0.76 | 0.3591 | +35 | 0.2600 | -94 | 0.0258 | +50 | 0.0212 | +6.0 | -91 | 0.0257 | +85 |
| 0.72 | 0.3923 | +30 | 0.2738 | -94 | 0.0283 | +44 | 0.0223 | +6.4 | -91 | 0.0278 | +80 |
| 0.68 | 0.4279 | +25 | 0.2890 | -94 | 0.0311 | +38 | 0.0236 | +6.7 | -91 | 0.0301 | +75 |
| 0.64 | 0.4664 | +21 | 0.3057 | -94 | 0.0341 | +33 | 0.0251 | +7.0 | -91 | 0.0328 | +70 |
| 0.60 | 0.5082 | +17 | 0.3243 | -94 | 0.0375 | +28 | 0.0267 | +7.2 | -92 | 0.0359 | +65 |
| 0.56 | 0.5538 | +14 | 0.3449 | -94 | 0.0412 | +24 | 0.0285 | +7.4 | -92 | 0.0395 | +60 |
| 0.52 | 0.6040 | +11 | 0.3680 | -94 | 0.0455 | +21 | 0.0306 | +7.6 | -92 | 0.0437 | +54 |
| 0.48 | 0.6597 | + 8 | 0.3941 | -94 | 0.0502 | +17 | 0.0329 | +7.8 | -92 | 0.0484 | +50 |
| 0.44 | 0.7219 | + 6 | 0.4238 | -94 | 0.0557 | +14 | 0.0357 | +8.0 | -92 | 0.0540 | +45 |
| 0.40 | 0.7920 | + 4 | 0.4578 | -94 | 0.0620 | +11 | 0.0389 | +8.1 | -93 | 0.0606 | +41 |
| 0.36 | 0.8718 | + 2 | 0.4972 | -93 | 0.0695 | + 9 | 0.0426 | +8.2 | -93 | 0.0684 | +36 |
| 0.32 | 0.9685 | + 1 | 0.5432 | -93 | 0.0783 | + 7 | 0.0472 | +8.3 | -93 | 0.0779 | +33 |
| 0.28 | 1.0699 | 0 | 0.5974 | -93 | 0.0891 | + 5 | 0.0528 | +8.4 | -93 | 0.0896 | +29 |
| 0.24 | 1.1941 | - 2 | 0.6617 | -93 | 0.1026 | + 3 | 0.0598 | +8.5 | -93 | 0.1044 | +25 |
| 0.20 | 1.3389 | - 3 | 0.7379 | -93 | 0.1197 | + 1 | 0.0689 | +8.6 | -93 | 0.1235 | +22 |
| 0.16 | 1.5082 | - 4 | 0.8264 | -93 | .. | .. | .. | .. | .. | 0.1493 | +19 |
| 0.12 | 1.6675 | - 4 | 0.9184 | -92 | 0.1738 | - 1 | 0.0978 | +8.6 | -93 | 0.1854 | +16 |
| 0.08 | 1.7316 | - 6 | 0.9652 | -91 | 0.2175 | - 2 | 0.1213 | +8.7 | -93 | 0.2377 | +14 |
| 0.04 | 1.2394 | -10 | 0.7843 | -85 | 0.2682 | - 3 | 0.1487 | +8.7 | -93 | 0.3071 | +12 |
| 0 | .. | .. | .. | .. | 0.2405 | - 3 | 0.1322 | +8.7 | -92 | 0.3156 | +10 |

TABLE 6 (continued)

| $\mu$ | $\mu_o = 0.50; \varphi = 90°$ | | | $\mu_o = 0.86; \varphi = 0°$ | | $\mu_o = 0.86; \varphi = 90°$ | | | $\mu_o = 1.00$ | |
|---|---|---|---|---|---|---|---|---|---|---|
| | $I_1+I_r$ | $\chi$ | $\delta$ | $I_1+I_r$ | $\delta$ | $I_1+I_r$ | $\chi$ | $\delta$ | $I_1+I_r$ | $\delta$ |
| 0 | 0.1903 | +29.0 | -93 | 0.2450 | +54 | 0.1978 | -31.2 | +94 | 0.1968 | +96 |
| 0.04 | 0.1902 | +28.8 | -92 | 0.2434 | +50 | 0.1920 | -31.3 | +93 | 0.1893 | +96 |
| 0.08 | 0.1504 | +28.8 | -92 | 0.1964 | +46 | 0.1512 | -31.3 | +93 | 0.1491 | +95 |
| 0.12 | 0.1200 | +28.7 | -92 | 0.1605 | +42 | 0.1209 | -31.4 | +92 | 0.1194 | +93 |
| 0.16 | 0.0991 | +28.6 | -91 | 0.1358 | +38 | 0.1002 | -31.6 | +90 | 0.0993 | +91 |
| 0.20 | 0.0842 | +28.4 | -91 | 0.1182 | +34 | 0.0857 | -31.8 | +88 | 0.0853 | +89 |
| 0.24 | 0.0732 | +28.2 | -90 | 0.1052 | +31 | 0.0752 | -32.0 | +86 | 0.0751 | +86 |
| 0.28 | 0.0648 | +28.0 | -89 | 0.0952 | +28 | 0.0671 | -32.3 | +84 | 0.0674 | +82 |
| 0.32 | 0.0581 | +27.7 | -88 | 0.0874 | +24 | 0.0609 | -32.6 | +81 | 0.0616 | +79 |
| 0.36 | 0.0528 | +27.4 | -87 | 0.0810 | +21 | 0.0560 | -33.0 | +78 | 0.0570 | +74 |
| 0.40 | 0.0484 | +27.1 | -86 | 0.0757 | +18 | 0.0520 | -33.4 | +75 | 0.0533 | +70 |
| 0.44 | 0.0447 | +26.7 | -85 | 0.0712 | +16 | 0.0488 | -33.9 | +71 | 0.0504 | +65 |
| 0.48 | 0.0416 | +26.2 | -84 | 0.0674 | +13 | 0.0461 | -34.5 | +68 | 0.0480 | +61 |
| 0.52 | 0.0390 | +25.6 | -82 | 0.0640 | +11 | 0.0439 | -35.2 | +64 | 0.0460 | +56 |
| 0.56 | 0.0366 | +25.1 | -81 | 0.0611 | + 8 | 0.0420 | -36.0 | +60 | 0.0444 | +51 |
| 0.60 | 0.0346 | +24.4 | -79 | 0.0584 | + 6 | 0.0405 | -36.9 | +56 | 0.0431 | +46 |
| 0.64 | 0.0329 | +23.6 | -77 | 0.0560 | + 5 | 0.0391 | -37.9 | +52 | 0.0421 | +41 |
| 0.68 | 0.0313 | +22.7 | -75 | 0.0538 | + 3 | 0.0380 | -39.2 | +48 | 0.0412 | +36 |
| 0.72 | 0.0299 | +21.7 | -73 | 0.0517 | + 2 | 0.0371 | -40.7 | +43 | 0.0405 | +31 |
| 0.76 | 0.0287 | +20.6 | -71 | 0.0497 | + 1 | 0.0363 | -42.4 | +39 | 0.0400 | +26 |
| 0.80 | 0.0276 | +19.2 | -69 | 0.0477 | 0 | 0.0356 | -44.5 | +35 | 0.0396 | +22 |
| 0.84 | 0.0266 | +17.6 | -66 | 0.0459 | - 1 | 0.0352 | +42.6 | -31 | 0.0393 | +17 |
| 0.88 | 0.0257 | +15.6 | -64 | 0.0437 | - 1 | 0.0346 | +39.1 | -27 | 0.0391 | +13 |
| 0.92 | 0.0249 | +13.1 | -62 | 0.0416 | 0 | 0.0342 | +34.1 | -22 | 0.0389 | + 8 |
| 0.96 | 0.0241 | + 9.5 | -59 | 0.0390 | + 3 | 0.0339 | +26.2 | -18 | 0.0388 | + 4 |
| 1.00 | 0.0235 | 0 | -56 | 0.0336 | +14 | 0.0336 | 0 | -14 | 0 | 0 |
| 0.96 | 0.0241 | + 9.5 | -59 | 0.0296 | +35 | 0.0339 | +26.2 | -18 | 0.0388 | + 4 |
| 0.92 | 0.0249 | +13.1 | -62 | 0.0284 | +47 | 0.0342 | +34.1 | -22 | 0.0389 | + 8 |
| 0.88 | 0.0257 | +15.6 | -64 | 0.0279 | +56 | 0.0346 | +39.1 | -27 | 0.0391 | +13 |
| 0.84 | 0.0266 | +17.6 | -66 | 0.0278 | +64 | 0.0352 | +42.6 | -31 | 0.0393 | +17 |
| 0.80 | 0.0276 | +19.2 | -69 | 0.0278 | +71 | 0.0356 | -44.5 | +35 | 0.0396 | +22 |
| 0.76 | 0.0287 | +20.6 | -71 | 0.0281 | +78 | 0.0363 | -42.4 | +39 | 0.0400 | +26 |
| 0.72 | 0.0299 | +21.7 | -73 | 0.0288 | +83 | 0.0371 | -40.7 | +43 | 0.0405 | +31 |
| 0.68 | 0.0313 | +22.7 | -75 | 0.0297 | +87 | 0.0380 | -39.2 | +48 | 0.0412 | +36 |
| 0.64 | 0.0329 | +23.6 | -77 | 0.0309 | +90 | 0.0391 | -37.9 | +52 | 0.0421 | +41 |
| 0.60 | 0.0346 | +24.4 | -79 | 0.0324 | +92 | 0.0405 | -36.9 | +56 | 0.0431 | +46 |
| 0.56 | 0.0366 | +25.1 | -81 | 0.0343 | +93 | 0.0420 | -36.0 | +60 | 0.0444 | +51 |
| 0.52 | 0.0390 | +25.6 | -82 | 0.0366 | +94 | 0.0439 | -35.2 | +64 | 0.0460 | +56 |
| 0.48 | 0.0416 | +26.2 | -84 | 0.0394 | +93 | 0.0461 | -34.5 | +68 | 0.0480 | +61 |
| 0.44 | 0.0447 | +26.7 | -85 | 0.0428 | +92 | 0.0488 | -33.9 | +71 | 0.0504 | +65 |
| 0.40 | 0.0484 | +27.1 | -86 | 0.0470 | +90 | 0.0520 | -33.4 | +75 | 0.0533 | +70 |
| 0.36 | 0.0528 | +27.4 | -87 | 0.0522 | +88 | 0.0560 | -33.0 | +78 | 0.0570 | +74 |
| 0.32 | 0.0581 | +27.7 | -88 | 0.0587 | +85 | 0.0609 | -32.6 | +81 | 0.0616 | +79 |
| 0.28 | 0.0648 | +28.0 | -89 | 0.0668 | +82 | 0.0671 | -32.3 | +84 | 0.0674 | +82 |
| 0.24 | 0.0732 | +28.2 | -90 | 0.0772 | +78 | 0.0752 | -32.0 | +86 | 0.0751 | +86 |
| 0.20 | 0.0842 | +28.4 | -91 | 0.0911 | +75 | 0.0857 | -31.8 | +88 | 0.0853 | +89 |
| 0.16 | 0.0991 | +28.6 | -91 | 0.1100 | +71 | 0.1002 | -31.6 | +90 | 0.0993 | +91 |
| 0.12 | 0.1200 | +28.7 | -92 | 0.1368 | +66 | 0.1209 | -31.4 | +92 | 0.1194 | +93 |
| 0.08 | 0.1504 | +28.8 | -92 | 0.1765 | +62 | 0.1512 | -31.3 | +93 | 0.1491 | +95 |
| 0.04 | 0.1902 | +28.8 | -92 | 0.2307 | +58 | 0.1920 | -31.3 | +93 | 0.1893 | +96 |
| 0 | 0.1903 | +29.0 | -93 | 0.2450 | +54 | 0.1978 | -31.2 | +94 | 0.1968 | +96 |

## TABLE 7

### THE ILLUMINATION AND POLARIZATION OF THE SKY FOR VARIOUS ZENITH DISTANCES OF THE SUN AND FOR AN ATMOSPHERE OF OPTICAL THICKNESS $\tau_1 = 0.20$

| μ | $\mu_0 = 0; \varphi = 0°$ | | $\mu_0 = 0; \varphi = 90°$ | | $\mu_0 = 0.16; \varphi = 0°$ | | $\mu_0 = 0.16; \varphi = 90°$ | | | $\mu_0 = 0.50; \varphi = 0°$ | |
|---|---|---|---|---|---|---|---|---|---|---|---|
| | $I_1+I_r$ | δ | $I_1+I_r$ | δ | $I_1+I_r$ | δ | $I_1+I_r$ | α° | δ | $I_1+I_r$ | δ |
| 0    | . .    | . .  | . .    | . .  | 0.1587 | − 7 | 0.0931 | +8.4 | −85 | 0.2920 | + 8 |
| 0.04 | 0.4489 | −18  | 0.3158 | −68  | 0.2015 | − 7 | 0.1173 | +8.3 | −86 | 0.3267 | + 5 |
| 0.08 | 0.7767 | −11  | 0.4839 | −79  | 0.2143 | − 7 | 0.1231 | +8.4 | −87 | 0.3213 | + 3 |
| 0.12 | 0.9510 | − 8  | 0.5670 | −84  | 0.2019 | − 7 | 0.1151 | +8.4 | −88 | 0.2941 | + 2 |
| 0.16 | 0.9920 | − 7  | 0.5817 | −86  | . .    | . . | . .    | . .  | . . | 0.2651 |  0  |
| 0.20 | 0.9710 | − 5  | 0.5662 | −87  | 0.1658 | − 7 | 0.0940 | +8.3 | −88 | 0.2393 | − 1 |
| 0.24 | 0.9245 | − 4  | 0.5391 | −87  | 0.1499 | − 7 | 0.0850 | +8.3 | −88 | 0.2173 | − 2 |
| 0.28 | 0.8689 | − 3  | 0.5085 | −88  | 0.1360 | − 6 | 0.0773 | +8.2 | −88 | 0.1986 | − 3 |
| 0.32 | 0.8115 | − 1  | 0.4781 | −88  | 0.1239 | − 5 | 0.0707 | +8.1 | −88 | 0.1826 | − 3 |
| 0.36 | 0.7556 |  0   | 0.4491 | −88  | 0.1134 | − 5 | 0.0650 | +8.0 | −88 | 0.1688 | − 4 |
| 0.40 | 0.7026 | + 2  | 0.4222 | −88  | 0.1040 | − 4 | 0.0600 | +7.9 | −88 | 0.1567 | − 4 |
| 0.44 | 0.6527 | + 4  | 0.3975 | −89  | 0.0957 | − 2 | 0.0557 | +7.8 | −88 | 0.1460 | − 5 |
| 0.48 | 0.6062 | + 6  | 0.3748 | −89  | 0.0883 | − 1 | 0.0519 | +7.7 | −88 | 0.1364 | − 5 |
| 0.52 | 0.5627 | + 9  | 0.3541 | −89  | 0.0815 | + 1 | 0.0486 | +7.5 | −87 | 0.1278 | − 5 |
| 0.56 | 0.5220 | +12  | 0.3351 | −89  | 0.0754 | + 3 | 0.0456 | +7.3 | −87 | 0.1198 | − 4 |
| 0.60 | 0.4840 | +15  | 0.3176 | −89  | 0.0697 | + 5 | 0.0429 | +7.1 | −87 | 0.1124 | − 4 |
| 0.64 | 0.4482 | +18  | 0.3015 | −89  | 0.0645 | + 8 | 0.0404 | +6.9 | −87 | 0.1055 | − 3 |
| 0.68 | 0.4146 | +22  | 0.2867 | −89  | 0.0596 | +11 | 0.0382 | +6.7 | −87 | 0.0991 | − 2 |
| 0.72 | 0.3829 | +27  | 0.2730 | −89  | 0.0550 | +15 | 0.0362 | +6.4 | −86 | 0.0929 |  0  |
| 0.76 | 0.3528 | +32  | 0.2602 | −89  | 0.0506 | +19 | 0.0344 | +6.0 | −86 | 0.0869 | + 2 |
| 0.80 | 0.3242 | +39  | 0.2483 | −89  | 0.0464 | +24 | 0.0327 | +5.6 | −86 | 0.0811 | + 5 |
| 0.84 | 0.2970 | +46  | 0.2373 | −89  | 0.0424 | +30 | 0.0311 | +5.1 | −85 | 0.0753 | + 8 |
| 0.88 | 0.2711 | +54  | 0.2269 | −89  | 0.0385 | +38 | 0.0296 | +4.5 | −85 | 0.0695 | +12 |
| 0.92 | 0.2462 | +64  | 0.2172 | −89  | 0.0347 | +47 | 0.0283 | +3.8 | −84 | 0.0634 | +18 |
| 0.96 | 0.2223 | +75  | 0.2080 | −89  | 0.0308 | +59 | 0.0270 | +2.7 | −84 | 0.0566 | +27 |
| 1.00 | 0.1994 | +89  | 0.1994 | −89  | 0.0258 | +84 | 0.0258 |  0   | −83 | 0.0450 | +55 |
| 0.96 | 0.2223 | +75  | 0.2080 | −89  | 0.0267 | +84 | 0.0270 | +2.7 | −84 | 0.0400 | +80 |
| 0.92 | 0.2462 | +64  | 0.2172 | −89  | 0.0289 | +76 | 0.0283 | +3.8 | −84 | 0.0402 | +86 |
| 0.88 | 0.2711 | +54  | 0.2269 | −89  | 0.0316 | +68 | 0.0296 | +4.5 | −85 | 0.0415 | +88 |
| 0.84 | 0.2970 | +46  | 0.2373 | −89  | 0.0345 | +60 | 0.0311 | +5.1 | −85 | 0.0435 | +87 |
| 0.80 | 0.3242 | +39  | 0.2483 | −89  | 0.0377 | +53 | 0.0327 | +5.6 | −86 | 0.0461 | +84 |
| 0.76 | 0.3528 | +32  | 0.2602 | −89  | 0.0412 | +46 | 0.0344 | +6.0 | −86 | 0.0493 | +80 |
| 0.72 | 0.3829 | +27  | 0.2730 | −89  | 0.0451 | +40 | 0.0362 | +6.4 | −86 | 0.0530 | +75 |
| 0.68 | 0.4146 | +22  | 0.2867 | −89  | 0.0492 | +34 | 0.0382 | +6.7 | −87 | 0.0573 | +70 |
| 0.64 | 0.4482 | +18  | 0.3015 | −89  | 0.0537 | +29 | 0.0404 | +6.9 | −87 | 0.0622 | +65 |
| 0.60 | 0.4840 | +15  | 0.3176 | −89  | 0.0587 | +25 | 0.0429 | +7.1 | −87 | 0.0678 | +60 |
| 0.56 | 0.5220 | +12  | 0.3351 | −89  | 0.0641 | +21 | 0.0456 | +7.3 | −87 | 0.0740 | +55 |
| 0.52 | 0.5627 | + 9  | 0.3541 | −89  | 0.0701 | +17 | 0.0486 | +7.5 | −87 | 0.0813 | +50 |
| 0.48 | 0.6062 | + 6  | 0.3748 | −89  | 0.0767 | +14 | 0.0519 | +7.7 | −88 | 0.0894 | +45 |
| 0.44 | 0.6527 | + 4  | 0.3975 | −89  | 0.0841 | +11 | 0.0557 | +7.8 | −88 | 0.0988 | +41 |
| 0.40 | 0.7026 | + 2  | 0.4222 | −88  | 0.0925 | + 8 | 0.0600 | +7.9 | −88 | 0.1096 | +37 |
| 0.36 | 0.7556 |  0   | 0.4491 | −88  | 0.1020 | + 6 | 0.0650 | +8.0 | −88 | 0.1221 | +33 |
| 0.32 | 0.8115 | − 1  | 0.4781 | −88  | 0.1128 | + 4 | 0.0707 | +8.1 | −88 | 0.1368 | +29 |
| 0.28 | 0.8689 | − 3  | 0.5085 | −88  | 0.1253 | + 2 | 0.0773 | +8.2 | −88 | 0.1540 | +25 |
| 0.24 | 0.9245 | − 4  | 0.5391 | −87  | 0.1397 |  0  | 0.0850 | +8.3 | −88 | 0.1747 | +22 |
| 0.20 | 0.9710 | − 5  | 0.5662 | −87  | 0.1563 | − 1 | 0.0940 | +8.3 | −88 | 0.1994 | +19 |
| 0.16 | 0.9920 | − 7  | 0.5817 | −86  | . .    | . . | . .    | . .  | . . | 0.2290 | +16 |
| 0.12 | 0.9510 | − 8  | 0.5670 | −84  | 0.1949 | − 4 | 0.1151 | +8.4 | −88 | 0.2635 | +14 |
| 0.08 | 0.7767 | −11  | 0.4839 | −79  | 0.2093 | − 5 | 0.1231 | +8.4 | −87 | 0.2986 | +11 |
| 0.04 | 0.4489 | −18  | 0.3158 | −68  | 0.1992 | − 6 | 0.1173 | +8.3 | −86 | 0.3150 | + 9 |
| 0    | . .    | . .  | . .    | . .  | 0.1587 | − 7 | 0.0931 | +8.4 | −85 | 0.2920 | + 8 |

TABLE 7 (continued)

| μ | $\mu_0 = 0.50; \varphi = 90°$ | | | $\mu_0 = 0.86; \varphi = 0°$ | | $\mu_0 = 0.86; \varphi = 90°$ | | | $\mu_0 = 1.00$ | |
|---|---|---|---|---|---|---|---|---|---|---|
| | $I_1+I_r$ | $\chi°$ | $\delta$ | $I_1+I_r$ | $\delta$ | $I_1+I_r$ | $\chi°$ | $\delta$ | $I_1+I_r$ | $\delta$ |
| 0 | 0.1821 | +28.5 | −88 | 0.2442 | +51 | 0.1995 | −31.5 | +89 | 0.1994 | +92 |
| 0.04 | 0.2020 | +28.2 | −87 | 0.2671 | +47 | 0.2136 | −31.7 | +88 | 0.2111 | +91 |
| 0.08 | 0.1955 | +28.2 | −87 | 0.2608 | +43 | 0.2037 | −31.7 | +88 | 0.2008 | +90 |
| 0.12 | 0.1763 | +28.1 | −87 | 0.2394 | +39 | 0.1830 | −31.8 | +87 | 0.1806 | +89 |
| 0.16 | 0.1569 | +28.0 | −87 | 0.2176 | +36 | 0.1631 | −32.0 | +86 | 0.1612 | +87 |
| 0.20 | 0.1400 | +27.9 | −86 | 0.1985 | +32 | 0.1462 | −32.1 | +84 | 0.1451 | +85 |
| 0.24 | 0.1260 | +27.7 | −86 | 0.1826 | +29 | 0.1324 | −32.3 | +82 | 0.1320 | +82 |
| 0.28 | 0.1143 | +27.5 | −85 | 0.1694 | +26 | 0.1212 | −32.6 | +80 | 0.1214 | +79 |
| 0.32 | 0.1046 | +27.3 | −84 | 0.1582 | +23 | 0.1120 | −32.9 | +78 | 0.1129 | +76 |
| 0.36 | 0.0963 | +27.0 | −83 | 0.1488 | +20 | 0.1044 | −33.3 | +75 | 0.1059 | +72 |
| 0.40 | 0.0893 | +26.7 | −82 | 0.1407 | +17 | 0.0981 | −33.7 | +72 | 0.1002 | +68 |
| 0.44 | 0.0833 | +26.3 | −81 | 0.1336 | +14 | 0.0928 | −34.2 | +69 | 0.0955 | +64 |
| 0.48 | 0.0780 | +25.9 | −80 | 0.1274 | +12 | 0.0883 | −34.7 | +65 | 0.0917 | +59 |
| 0.52 | 0.0736 | +25.3 | −79 | 0.1218 | +10 | 0.0845 | −35.4 | +62 | 0.0885 | +55 |
| 0.56 | 0.0694 | +24.8 | −77 | 0.1169 | + 8 | 0.0813 | −36.1 | +58 | 0.0858 | +50 |
| 0.60 | 0.0659 | +24.2 | −76 | 0.1123 | + 6 | 0.0786 | −37.0 | +54 | 0.0837 | +45 |
| 0.64 | 0.0627 | +23.5 | −74 | 0.1081 | + 4 | 0.0762 | −38.0 | +50 | 0.0819 | +40 |
| 0.68 | 0.0599 | +22.7 | −72 | 0.1041 | + 3 | 0.0743 | −39.1 | +46 | 0.0805 | +36 |
| 0.72 | 0.0573 | +21.7 | −70 | 0.1004 | + 1 | 0.0726 | −40.5 | +42 | 0.0793 | +31 |
| 0.76 | 0.0551 | +20.6 | −68 | 0.0967 | 0 | 0.0711 | −42.2 | +38 | 0.0785 | +26 |
| 0.80 | 0.0530 | +19.3 | −66 | 0.0932 | 0 | 0.0699 | −44.2 | +34 | 0.0778 | +22 |
| 0.84 | 0.0511 | +17.8 | −64 | 0.0896 | − 1 | 0.0690 | +43.1 | −30 | 0.0773 | +17 |
| 0.88 | 0.0493 | +15.8 | −61 | 0.0856 | 0 | 0.0679 | +39.8 | −26 | 0.0770 | +13 |
| 0.92 | 0.0478 | +13.3 | −59 | 0.0816 | 0 | 0.0673 | +35.0 | −22 | 0.0768 | + 9 |
| 0.96 | 0.0463 | + 9.7 | −56 | 0.0767 | + 3 | 0.0668 | +27.3 | −18 | 0.0768 | + 5 |
| 1.00 | 0.0450 | 0 | −53 | 0.0663 | +15 | 0.0663 | 0 | −13 | 0 | 0 |
| 0.96 | 0.0463 | + 9.7 | −56 | 0.0583 | +35 | 0.0668 | +27.3 | −18 | 0.0768 | + 5 |
| 0.92 | 0.0478 | +13.3 | −59 | 0.0561 | +46 | 0.0673 | +35.0 | −22 | 0.0768 | + 9 |
| 0.88 | 0.0493 | +15.8 | −61 | 0.0549 | +55 | 0.0679 | +39.8 | −26 | 0.0770 | +13 |
| 0.84 | 0.0511 | +17.8 | −64 | 0.0546 | +63 | 0.0690 | +43.1 | −30 | 0.0773 | +17 |
| 0.80 | 0.0530 | +19.3 | −66 | 0.0547 | +70 | 0.0699 | −44.2 | +34 | 0.0778 | +22 |
| 0.76 | 0.0551 | +20.6 | −68 | 0.0554 | +75 | 0.0711 | −42.2 | +38 | 0.0785 | +26 |
| 0.72 | 0.0573 | +21.7 | −70 | 0.0565 | +80 | 0.0726 | −40.5 | +42 | 0.0793 | +31 |
| 0.68 | 0.0599 | +22.7 | −72 | 0.0582 | +84 | 0.0743 | −39.1 | +46 | 0.0805 | +36 |
| 0.64 | 0.0627 | +23.5 | −74 | 0.0603 | +87 | 0.0762 | −38.0 | +50 | 0.0819 | +40 |
| 0.60 | 0.0659 | +24.2 | −76 | 0.0631 | +88 | 0.0786 | −37.0 | +54 | 0.0837 | +45 |
| 0.56 | 0.0694 | +24.8 | −77 | 0.0665 | +89 | 0.0813 | −36.1 | +58 | 0.0858 | +50 |
| 0.52 | 0.0736 | +25.3 | −79 | 0.0706 | +90 | 0.0845 | −35.4 | +62 | 0.0885 | +55 |
| 0.48 | 0.0780 | +25.9 | −80 | 0.0756 | +89 | 0.0883 | −34.7 | +65 | 0.0917 | +59 |
| 0.44 | 0.0833 | +26.3 | −81 | 0.0815 | +88 | 0.0928 | −34.2 | +69 | 0.0955 | +64 |
| 0.40 | 0.0893 | +26.7 | −82 | 0.0886 | +86 | 0.0981 | −33.7 | +72 | 0.1002 | +68 |
| 0.36 | 0.0963 | +27.0 | −83 | 0.0972 | +83 | 0.1044 | −33.3 | +75 | 0.1059 | +72 |
| 0.32 | 0.1046 | +27.3 | −84 | 0.1075 | +81 | 0.1120 | −32.9 | +78 | 0.1129 | +76 |
| 0.28 | 0.1143 | +27.5 | −85 | 0.1201 | +77 | 0.1212 | −32.6 | +80 | 0.1214 | +79 |
| 0.24 | 0.1260 | +27.7 | −86 | 0.1354 | +74 | 0.1324 | −32.3 | +82 | 0.1320 | +82 |
| 0.20 | 0.1400 | +27.9 | −86 | 0.1542 | +70 | 0.1462 | −32.1 | +84 | 0.1451 | +85 |
| 0.16 | 0.1569 | +28.0 | −87 | 0.1774 | +66 | 0.1631 | −32.0 | +86 | 0.1612 | +87 |
| 0.12 | 0.1763 | +28.1 | −87 | 0.2053 | +62 | 0.1830 | −31.8 | +87 | 0.1806 | +89 |
| 0.08 | 0.1955 | +28.2 | −87 | 0.2352 | +58 | 0.2037 | −31.7 | +88 | 0.2008 | +90 |
| 0.04 | 0.2020 | +28.2 | −87 | 0.2537 | +54 | 0.2136 | −31.7 | +88 | 0.2111 | +91 |
| 0 | 0.1821 | +28.5 | −88 | 0.2442 | +51 | 0.1995 | −31.5 | +89 | 0.1994 | +92 |

TABLE 8

THE EFFECT OF A GROUND REFLECTING ACCORDING TO LAMBERT'S LAW WITH AN ALBEDO $\lambda_0$

ON THE ILLUMINATION AND POLARIZATION OF THE SKY ( $\tau_1 = 0.15$ )

$\varphi = 0°$

| μ | $\mu_0=0; \lambda_*=0.10$ $I_l+I_r$ | δ | $\mu_0=0; \lambda_*=0.25$ $I_l+I_r$ | δ | $\mu_0=0.24;\lambda_*=0.10$ $I_l+I_r$ | δ | $\mu_0=0.24;\lambda_*=0.25$ $I_l+I_r$ | δ | $\mu_0=0.52;\lambda_*=0.10$ $I_l+I_r$ | δ | $\mu_0=0.52;\lambda_*=0.25$ $I_l+I_r$ | δ | $\mu_0=0.64;\lambda_*=0.10$ $I_l+I_r$ | δ |
|---|---|---|---|---|---|---|---|---|---|---|---|---|---|---|
| 0 | | | | | 0.2629 | - 3 | 0.2801 | - 3 | 0.3311 | + 9 | 0.3737 | + 8 | 0.3278 | +18 |
| 0.04 | 0.6682 | -15 | 0.7018 | -14 | 0.3021 | - 4 | 0.3175 | - 4 | 0.3554 | + 7 | 0.3935 | + 6 | 0.3479 | +15 |
| 0.08 | 1.1399 | - 8 | 1.1680 | - 8 | 0.2801 | - 5 | 0.2929 | - 5 | 0.3215 | + 5 | 0.3532 | + 4 | 0.3139 | +13 |
| 0.12 | 1.2589 | - 6 | 1.2821 | - 6 | 0.2431 | - 5 | 0.2537 | - 5 | 0.2778 | + 3 | 0.3042 | + 3 | 0.2717 | +11 |
| 0.16 | 1.2267 | - 5 | 1.2464 | - 5 | 0.2107 | - 5 | 0.2196 | - 5 | 0.2415 | + 2 | 0.2637 | + 2 | 0.2368 | + 9 |
| 0.20 | 1.1478 | - 4 | 1.1648 | - 4 | 0.1844 | - 6 | 0.1921 | - 5 | 0.2127 | + 1 | 0.2319 | + 1 | 0.2094 | + 7 |
| 0.24 | 1.0588 | - 3 | 1.0737 | - 3 | | | | | 0.1899 | 0 | 0.2067 | 0 | 0.1876 | + 5 |
| 0.28 | 0.9722 | - 2 | 0.9855 | - 2 | 0.1461 | - 5 | 0.1521 | - 5 | 0.1714 | - 1 | 0.1863 | - 1 | 0.1701 | + 4 |
| 0.32 | 0.8921 | 0 | 0.9040 | 0 | 0.1318 | - 5 | 0.1372 | - 5 | 0.1561 | - 2 | 0.1696 | - 2 | 0.1557 | + 3 |
| 0.36 | 0.8191 | + 1 | 0.8299 | + 1 | 0.1196 | - 5 | 0.1246 | - 4 | 0.1432 | - 3 | 0.1555 | - 2 | 0.1436 | + 1 |
| 0.40 | 0.7530 | + 3 | 0.7629 | + 3 | 0.1092 | - 4 | 0.1137 | - 4 | 0.1322 | - 3 | 0.1484 | - 3 | 0.1332 | 0 |
| 0.44 | 0.6931 | + 5 | 0.7023 | + 5 | 0.1001 | - 3 | 0.1043 | - 3 | 0.1226 | - 4 | 0.1330 | - 3 | 0.1243 | - 1 |
| 0.48 | 0.6387 | + 7 | 0.6472 | + 7 | 0.0921 | - 2 | 0.0960 | - 2 | 0.1142 | - 4 | 0.1238 | - 3 | 0.1164 | - 1 |
| 0.52 | 0.5890 | +10 | 0.5969 | +10 | 0.0850 | - 1 | 0.0886 | - 1 | | | | | 0.1094 | - 2 |
| 0.56 | 0.5433 | +12 | 0.5508 | +12 | 0.0785 | + 1 | 0.0819 | + 1 | 0.0998 | - 3 | 0.1082 | - 3 | 0.1030 | - 2 |
| 0.60 | 0.5013 | +16 | 0.5082 | +15 | 0.0726 | + 3 | 0.0758 | + 3 | 0.0935 | - 3 | 0.1014 | - 3 | 0.0972 | - 3 |
| 0.64 | 0.4623 | +19 | 0.4689 | +19 | 0.0672 | + 5 | 0.0702 | + 5 | 0.0877 | - 2 | 0.0952 | - 2 | | |
| 0.68 | 0.4260 | +23 | 0.4322 | +23 | 0.0621 | + 8 | 0.0650 | + 7 | 0.0823 | - 1 | 0.0894 | - 1 | 0.0868 | - 2 |
| 0.72 | 0.3921 | +28 | 0.3980 | +28 | 0.0574 | +11 | 0.0601 | +10 | 0.0772 | 0 | 0.0889 | 0 | 0.0820 | - 2 |
| 0.76 | 0.3602 | +33 | 0.3659 | +33 | 0.0529 | +14 | 0.0555 | +14 | 0.0723 | + 2 | 0.0787 | + 2 | 0.0774 | - 1 |
| 0.80 | 0.3302 | +40 | 0.3356 | +39 | 0.0487 | +19 | 0.0511 | +18 | 0.0675 | + 4 | 0.0736 | + 4 | 0.0729 | 0 |
| 0.84 | 0.3018 | +47 | 0.3070 | +46 | 0.0446 | +24 | 0.0469 | +23 | 0.0628 | + 7 | 0.0686 | + 6 | 0.0685 | + 2 |
| 0.88 | 0.2749 | +55 | 0.2798 | +54 | 0.0406 | +31 | 0.0428 | +29 | 0.0581 | +11 | 0.0637 | +10 | 0.0689 | + 5 |
| 0.92 | 0.2492 | +65 | 0.2539 | +63 | 0.0366 | +39 | 0.0388 | +37 | 0.0532 | +16 | 0.0586 | +15 | 0.0592 | + 9 |
| 0.96 | 0.2247 | +76 | 0.2292 | +74 | 0.0325 | +51 | 0.0345 | +48 | 0.0478 | +24 | 0.0529 | +22 | 0.0538 | +15 |
| 1.00 | 0.2012 | +90 | 0.2056 | +88 | 0.0267 | +77 | 0.0287 | +72 | 0.0385 | +49 | 0.0434 | +43 | 0.0440 | +36 |
| 0.96 | 0.2247 | +76 | 0.2292 | +74 | 0.0265 | +85 | 0.0286 | +79 | 0.0344 | +73 | 0.0396 | +63 | 0.0386 | +60 |
| 0.92 | 0.2492 | +65 | 0.2539 | +63 | 0.0283 | +80 | 0.0304 | +74 | 0.0346 | +79 | 0.0399 | +68 | 0.0380 | +69 |
| 0.88 | 0.2749 | +55 | 0.2798 | +54 | 0.0305 | +74 | 0.0328 | +69 | 0.0356 | +81 | 0.0412 | +70 | 0.0364 | +75 |
| 0.84 | 0.3018 | +47 | 0.3070 | +46 | 0.0332 | +67 | 0.0355 | +62 | 0.0372 | +80 | 0.0431 | +69 | 0.0394 | +78 |
| 0.80 | 0.3302 | +40 | 0.3356 | +39 | 0.0361 | +60 | 0.0386 | +56 | 0.0394 | +78 | 0.0454 | +68 | 0.0409 | +79 |
| 0.76 | 0.3602 | +33 | 0.3659 | +33 | 0.0394 | +54 | 0.0420 | +50 | 0.0419 | +75 | 0.0483 | +65 | 0.0429 | +79 |
| 0.72 | 0.3921 | +28 | 0.3980 | +28 | 0.0430 | +48 | 0.0457 | +45 | 0.0450 | +72 | 0.0517 | +62 | 0.0453 | +78 |
| 0.68 | 0.4260 | +23 | 0.4322 | +23 | 0.0470 | +42 | 0.0499 | +40 | 0.0485 | +68 | 0.0555 | +59 | 0.0483 | +76 |
| 0.64 | 0.4623 | +19 | 0.4689 | +19 | 0.0514 | +37 | 0.0545 | +35 | 0.0525 | +63 | 0.0600 | +55 | | |
| 0.60 | 0.5013 | +16 | 0.5082 | +15 | 0.0564 | +32 | 0.0595 | +31 | 0.0572 | +59 | 0.0651 | +52 | 0.0558 | +70 |
| 0.56 | 0.5433 | +12 | 0.5508 | +12 | 0.0618 | +28 | 0.0652 | +27 | 0.0625 | +54 | 0.0709 | +48 | 0.0605 | +66 |
| 0.52 | 0.5890 | +10 | 0.5969 | +10 | 0.0680 | +24 | 0.0716 | +23 | | | | | 0.0660 | +62 |
| 0.48 | 0.6387 | + 7 | 0.6472 | + 7 | 0.0749 | +21 | 0.0788 | +20 | 0.0755 | +46 | 0.0851 | +41 | 0.0724 | +58 |
| 0.44 | 0.6931 | + 5 | 0.7023 | + 5 | 0.0827 | +17 | 0.0869 | +17 | 0.0836 | +42 | 0.0939 | +37 | 0.0799 | +55 |
| 0.40 | 0.7530 | + 3 | 0.7629 | + 3 | 0.0918 | +14 | 0.0963 | +14 | 0.0930 | +38 | 0.1042 | +34 | 0.0887 | +51 |
| 0.36 | 0.8191 | + 1 | 0.8299 | + 1 | 0.1023 | +12 | 0.1072 | +11 | 0.1041 | +34 | 0.1164 | +30 | 0.0991 | +47 |
| 0.32 | 0.8921 | 0 | 0.9040 | 0 | 0.1146 | + 9 | 0.1200 | + 9 | 0.1174 | +30 | 0.1309 | +27 | 0.1116 | +43 |
| 0.28 | 0.9722 | - 2 | 0.9855 | - 2 | 0.1293 | + 7 | 0.1353 | + 7 | 0.1324 | +27 | 0.1484 | +24 | 0.1269 | +39 |
| 0.24 | 1.0588 | - 3 | 1.0737 | - 3 | | | | | 0.1531 | +24 | 0.1699 | +21 | 0.1457 | +36 |
| 0.20 | 1.1478 | - 4 | 1.1648 | - 4 | 0.1689 | + 3 | 0.1767 | + 3 | 0.1777 | +21 | 0.1966 | +19 | 0.1694 | +32 |
| 0.16 | 1.2267 | - 5 | 1.2464 | - 5 | 0.1964 | + 1 | 0.2054 | + 1 | 0.2090 | +18 | 0.2313 | +16 | 0.1998 | +29 |
| 0.12 | 1.2589 | - 6 | 1.2821 | - 6 | 0.2207 | 0 | 0.2413 | 0 | 0.2493 | +15 | 0.2756 | +14 | 0.2391 | +26 |
| 0.08 | 1.1399 | - 8 | 1.1680 | - 8 | 0.2705 | - 1 | 0.2833 | - 1 | 0.2991 | +13 | 0.3308 | +12 | 0.2883 | +23 |
| 0.04 | 0.6682 | -15 | 0.7018 | -14 | 0.2969 | - 3 | 0.3123 | - 2 | 0.3429 | +11 | 0.3809 | + 9 | 0.3334 | +20 |
| 0 | | | | | 0.2629 | - 3 | 0.2801 | - 3 | 0.3311 | + 9 | 0.3737 | + 8 | 0.3278 | +18 |

TABLE 8 (continued)

| μ | $\varphi = 0°$ $\mu_o=0.64;\lambda=0.25$ | | $\mu_o = 0.64;\ \lambda_o = 0.10$ | | | | | | | | | |
|---|---|---|---|---|---|---|---|---|---|---|---|---|
| | | | $\varphi = 10°$ | $\varphi = 20°$ | $\varphi = 30°$ | $\varphi = 40°$ | $\varphi = 50°$ | $\varphi = 60°$ | $\varphi = 70°$ | $\varphi = 80°$ | $\varphi = 90°$ |
| | $I_1+I_r$ | $\delta$ | $I_1+I_r$ | $I_1+I_r$ | $I_1+I_r$ | $I_1+I_r$ | $I_1+I_r$ | $I_1+I_r$ | $I_1+I_r$ | $I_1+I_r$ | $I_1+I_r$ |
| 0 | 0.3815 | +15 | 0.3249 | 0.3163 | 0.3033 | 0.2872 | 0.2701 | 0.2541 | 0.2410 | 0.2325 | 0.2295 |
| 0.04 | 0.3958 | +13 | 0.3447 | 0.3355 | 0.3213 | 0.3039 | 0.2852 | 0.2675 | 0.2527 | 0.2426 | 0.2383 |
| 0.08 | 0.3839 | +11 | 0.3109 | 0.3025 | 0.2895 | 0.2734 | 0.2561 | 0.2395 | 0.2254 | 0.2154 | 0.2105 |
| 0.12 | 0.3048 | +10 | 0.2691 | 0.2617 | 0.2504 | 0.2362 | 0.2209 | 0.2061 | 0.1934 | 0.1840 | 0.1789 |
| 0.16 | 0.2648 | + 8 | 0.2345 | 0.2281 | 0.2181 | 0.2057 | 0.1921 | 0.1789 | 0.1674 | 0.1586 | 0.1534 |
| 0.20 | 0.2335 | + 6 | 0.2074 | 0.2017 | 0.1928 | 0.1817 | 0.1696 | 0.1577 | 0.1471 | 0.1389 | 0.1338 |
| 0.24 | 0.2088 | + 5 | 0.1859 | 0.1807 | 0.1728 | 0.1628 | 0.1519 | 0.1410 | 0.1313 | 0.1236 | 0.1185 |
| 0.28 | 0.1890 | + 3 | 0.1685 | 0.1639 | 0.1567 | 0.1477 | 0.1377 | 0.1277 | 0.1187 | 0.1114 | 0.1064 |
| 0.32 | 0.1727 | + 2 | 0.1542 | 0.1500 | 0.1435 | 0.1352 | 0.1261 | 0.1169 | 0.1085 | 0.1016 | 0.0966 |
| 0.36 | 0.1590 | + 1 | 0.1423 | 0.1384 | 0.1324 | 0.1248 | 0.1164 | 0.1079 | 0.1000 | 0.0934 | 0.0886 |
| 0.40 | 0.1474 | 0 | 0.1320 | 0.1285 | 0.1230 | 0.1160 | 0.1082 | 0.1003 | 0.0930 | 0.0867 | 0.0819 |
| 0.44 | 0.1373 | - 1 | 0.1232 | 0.1199 | 0.1149 | 0.1084 | 0.1012 | 0.0939 | 0.0869 | 0.0810 | 0.0763 |
| 0.48 | 0.1285 | - 1 | 0.1154 | 0.1124 | 0.1077 | 0.1018 | 0.0951 | 0.0883 | 0.0817 | 0.0761 | 0.0715 |
| 0.52 | 0.1207 | - 2 | 0.1084 | 0.1057 | 0.1014 | 0.0959 | 0.0897 | 0.0834 | 0.0772 | 0.0718 | 0.0675 |
| 0.56 | 0.1136 | - 2 | 0.1021 | 0.0996 | 0.0957 | 0.0906 | 0.0849 | 0.0790 | 0.0732 | 0.0681 | 0.0639 |
| 0.60 | 0.1071 | - 2 | 0.0964 | 0.0941 | 0.0905 | 0.0859 | 0.0806 | 0.0751 | 0.0697 | 0.0649 | 0.0608 |
| 0.64 | .. | .. | .. | .. | .. | .. | .. | .. | .. | .. | .. |
| 0.68 | 0.0957 | - 2 | 0.0861 | 0.0842 | 0.0812 | 0.0774 | 0.0730 | 0.0683 | 0.0637 | 0.0594 | 0.0557 |
| 0.72 | 0.0904 | - 2 | 0.0814 | 0.0797 | 0.0770 | 0.0735 | 0.0695 | 0.0651 | 0.0611 | 0.0571 | 0.0536 |
| 0.76 | 0.0854 | - 1 | 0.0769 | 0.0754 | 0.0730 | 0.0699 | 0.0663 | 0.0625 | 0.0586 | 0.0550 | 0.0517 |
| 0.80 | 0.0806 | 0 | 0.0724 | 0.0711 | 0.0690 | 0.0663 | 0.0632 | 0.0598 | 0.0563 | 0.0530 | 0.0501 |
| 0.84 | 0.0758 | + 2 | 0.0681 | 0.0670 | 0.0652 | 0.0628 | 0.0601 | 0.0572 | 0.0542 | 0.0513 | 0.0486 |
| 0.88 | 0.0710 | + 4 | 0.0636 | 0.0627 | 0.0612 | 0.0593 | 0.0571 | 0.0546 | 0.0521 | 0.0496 | 0.0472 |
| 0.92 | 0.0659 | + 8 | 0.0589 | 0.0582 | 0.0571 | 0.0557 | 0.0539 | 0.0520 | 0.0500 | 0.0480 | 0.0460 |
| 0.96 | 0.0603 | +13 | 0.0537 | 0.0532 | 0.0525 | 0.0515 | 0.0504 | 0.0491 | 0.0477 | 0.0463 | 0.0450 |
| 1.00 | 0.0502 | +31 | 0.0440 | 0.0440 | 0.0440 | 0.0440 | 0.0440 | 0.0440 | 0.0440 | 0.0440 | 0.0440 |
| 0.96 | 0.0451 | +52 | 0.0387 | 0.0389 | 0.0393 | 0.0399 | 0.0406 | 0.0415 | 0.0425 | 0.0437 | 0.0450 |
| 0.92 | 0.0448 | +59 | 0.0381 | 0.0384 | 0.0388 | 0.0394 | 0.0403 | 0.0414 | 0.0427 | 0.0443 | 0.0460 |
| 0.88 | 0.0454 | +63 | 0.0385 | 0.0387 | 0.0391 | 0.0397 | 0.0406 | 0.0418 | 0.0433 | 0.0451 | 0.0472 |
| 0.84 | 0.0467 | +66 | 0.0394 | 0.0396 | 0.0400 | 0.0406 | 0.0414 | 0.0426 | 0.0442 | 0.0462 | 0.0486 |
| 0.80 | 0.0486 | +67 | 0.0409 | 0.0410 | 0.0413 | 0.0418 | 0.0426 | 0.0438 | 0.0454 | 0.0475 | 0.0501 |
| 0.76 | 0.0509 | +66 | 0.0429 | 0.0429 | 0.0431 | 0.0434 | 0.0441 | 0.0452 | 0.0468 | 0.0490 | 0.0517 |
| 0.72 | 0.0538 | +65 | 0.0453 | 0.0453 | 0.0453 | 0.0455 | 0.0460 | 0.0470 | 0.0485 | 0.0507 | 0.0536 |
| 0.68 | 0.0572 | +64 | 0.0482 | 0.0480 | 0.0479 | 0.0479 | 0.0482 | 0.0490 | 0.0505 | 0.0527 | 0.0557 |
| 0.64 | .. | .. | .. | .. | .. | .. | .. | .. | .. | .. | .. |
| 0.60 | 0.0657 | +59 | 0.0556 | 0.0552 | 0.0546 | 0.0541 | 0.0540 | 0.0544 | 0.0556 | 0.0577 | 0.0608 |
| 0.56 | 0.0711 | +56 | 0.0603 | 0.0597 | 0.0589 | 0.0581 | 0.0576 | 0.0577 | 0.0587 | 0.0607 | 0.0639 |
| 0.52 | 0.0773 | +53 | 0.0657 | 0.0650 | 0.0639 | 0.0627 | 0.0619 | 0.0617 | 0.0624 | 0.0643 | 0.0675 |
| 0.48 | 0.0845 | +50 | 0.0720 | 0.0710 | 0.0696 | 0.0681 | 0.0668 | 0.0663 | 0.0667 | 0.0684 | 0.0715 |
| 0.44 | 0.0929 | +47 | 0.0794 | 0.0782 | 0.0764 | 0.0744 | 0.0727 | 0.0717 | 0.0717 | 0.0732 | 0.0763 |
| 0.40 | 0.1028 | +44 | 0.0881 | 0.0866 | 0.0844 | 0.0819 | 0.0796 | 0.0780 | 0.0777 | 0.0789 | 0.0819 |
| 0.36 | 0.1145 | +40 | 0.0985 | 0.0966 | 0.0939 | 0.0908 | 0.0878 | 0.0857 | 0.0848 | 0.0857 | 0.0886 |
| 0.32 | 0.1286 | +37 | 0.1109 | 0.1086 | 0.1053 | 0.1015 | 0.0978 | 0.0949 | 0.0934 | 0.0939 | 0.0966 |
| 0.28 | 0.1457 | +34 | 0.1259 | 0.1233 | 0.1192 | 0.1145 | 0.1099 | 0.1061 | 0.1039 | 0.1039 | 0.1064 |
| 0.24 | 0.1669 | +31 | 0.1446 | 0.1413 | 0.1365 | 0.1307 | 0.1249 | 0.1201 | 0.1170 | 0.1163 | 0.1185 |
| 0.20 | 0.1936 | +28 | 0.1680 | 0.1641 | 0.1582 | 0.1511 | 0.1439 | 0.1377 | 0.1335 | 0.1320 | 0.1338 |
| 0.16 | 0.2278 | +25 | 0.1981 | 0.1933 | 0.1860 | 0.1773 | 0.1683 | 0.1604 | 0.1547 | 0.1522 | 0.1534 |
| 0.12 | 0.2723 | +23 | 0.2370 | 0.2311 | 0.2222 | 0.2113 | 0.2000 | 0.1899 | 0.1823 | 0.1784 | 0.1789 |
| 0.08 | 0.3283 | +20 | 0.2857 | 0.2784 | 0.2673 | 0.2538 | 0.2397 | 0.2267 | 0.2167 | 0.2110 | 0.2105 |
| 0.04 | 0.3814 | +17 | 0.3305 | 0.3219 | 0.3088 | 0.2928 | 0.2760 | 0.2603 | 0.2478 | 0.2401 | 0.2383 |
| 0 | 0.3815 | +15 | 0.3249 | 0.3163 | 0.3033 | 0.2872 | 0.2701 | 0.2541 | 0.2410 | 0.2325 | 0.2295 |

TABLE 8 (continued)

| μ | $\mu_o=0.64$ $\lambda_o=0.25$ $\varphi=90°$ $I_1+I_r$ | $\mu_o=0.72;\ \varphi=0°$ $\lambda_o=0.10$ $I_1+I_r$ | $\delta$ | $\lambda_o=0.25$ $I_1+I_r$ | $\delta$ | $\mu_o=0.80;\ \varphi=0°$ $\lambda_o=0.10$ $I_1+I_r$ | $\delta$ | $\lambda_o=0.25$ $I_1+I_r$ | $\delta$ | $\mu_o=1.00$ $\lambda_o=0.10$ $I_1+I_r$ | $\delta$ | $\lambda_o=0.25$ $I_1+I_r$ | $\delta$ |
|---|---|---|---|---|---|---|---|---|---|---|---|---|---|
| 0 | 0.2832 | 0.3190 | +25 | 0.3801 | +21 | 0.3057 | +35 | 0.3742 | +28 | 0.2548 | +72 | 0.3419 | +53 |
| 0.04 | 0.2863 | 0.3363 | +23 | 0.3909 | +19 | 0.3200 | +32 | 0.3612 | +27 | 0.2556 | +75 | 0.3335 | +57 |
| 0.08 | 0.2505 | 0.3030 | +20 | 0.3486 | +17 | 0.2879 | +29 | 0.3390 | +25 | 0.2238 | +75 | 0.2888 | +58 |
| 0.12 | 0.2120 | 0.2625 | +18 | 0.3003 | +15 | 0.2495 | +27 | 0.2918 | +23 | 0.1901 | +74 | 0.2439 | +58 |
| 0.16 | 0.1815 | 0.2293 | +15 | 0.2612 | +13 | 0.2182 | +24 | 0.2540 | +20 | 0.1636 | +73 | 0.2091 | +57 |
| 0.20 | 0.1580 | 0.2032 | +13 | 0.2307 | +11 | 0.1937 | +21 | 0.2246 | +18 | 0.1434 | +71 | 0.1827 | +56 |
| 0.24 | 0.1397 | 0.1826 | +11 | 0.2067 | +10 | 0.1745 | +19 | 0.2016 | +16 | 0.1280 | +69 | 0.1624 | +55 |
| 0.28 | 0.1253 | 0.1660 | + 9 | 0.1875 | + 8 | 0.1591 | +16 | 0.1832 | +14 | 0.1160 | +67 | 0.1466 | +53 |
| 0.32 | 0.1136 | 0.1524 | + 7 | 0.1718 | + 6 | 0.1466 | +14 | 0.1683 | +12 | 0.1064 | +64 | 0.1340 | +51 |
| 0.36 | 0.1040 | 0.1411 | + 6 | 0.1587 | + 5 | 0.1361 | +12 | 0.1558 | +10 | 0.0988 | +61 | 0.1239 | +49 |
| 0.40 | 0.0961 | 0.1314 | + 4 | 0.1475 | + 4 | 0.1273 | +10 | 0.1453 | + 8 | 0.0926 | +58 | 0.1155 | +46 |
| 0.44 | 0.0894 | 0.1231 | + 3 | 0.1379 | + 2 | 0.1197 | + 8 | 0.1363 | + 7 | 0.0874 | +54 | 0.1087 | +44 |
| 0.48 | 0.0837 | 0.1157 | + 1 | 0.1295 | + 1 | 0.1130 | + 6 | 0.1285 | + 5 | 0.0832 | +51 | 0.1029 | +41 |
| 0.52 | 0.0788 | 0.1092 | 0 | 0.1221 | 0 | 0.1071 | + 4 | 0.1216 | + 4 | 0.0797 | +47 | 0.0981 | +38 |
| 0.56 | 0.0745 | 0.1033 | 0 | 0.1154 | 0 | 0.1019 | + 3 | 0.1154 | + 3 | 0.0768 | +43 | 0.0940 | +35 |
| 0.60 | 0.0708 | 0.0980 | - 1 | 0.1093 | - 1 | 0.0971 | + 2 | 0.1098 | + 2 | 0.0744 | +39 | 0.0905 | +32 |
| 0.64 | . . | 0.0930 | - 2 | 0.1037 | - 1 | 0.0927 | + 1 | 0.1047 | + 1 | 0.0723 | +35 | 0.0876 | +29 |
| 0.68 | 0.0646 | 0.0884 | - 2 | 0.0985 | - 2 | 0.0886 | 0 | 0.0999 | 0 | 0.0706 | +31 | 0.0851 | +26 |
| 0.72 | 0.0621 | . . | . . | . . | . . | 0.0847 | - 1 | 0.0955 | - 1 | 0.0692 | +27 | 0.0829 | +23 |
| 0.76 | 0.0598 | 0.0797 | - 2 | 0.0889 | - 1 | 0.0810 | - 1 | 0.0912 | - 1 | 0.0680 | +23 | 0.0811 | +19 |
| 0.80 | 0.0577 | 0.0756 | - 1 | 0.0843 | - 1 | . . | . . | . . | . . | 0.0671 | +19 | 0.0796 | +16 |
| 0.84 | 0.0559 | 0.0715 | 0 | 0.0798 | 0 | 0.0737 | - 1 | 0.0831 | - 1 | 0.0663 | +15 | 0.0782 | +13 |
| 0.88 | 0.0543 | 0.0673 | + 2 | 0.0753 | + 2 | 0.0700 | 0 | 0.0790 | 0 | 0.0657 | +11 | 0.0772 | +10 |
| 0.92 | 0.0528 | 0.0628 | + 5 | 0.0705 | + 4 | 0.0660 | + 2 | 0.0747 | + 2 | 0.0653 | + 6 | 0.0762 | + 7 |
| 0.96 | 0.0514 | 0.0577 | +10 | 0.0651 | + 9 | 0.0614 | + 5 | 0.0697 | + 5 | 0.0649 | + 4 | 0.0755 | + 3 |
| 1.00 | 0.0502 | 0.0480 | +27 | 0.0551 | +24 | 0.0524 | +19 | 0.0603 | +16 | 0.0066 | 0 | 0.0167 | 0 |
| 0.96 | 0.0514 | 0.0421 | +51 | 0.0495 | +43 | 0.0463 | +40 | 0.0546 | +84 | 0.0649 | + 4 | 0.0755 | + 3 |
| 0.92 | 0.0528 | 0.0410 | +61 | 0.0487 | +51 | 0.0449 | +50 | 0.0535 | +42 | 0.0653 | + 8 | 0.0762 | + 7 |
| 0.88 | 0.0543 | 0.0410 | +67 | 0.0490 | +56 | 0.0445 | +58 | 0.0535 | +48 | 0.0657 | +11 | 0.0772 | +10 |
| 0.84 | 0.0559 | 0.0415 | +72 | 0.0499 | +60 | 0.0447 | +64 | 0.0540 | +53 | 0.0663 | +15 | 0.0782 | +13 |
| 0.80 | 0.0577 | 0.0426 | +76 | 0.0514 | +63 | . . | . . | . . | . . | 0.0671 | +19 | 0.0796 | +16 |
| 0.76 | 0.0598 | 0.0442 | +77 | 0.0533 | +64 | 0.0465 | +72 | 0.0567 | +59 | 0.0680 | +23 | 0.0811 | +19 |
| 0.72 | 0.0621 | . . | . . | . . | . . | 0.0481 | +75 | 0.0589 | +61 | 0.0692 | +27 | 0.0829 | +23 |
| 0.68 | 0.0646 | 0.0488 | +78 | 0.0589 | +64 | 0.0501 | +77 | 0.0615 | +62 | 0.0706 | +31 | 0.0851 | +26 |
| 0.64 | . . | 0.0518 | +77 | 0.0625 | +64 | 0.0527 | +77 | 0.0647 | +63 | 0.0723 | +35 | 0.0876 | +29 |
| 0.60 | 0.0708 | 0.0554 | +75 | 0.0667 | +62 | 0.0557 | +77 | 0.0685 | +63 | 0.0744 | +39 | 0.0905 | +32 |
| 0.56 | 0.0745 | 0.0596 | +73 | 0.0716 | +60 | 0.0594 | +77 | 0.0729 | +62 | 0.0768 | +43 | 0.0940 | +35 |
| 0.52 | 0.0788 | 0.0646 | +70 | 0.0774 | +58 | 0.0638 | +75 | 0.0783 | +62 | 0.0797 | +47 | 0.0981 | +38 |
| 0.48 | 0.0837 | 0.0704 | +67 | 0.0841 | +56 | 0.0690 | +74 | 0.0845 | +60 | 0.0832 | +51 | 0.1029 | +41 |
| 0.44 | 0.0894 | 0.0773 | +63 | 0.0922 | +53 | 0.0753 | +71 | 0.0919 | +58 | 0.0874 | +54 | 0.1087 | +44 |
| 0.40 | 0.0961 | 0.0855 | +60 | 0.1016 | +50 | 0.0827 | +69 | 0.1008 | +56 | 0.0926 | +58 | 0.1155 | +46 |
| 0.36 | 0.1040 | 0.0953 | +56 | 0.1129 | +48 | 0.0917 | +66 | 0.1114 | +54 | 0.0988 | +61 | 0.1239 | +49 |
| 0.32 | 0.1136 | 0.1071 | +53 | 0.1264 | +45 | 0.1025 | +63 | 0.1242 | +52 | 0.1064 | +64 | 0.1340 | +51 |
| 0.28 | 0.1253 | 0.1215 | +49 | 0.1430 | +42 | 0.1159 | +60 | 0.1400 | +49 | 0.1160 | +67 | 0.1466 | +53 |
| 0.24 | 0.1397 | 0.1394 | +45 | 0.1635 | +39 | 0.1326 | +56 | 0.1597 | +47 | 0.1280 | +69 | 0.1624 | +55 |
| 0.20 | 0.1580 | 0.1620 | +42 | 0.1895 | +36 | 0.1537 | +53 | 0.1846 | +44 | 0.1434 | +71 | 0.1827 | +56 |
| 0.16 | 0.1815 | 0.1911 | +38 | 0.2230 | +33 | 0.1811 | +49 | 0.2169 | +41 | 0.1636 | +73 | 0.2091 | +57 |
| 0.12 | 0.2120 | 0.2290 | +35 | 0.2667 | +30 | 0.2169 | +46 | 0.2592 | +38 | 0.1901 | +74 | 0.2439 | +58 |
| 0.08 | 0.2505 | 0.2766 | +32 | 0.3222 | +27 | 0.2621 | +42 | 0.3132 | +35 | 0.2238 | +75 | 0.2888 | +58 |
| 0.04 | 0.2863 | 0.3214 | +28 | 0.3760 | +24 | 0.3055 | +38 | 0.3667 | +32 | 0.2556 | +75 | 0.3335 | +57 |
| 0 | 0.2832 | 0.3190 | +25 | 0.3801 | +21 | 0.3057 | +35 | 0.3742 | +28 | 0.2548 | +72 | 0.3419 | +53 |

TABLE 9

THE EFFECT OF A GROUND REFLECTING ACCORDING TO LAMBERT'S LAW WITH AN ALBEDO $\lambda_o$

ON THE ILLUMINATION AND POLARIZATION OF THE SKY ( $\tau_1 = 0.10$ )

| $\mu$ | $\mu_o = 0$ | | | | $\mu_o = 0.16$ | | | | $\mu_o = 0.50$ | |
| | $\lambda_o = 0.10$ | | $\lambda_o = 0.20$ | | $\lambda_o = 0.10$ | | $\lambda_o = 0.20$ | | $\lambda_o = 0.10$ | |
| | $I_1+I_r$ | $\delta$ | $I_1+I_r$ | $\delta$ | $I_1+I_r$ | $\delta$ | $I_1+I_r$ | $\delta$ | $I_1+I_r$ | $\delta$ |
|---|---|---|---|---|---|---|---|---|---|---|
| 0 | . . | . . | . . | . . | 0.2476 | − 3 | 0.2548 | − 3 | 0.3419 | + 9 |
| 0.04 | 1.2601 | − 9 | 1.2812 | − 9 | 0.2775 | − 4 | 0.2835 | − 4 | 0.3411 | + 7 |
| 0.08 | 1.7473 | − 6 | 1.7633 | − 6 | 0.2275 | − 4 | 0.2321 | − 4 | 0.2733 | + 5 |
| 0.12 | 1.6798 | − 4 | 1.6923 | − 4 | 0.1839 | − 5 | 0.1875 | − 5 | 0.2208 | + 4 |
| 0.16 | 1.5132 | − 4 | 1.5235 | − 3 | . . | . . | . . | . . | 0.1844 | + 2 |
| 0.20 | 1.3474 | − 3 | 1.3561 | − 3 | 0.1297 | − 4 | 0.1322 | − 4 | 0.1583 | + 1 |
| 0.24 | 1.2015 | − 2 | 1.2090 | − 2 | 0.1125 | − 4 | 0.1146 | − 4 | 0.1387 | 0 |
| 0.28 | 1.0764 | 0 | 1.0829 | 0 | 0.0989 | − 4 | 0.1008 | − 4 | 0.1235 | − 1 |
| 0.32 | 0.9693 | + 1 | 0.9752 | + 1 | 0.0880 | − 3 | 0.0897 | − 3 | 0.1112 | − 2 |
| 0.36 | 0.8770 | + 2 | 0.8823 | + 2 | 0.0790 | − 2 | 0.0806 | − 2 | 0.1012 | − 2 |
| 0.40 | 0.7967 | + 4 | 0.8016 | + 4 | 0.0714 | − 1 | 0.0728 | − 1 | 0.0927 | − 3 |
| 0.44 | 0.7262 | + 6 | 0.7307 | + 6 | 0.0649 | 0 | 0.0662 | 0 | 0.0855 | − 3 |
| 0.48 | 0.6637 | + 8 | 0.6678 | + 8 | 0.0593 | + 1 | 0.0604 | + 1 | 0.0792 | − 3 |
| 0.52 | 0.6078 | +11 | 0.6116 | +11 | 0.0543 | + 3 | 0.0554 | + 3 | 0.0736 | − 3 |
| 0.56 | 0.5573 | +14 | 0.5609 | +14 | 0.0498 | + 5 | 0.0508 | + 5 | 0.0686 | − 3 |
| 0.60 | 0.5115 | +17 | 0.5148 | +17 | 0.0458 | + 7 | 0.0467 | + 7 | 0.0641 | − 2 |
| 0.64 | 0.4695 | +21 | 0.4727 | +21 | 0.0421 | +10 | 0.0430 | +10 | 0.0599 | − 1 |
| 0.68 | 0.4308 | +25 | 0.4338 | +25 | 0.0387 | +13 | 0.0396 | +13 | 0.0560 | 0 |
| 0.72 | 0.3951 | +30 | 0.3979 | +29 | 0.0356 | +17 | 0.0364 | +16 | 0.0524 | + 1 |
| 0.76 | 0.3617 | +35 | 0.3644 | +35 | 0.0326 | +21 | 0.0334 | +21 | 0.0489 | + 3 |
| 0.80 | 0.3305 | +41 | 0.3331 | +41 | 0.0299 | +26 | 0.0306 | +25 | 0.0456 | + 5 |
| 0.84 | 0.3012 | +49 | 0.3037 | +48 | 0.0272 | +32 | 0.0279 | +31 | 0.0423 | + 8 |
| 0.88 | 0.2736 | +57 | 0.2759 | +56 | 0.0247 | +39 | 0.0254 | +38 | 0.0390 | +12 |
| 0.92 | 0.2474 | +67 | 0.2496 | +66 | 0.0222 | +49 | 0.0228 | +47 | 0.0356 | +18 |
| 0.96 | 0.2225 | +78 | 0.2247 | +78 | 0.0197 | +61 | 0.0203 | +59 | 0.0319 | +27 |
| 1.00 | 0.1988 | +92 | 0.2009 | +92 | 0.0165 | +85 | 0.0171 | +82 | 0.0256 | +52 |
| 0.96 | 0.2225 | +78 | 0.2247 | +78 | 0.0170 | +86 | 0.0177 | +83 | 0.0230 | +76 |
| 0.92 | 0.2474 | +67 | 0.2496 | +66 | 0.0185 | +79 | 0.0191 | +76 | 0.0232 | +81 |
| 0.88 | 0.2736 | +57 | 0.2759 | +56 | 0.0202 | +70 | 0.0209 | +68 | 0.0240 | +83 |
| 0.84 | 0.3012 | +49 | 0.3037 | +48 | 0.0221 | +63 | 0.0228 | +61 | 0.0252 | +82 |
| 0.80 | 0.3305 | +41 | 0.3331 | +41 | 0.0242 | +55 | 0.0250 | +54 | 0.0267 | +80 |
| 0.76 | 0.3617 | +35 | 0.3644 | +35 | 0.0266 | +49 | 0.0273 | +47 | 0.0286 | +76 |
| 0.72 | 0.3951 | +30 | 0.3979 | +29 | 0.0291 | +43 | 0.0299 | +41 | 0.0307 | +72 |
| 0.68 | 0.4308 | +25 | 0.4338 | +25 | 0.0319 | +37 | 0.0328 | +36 | 0.0332 | +68 |
| 0.64 | 0.4695 | +21 | 0.4727 | +21 | 0.0350 | +32 | 0.0359 | +31 | 0.0361 | +64 |
| 0.60 | 0.5115 | +17 | 0.5148 | +17 | 0.0384 | +28 | 0.0394 | +27 | 0.0395 | +59 |
| 0.56 | 0.5573 | +14 | 0.5609 | +14 | 0.0422 | +24 | 0.0433 | +23 | 0.0433 | +54 |
| 0.52 | 0.6078 | +11 | 0.6116 | +11 | 0.0465 | +20 | 0.0476 | +20 | 0.0477 | +50 |
| 0.48 | 0.6637 | + 8 | 0.6678 | + 8 | 0.0514 | +17 | 0.0526 | +16 | 0.0527 | +46 |
| 0.44 | 0.7262 | + 6 | 0.7307 | + 6 | 0.0570 | +14 | 0.0582 | +13 | 0.0587 | +41 |
| 0.40 | 0.7967 | + 4 | 0.8016 | + 4 | 0.0634 | +11 | 0.0648 | +11 | 0.0657 | +37 |
| 0.36 | 0.8770 | + 2 | 0.8823 | + 2 | 0.0710 | + 9 | 0.0725 | + 9 | 0.0740 | +34 |
| 0.32 | 0.9693 | + 1 | 0.9752 | + 1 | 0.0800 | + 7 | 0.0817 | + 6 | 0.0841 | +30 |
| 0.28 | 1.0764 | 0 | 1.0829 | 0 | 0.0910 | + 5 | 0.0929 | + 5 | 0.0966 | +27 |
| 0.24 | 1.2015 | − 2 | 1.2090 | − 2 | 0.1047 | + 3 | 0.1068 | + 3 | 0.1123 | +24 |
| 0.20 | 1.3474 | − 3 | 1.3561 | − 3 | 0.1222 | + 1 | 0.1247 | + 1 | 0.1326 | +21 |
| 0.16 | 1.5132 | − 4 | 1.5235 | − 3 | . . | . . | . . | . . | 0.1601 | +18 |
| 0.12 | 1.6798 | − 4 | 1.6923 | − 4 | 0.1774 | − 1 | 0.1810 | − 1 | 0.1986 | +15 |
| 0.08 | 1.7473 | − 6 | 1.7633 | − 6 | 0.2221 | − 2 | 0.2267 | − 2 | 0.2545 | +13 |
| 0.04 | 1.2601 | − 9 | 1.2812 | − 9 | 0.2742 | − 3 | 0.2803 | − 3 | 0.3293 | +11 |
| 0 | . . | . . | . . | . . | 0.2476 | − 3 | 0.2548 | − 3 | 0.3419 | + 9 |

TABLE 9 (continued)

| μ | μ₀ = 0.50 λ = 0.20 $I_1+I_r$ | δ | μ₀ = 0.86 λ = 0.10 $I_1+I_r$ | δ | μ₀ = 0.86 λ = 0.20 $I_1+I_r$ | δ | μ₀ = 1.00 λ = 0.10 $I_1+I_r$ | δ | μ₀ = 1.00 λ = 0.20 $I_1+I_r$ | δ |
|---|---|---|---|---|---|---|---|---|---|---|
| 0 | 0.3688 | + 8 | 0.2922 | +45 | 0.3402 | +38 | 0.2521 | +74 | 0.3083 | +60 |
| 0.04 | 0.3637 | + 6 | 0.2831 | +43 | 0.3234 | +37 | 0.2358 | +76 | 0.2831 | +64 |
| 0.08 | 0.2904 | + 5 | 0.2265 | +40 | 0.2571 | +35 | 0.1844 | +76 | 0.2202 | +64 |
| 0.12 | 0.2343 | + 3 | 0.1841 | +36 | 0.2081 | +32 | 0.1471 | +76 | 0.1752 | +63 |
| 0.16 | 0.1954 | + 2 | 0.1551 | +33 | 0.1748 | +30 | 0.1220 | +74 | 0.1450 | +62 |
| 0.20 | 0.1676 | + 1 | 0.1345 | +30 | 0.1511 | +27 | 0.1044 | +72 | 0.1238 | +61 |
| 0.24 | 0.1467 | 0 | 0.1193 | +27 | 0.1337 | +24 | 0.0916 | +70 | 0.1084 | +59 |
| 0.28 | 0.1305 | - 1 | 0.1076 | +24 | 0.1203 | +22 | 0.0820 | +68 | 0.0968 | +57 |
| 0.32 | 0.1175 | - 2 | 0.0984 | +22 | 0.1097 | +19 | 0.0745 | +65 | 0.0877 | +55 |
| 0.36 | 0.1069 | - 2 | 0.0910 | +19 | 0.1011 | +17 | 0.0687 | +62 | 0.0806 | +53 |
| 0.40 | 0.0979 | - 3 | 0.0848 | +16 | 0.0940 | +15 | 0.0640 | +58 | 0.0748 | +50 |
| 0.44 | 0.0903 | - 3 | 0.0796 | +14 | 0.0881 | +13 | 0.0602 | +55 | 0.0701 | +47 |
| 0.48 | 0.0836 | - 3 | 0.0751 | +12 | 0.0830 | +11 | 0.0570 | +51 | 0.0662 | +44 |
| 0.52 | 0.0777 | - 3 | 0.0712 | +10 | 0.0765 | + 9 | 0.0544 | +47 | 0.0630 | +41 |
| 0.56 | 0.0724 | - 3 | 0.0678 | + 8 | 0.0746 | + 7 | 0.0523 | +43 | 0.0603 | +37 |
| 0.60 | 0.0677 | - 2 | 0.0647 | + 6 | 0.0711 | + 5 | 0.0505 | +39 | 0.0580 | +34 |
| 0.64 | 0.0633 | - 1 | 0.0619 | + 4 | 0.0680 | + 4 | 0.0490 | +35 | 0.0561 | +31 |
| 0.68 | 0.0592 | 0 | 0.0594 | + 3 | 0.0651 | + 3 | 0.0478 | +31 | 0.0545 | +27 |
| 0.72 | 0.0554 | + 1 | 0.0570 | + 2 | 0.0624 | + 1 | 0.0468 | +27 | 0.0531 | +24 |
| 0.76 | 0.0518 | + 3 | 0.0547 | + 1 | 0.0599 | + 1 | 0.0459 | +23 | 0.0520 | +20 |
| 0.80 | 0.0483 | + 5 | 0.0525 | 0 | 0.0574 | 0 | 0.0452 | +19 | 0.0510 | +17 |
| 0.84 | 0.0449 | + 8 | 0.0505 | - 1 | 0.0552 | - 1 | 0.0447 | +15 | 0.0502 | +13 |
| 0.88 | 0.0415 | +12 | 0.0481 | 0 | 0.0526 | 0 | 0.0442 | +11 | 0.0495 | +10 |
| 0.92 | 0.0381 | +17 | 0.0458 | 0 | 0.0501 | 0 | 0.0439 | + 7 | 0.0489 | + 7 |
| 0.96 | 0.0342 | +25 | 0.0430 | + 2 | 0.0472 | + 2 | 0.0436 | + 4 | 0.0484 | + 3 |
| 1.00 | 0.0279 | +48 | 0.0375 | +13 | 0.0414 | +11 | 0.0046 | 0 | 0.0092 | 0 |
| 0.96 | 0.0253 | +69 | 0.0336 | +31 | 0.0378 | +28 | 0.0436 | + 4 | 0.0484 | + 3 |
| 0.92 | 0.0256 | +74 | 0.0327 | +41 | 0.0370 | +36 | 0.0439 | + 7 | 0.0489 | + 7 |
| 0.88 | 0.0265 | +75 | 0.0323 | +48 | 0.0368 | +42 | 0.0442 | +11 | 0.0495 | +10 |
| 0.84 | 0.0278 | +74 | 0.0324 | +55 | 0.0371 | +48 | 0.0447 | +15 | 0.0502 | +13 |
| 0.80 | 0.0295 | +72 | 0.0326 | +61 | 0.0375 | +53 | 0.0452 | +19 | 0.0510 | +17 |
| 0.76 | 0.0314 | +69 | 0.0332 | +66 | 0.0383 | +57 | 0.0459 | +23 | 0.0520 | +20 |
| 0.72 | 0.0338 | +66 | 0.0341 | +70 | 0.0395 | +60 | 0.0468 | +27 | 0.0531 | +24 |
| 0.68 | 0.0364 | +62 | 0.0353 | +73 | 0.0410 | +68 | 0.0478 | +31 | 0.0545 | +27 |
| 0.64 | 0.0395 | +58 | 0.0368 | +75 | 0.0428 | +65 | 0.0490 | +35 | 0.0561 | +31 |
| 0.60 | 0.0431 | +54 | 0.0387 | +77 | 0.0451 | +66 | 0.0505 | +39 | 0.0580 | +34 |
| 0.56 | 0.0471 | +50 | 0.0410 | +78 | 0.0478 | +67 | 0.0523 | +43 | 0.0603 | +37 |
| 0.52 | 0.0518 | +46 | 0.0438 | +78 | 0.0511 | +67 | 0.0544 | +47 | 0.0630 | +41 |
| 0.48 | 0.0571 | +42 | 0.0471 | +78 | 0.0550 | +67 | 0.0570 | +51 | 0.0662 | +44 |
| 0.44 | 0.0634 | +38 | 0.0512 | +77 | 0.0597 | +66 | 0.0602 | +55 | 0.0701 | +47 |
| 0.40 | 0.0709 | +35 | 0.0561 | +76 | 0.0654 | +65 | 0.0640 | +58 | 0.0748 | +50 |
| 0.36 | 0.0797 | +31 | 0.0622 | +74 | 0.0724 | +63 | 0.0687 | +62 | 0.0806 | +53 |
| 0.32 | 0.0904 | +28 | 0.0697 | +72 | 0.0810 | +62 | 0.0745 | +65 | 0.0877 | +55 |
| 0.28 | 0.1036 | +25 | 0.0792 | +69 | 0.0918 | +60 | 0.0820 | +68 | 0.0968 | +57 |
| 0.24 | 0.1203 | +22 | 0.0913 | +66 | 0.1057 | +57 | 0.0916 | +70 | 0.1084 | +59 |
| 0.20 | 0.1419 | +19 | 0.1074 | +63 | 0.1240 | +55 | 0.1044 | +72 | 0.1238 | +61 |
| 0.16 | 0.1711 | +17 | 0.1293 | +60 | 0.1489 | +52 | 0.1220 | +74 | 0.1450 | +62 |
| 0.12 | 0.2120 | +14 | 0.1605 | +57 | 0.1845 | +49 | 0.1471 | +76 | 0.1752 | +63 |
| 0.08 | 0.2717 | +12 | 0.2066 | +53 | 0.2372 | +46 | 0.1844 | +76 | 0.2202 | +64 |
| 0.04 | 0.3518 | +10 | 0.2704 | +49 | 0.3107 | +43 | 0.2358 | +76 | 0.2831 | +64 |
| 0 | 0.3688 | + 8 | 0.2922 | +45 | 0.3402 | +38 | 0.2521 | +74 | 0.3083 | +60 |

TABLE 10

THE EFFECT OF A GROUND REFLECTING ACCORDING TO LAMBERT'S LAW WITH AN ALBEDO

$\lambda_o$ ON THE ILLUMINATION AND POLARIZATION OF THE SKY ( $\tau_1$ = 0.20)

| $\mu$ | $\mu_o = 0$ | | | | $\mu_o = 0.16$ | | | | $\mu_o = 0.50$ | |
| | $\lambda_o = 0.10$ | | $\lambda_o = 0.20$ | | $\lambda_o = 0.10$ | | $\lambda_o = 0.20$ | | $\lambda_o = 0.10$ | |
| | $I_1+I_r$ | $\delta$ | $I_1+I_r$ | $\delta$ | $I_1+I_r$ | $\delta$ | $I_1+I_r$ | $\delta$ | $I_1+I_r$ | $\delta$ |
|---|---|---|---|---|---|---|---|---|---|---|
| 0 | . . | . . | . . | . . | 0.1650 | - 6 | 0.1715 | - 6 | 0.3182 | + 7 |
| 0.04 | 0.4708 | -18 | 0.4934 | -17 | 0.2073 | - 7 | 0.2132 | - 7 | 0.3506 | + 4 |
| 0.08 | 0.7960 | -11 | 0.8159 | -11 | 0.2193 | - 7 | 0.2246 | - 7 | 0.3423 | + 3 |
| 0.12 | 0.9677 | - 8 | 0.9848 | - 8 | 0.2063 | - 7 | 0.2108 | - 7 | 0.3123 | + 2 |
| 0.16 | 1.0065 | - 7 | 1.0214 | - 6 | . . | . . | . . | . . | 0.2808 | 0 |
| 0.20 | 0.9837 | - 5 | 0.9968 | - 5 | 0.1692 | - 7 | 0.1726 | - 7 | 0.2531 | - 1 |
| 0.24 | 0.9358 | - 4 | 0.9474 | - 4 | 0.1529 | - 6 | 0.1559 | - 6 | 0.2296 | - 2 |
| 0.28 | 0.8790 | - 3 | 0.8895 | - 3 | 0.1387 | - 6 | 0.1415 | - 6 | 0.2097 | - 3 |
| 0.32 | 0.8207 | - 1 | 0.8302 | - 1 | 0.1264 | - 5 | 0.1289 | - 5 | 0.1927 | - 3 |
| 0.36 | 0.7641 | 0 | 0.7727 | 0 | 0.1156 | - 5 | 0.1179 | - 4 | 0.1780 | - 4 |
| 0.40 | 0.7103 | + 2 | 0.7183 | + 2 | 0.1061 | - 4 | 0.1082 | - 3 | 0.1652 | - 4 |
| 0.44 | 0.6599 | + 4 | 0.6673 | + 4 | 0.0976 | - 2 | 0.0996 | - 2 | 0.1538 | - 4 |
| 0.48 | 0.6128 | + 6 | 0.6197 | + 6 | 0.0900 | - 1 | 0.0919 | - 1 | 0.1437 | - 4 |
| 0.52 | 0.5689 | + 9 | 0.5754 | + 8 | 0.0832 | + 1 | 0.0849 | + 1 | 0.1346 | - 4 |
| 0.56 | 0.5279 | +11 | 0.5340 | +11 | 0.0769 | + 3 | 0.0785 | + 3 | 0.1262 | - 4 |
| 0.60 | 0.4895 | +15 | 0.4952 | +14 | 0.0712 | + 5 | 0.0727 | + 5 | 0.1185 | - 3 |
| 0.64 | 0.4535 | +18 | 0.4589 | +18 | 0.0658 | + 8 | 0.0673 | + 8 | 0.1113 | - 3 |
| 0.68 | 0.4196 | +22 | 0.4247 | +22 | 0.0609 | +11 | 0.0622 | +11 | 0.1045 | - 1 |
| 0.72 | 0.3876 | +27 | 0.3925 | +26 | 0.0562 | +14 | 0.0575 | +14 | 0.0980 | 0 |
| 0.76 | 0.3573 | +32 | 0.3619 | +32 | 0.0518 | +19 | 0.0530 | +18 | 0.0918 | + 2 |
| 0.80 | 0.3285 | +38 | 0.3330 | +38 | 0.0476 | +23 | 0.0487 | +23 | 0.0858 | + 4 |
| 0.84 | 0.3012 | +45 | 0.3054 | +44 | 0.0435 | +29 | 0.0446 | +29 | 0.0798 | + 7 |
| 0.88 | 0.2750 | +53 | 0.2791 | +52 | 0.0396 | +37 | 0.0407 | +36 | 0.0738 | +11 |
| 0.92 | 0.2500 | +63 | 0.2539 | +62 | 0.0357 | +46 | 0.0367 | +44 | 0.0675 | +17 |
| 0.96 | 0.2260 | +74 | 0.2297 | +73 | 0.0318 | +58 | 0.0328 | +56 | 0.0606 | +26 |
| 1.00 | 0.2029 | +87 | 0.2065 | +86 | 0.0267 | +81 | 0.0277 | +78 | 0.0488 | +50 |
| 0.96 | 0.2260 | +74 | 0.2297 | +73 | 0.0276 | +81 | 0.0286 | +79 | 0.0439 | +73 |
| 0.92 | 0.2500 | +63 | 0.2539 | +62 | 0.0299 | +74 | 0.0310 | +71 | 0.0443 | +78 |
| 0.88 | 0.2750 | +53 | 0.2791 | +52 | 0.0326 | +66 | 0.0337 | +64 | 0.0458 | +80 |
| 0.84 | 0.3012 | +45 | 0.3054 | +44 | 0.0356 | +58 | 0.0367 | +56 | 0.0480 | +78 |
| 0.80 | 0.3285 | +38 | 0.3330 | +38 | 0.0389 | +51 | 0.0400 | +50 | 0.0508 | +76 |
| 0.76 | 0.3573 | +32 | 0.3619 | +32 | 0.0424 | +45 | 0.0436 | +43 | 0.0542 | +72 |
| 0.72 | 0.3876 | +27 | 0.3925 | +26 | 0.0463 | +39 | 0.0476 | +38 | 0.0582 | +68 |
| 0.68 | 0.4196 | +22 | 0.4247 | +22 | 0.0505 | +33 | 0.0519 | +33 | 0.0627 | +64 |
| 0.64 | 0.4535 | +18 | 0.4589 | +18 | 0.0551 | +29 | 0.0565 | +28 | 0.0679 | +60 |
| 0.60 | 0.4895 | +15 | 0.4952 | +14 | 0.0601 | +24 | 0.0616 | +24 | 0.0738 | +55 |
| 0.56 | 0.5279 | +11 | 0.5340 | +11 | 0.0656 | +20 | 0.0672 | +20 | 0.0804 | +51 |
| 0.52 | 0.5689 | + 9 | 0.5754 | + 8 | 0.0717 | +17 | 0.0734 | +17 | 0.0881 | +46 |
| 0.48 | 0.6128 | + 6 | 0.6197 | + 6 | 0.0785 | +14 | 0.0803 | +13 | 0.0967 | +42 |
| 0.44 | 0.6599 | + 4 | 0.6673 | + 4 | 0.0860 | +11 | 0.0880 | +11 | 0.1066 | +38 |
| 0.40 | 0.7103 | + 2 | 0.7183 | + 2 | 0.0945 | + 8 | 0.0966 | + 8 | 0.1180 | +34 |
| 0.36 | 0.7641 | 0 | 0.7727 | 0 | 0.1042 | + 6 | 0.1064 | + 6 | 0.1313 | +30 |
| 0.32 | 0.8207 | - 1 | 0.8302 | - 1 | 0.1152 | + 4 | 0.1177 | + 4 | 0.1468 | +27 |
| 0.28 | 0.8790 | - 3 | 0.8895 | - 3 | 0.1279 | + 2 | 0.1307 | + 2 | 0.1651 | +24 |
| 0.24 | 0.9358 | - 4 | 0.9474 | - 4 | 0.1426 | 0 | 0.1457 | 0 | 0.1870 | +21 |
| 0.20 | 0.9837 | - 5 | 0.9968 | - 5 | 0.1597 | - 1 | 0.1631 | - 1 | 0.2132 | +18 |
| 0.16 | 1.0065 | - 7 | 1.0214 | - 6 | . . | . . | . . | . . | 0.2448 | +15 |
| 0.12 | 0.9677 | - 8 | 0.9848 | - 8 | 0.1993 | - 4 | 0.2038 | - 4 | 0.2817 | +13 |
| 0.08 | 0.7960 | -11 | 0.8159 | -11 | 0.2144 | - 5 | 0.2196 | - 5 | 0.3197 | +10 |
| 0.04 | 0.4708 | -18 | 0.4934 | -17 | 0.2049 | - 6 | 0.2109 | - 6 | 0.3388 | + 8 |
| 0 | . . | . . | . . | . . | 0.1650 | - 6 | 0.1715 | - 6 | 0.3182 | + 7 |

TABLE 10 (continued)

| μ | μ₀ = 0.50 λ=0.20 $I_1{+}I_r$ | δ | μ₀ = 0.86 λ=0.10 $I_1{+}I_r$ | δ | λ=0.20 $I_1{+}I_r$ | δ | μ₀ = 1.00 λ=0.10 $I_1{+}I_r$ | δ | λ=0.20 $I_1{+}I_r$ | δ |
|---|---|---|---|---|---|---|---|---|---|---|
| 0 | 0.3451 | + 6 | 0.2925 | +42 | 0.3423 | +36 | 0.2564 | +71 | 0.3152 | +57 |
| 0.04 | 0.3752 | + 4 | 0.3113 | +40 | 0.3568 | +35 | 0.2632 | +73 | 0.3169 | +60 |
| 0.08 | 0.3640 | + 3 | 0.2997 | +37 | 0.3398 | +33 | 0.2467 | +73 | 0.2941 | +61 |
| 0.12 | 0.3310 | + 1 | 0.2730 | +34 | 0.3076 | +30 | 0.2202 | +73 | 0.2610 | +61 |
| 0.16 | 0.2971 | 0 | 0.2467 | +31 | 0.2767 | +28 | 0.1956 | +72 | 0.2311 | +61 |
| 0.20 | 0.2674 | - 1 | 0.2241 | +28 | 0.2505 | +25 | 0.1753 | +70 | 0.2064 | +60 |
| 0.24 | 0.2423 | - 2 | 0.2054 | +26 | 0.2288 | +23 | 0.1589 | +68 | 0.1865 | +58 |
| 0.28 | 0.2211 | - 2 | 0.1898 | +23 | 0.2109 | +21 | 0.1456 | +66 | 0.1705 | +56 |
| 0.32 | 0.2030 | - 3 | 0.1768 | +20 | 0.1959 | +18 | 0.1348 | +63 | 0.1574 | +54 |
| 0.36 | 0.1874 | - 4 | 0.1657 | +18 | 0.1832 | +16 | 0.1260 | +61 | 0.1466 | +52 |
| 0.40 | 0.1739 | - 4 | 0.1563 | +15 | 0.1724 | +14 | 0.1187 | +57 | 0.1377 | +49 |
| 0.44 | 0.1619 | - 4 | 0.1481 | +13 | 0.1630 | +12 | 0.1126 | +54 | 0.1302 | +47 |
| 0.48 | 0.1512 | - 4 | 0.1409 | +11 | 0.1548 | +10 | 0.1076 | +50 | 0.1240 | +44 |
| 0.52 | 0.1417 | - 4 | 0.1345 | + 9 | 0.1475 | + 8 | 0.1034 | +47 | 0.1187 | +41 |
| 0.56 | 0.1328 | - 4 | 0.1287 | + 7 | 0.1409 | + 6 | 0.0998 | +43 | 0.1142 | +38 |
| 0.60 | 0.1247 | - 3 | 0.1235 | + 5 | 0.1350 | + 5 | 0.0969 | +39 | 0.1105 | +34 |
| 0.64 | 0.1172 | - 2 | 0.1187 | + 4 | 0.1296 | + 3 | 0.0944 | +35 | 0.1072 | +31 |
| 0.68 | 0.1101 | - 1 | 0.1142 | + 2 | 0.1245 | + 2 | 0.0923 | +31 | 0.1045 | +28 |
| 0.72 | 0.1033 | 0 | 0.1099 | + 1 | 0.1197 | + 1 | 0.0906 | +27 | 0.1022 | +24 |
| 0.76 | 0.0969 | + 2 | 0.1058 | 0 | 0.1152 | 0 | 0.0892 | +23 | 0.1002 | +21 |
| 0.80 | 0.0906 | + 4 | 0.1018 | 0 | 0.1108 | 0 | 0.0880 | +19 | 0.0986 | +17 |
| 0.84 | 0.0844 | + 7 | 0.0979 | - 1 | 0.1065 | - 1 | 0.0871 | +16 | 0.0972 | +14 |
| 0.86 | 0.0782 | +11 | 0.0936 | 0 | 0.1018 | 0 | 0.0864 | +12 | 0.0961 | +11 |
| 0.92 | 0.0718 | +16 | 0.0892 | 0 | 0.0971 | 0 | 0.0859 | + 8 | 0.0952 | + 7 |
| 0.96 | 0.0647 | +24 | 0.0840 | + 3 | 0.0916 | + 3 | 0.0855 | + 4 | 0.0945 | + 4 |
| 1.00 | 0.0528 | +47 | 0.0734 | +13 | 0.0807 | +12 | 0.0084 | 0 | 0.0170 | 0 |
| 0.96 | 0.0480 | +67 | 0.0657 | +31 | 0.0733 | +28 | 0.0855 | + 4 | 0.0945 | + 4 |
| 0.92 | 0.0486 | +72 | 0.0687 | +41 | 0.0716 | +36 | 0.0859 | + 8 | 0.0952 | + 7 |
| 0.88 | 0.0502 | +73 | 0.0628 | +48 | 0.0710 | +43 | 0.0864 | +12 | 0.0961 | +11 |
| 0.84 | 0.0526 | +72 | 0.0629 | +55 | 0.0715 | +48 | 0.0871 | +16 | 0.0972 | +14 |
| 0.80 | 0.0557 | +69 | 0.0634 | +60 | 0.0723 | +53 | 0.0880 | +19 | 0.0986 | +17 |
| 0.76 | 0.0593 | +66 | 0.0644 | +65 | 0.0738 | +57 | 0.0892 | +23 | 0.1002 | +21 |
| 0.72 | 0.0635 | +63 | 0.0660 | +69 | 0.0759 | +60 | 0.0906 | +27 | 0.1022 | +24 |
| 0.68 | 0.0683 | +59 | 0.0682 | +72 | 0.0785 | +62 | 0.0923 | +31 | 0.1045 | +28 |
| 0.64 | 0.0738 | +55 | 0.0709 | +74 | 0.0818 | +64 | 0.0944 | +35 | 0.1072 | +31 |
| 0.60 | 0.0800 | +51 | 0.0743 | +75 | 0.0858 | +65 | 0.0969 | +39 | 0.1105 | +34 |
| 0.56 | 0.0871 | +47 | 0.0783 | +76 | 0.0906 | +66 | 0.0998 | +43 | 0.1142 | +38 |
| 0.52 | 0.0952 | +43 | 0.0832 | +76 | 0.0962 | +66 | 0.1034 | +47 | 0.1187 | +41 |
| 0.48 | 0.1042 | +39 | 0.0890 | +75 | 0.1029 | +65 | 0.1076 | +50 | 0.1240 | +44 |
| 0.44 | 0.1147 | +35 | 0.0960 | +74 | 0.1109 | +64 | 0.1126 | +54 | 0.1302 | +47 |
| 0.40 | 0.1267 | +32 | 0.1043 | +73 | 0.1204 | +63 | 0.1187 | +57 | 0.1377 | +49 |
| 0.36 | 0.1407 | +28 | 0.1142 | +71 | 0.1317 | +62 | 0.1260 | +61 | 0.1466 | +52 |
| 0.32 | 0.1571 | +25 | 0.1261 | +69 | 0.1452 | +60 | 0.1348 | +63 | 0.1574 | +54 |
| 0.28 | 0.1765 | +22 | 0.1405 | +66 | 0.1616 | +57 | 0.1456 | +66 | 0.1705 | +56 |
| 0.24 | 0.1996 | +19 | 0.1581 | +63 | 0.1816 | +55 | 0.1589 | +68 | 0.1865 | +58 |
| 0.20 | 0.2275 | +17 | 0.1798 | +60 | 0.2062 | +52 | 0.1753 | +70 | 0.2064 | +60 |
| 0.16 | 0.2610 | +14 | 0.2066 | +57 | 0.2366 | +50 | 0.1956 | +72 | 0.2311 | +61 |
| 0.12 | 0.3004 | +12 | 0.2388 | +54 | 0.2734 | +47 | 0.2202 | +73 | 0.2610 | +61 |
| 0.08 | 0.3413 | +10 | 0.2741 | +50 | 0.3142 | +43 | 0.2467 | +73 | 0.2941 | +61 |
| 0.04 | 0.3635 | + 7 | 0.2978 | +46 | 0.3434 | +40 | 0.2632 | +73 | 0.3169 | +60 |
| 0 | 0.3451 | + 6 | 0.2925 | +42 | 0.3423 | +36 | 0.2564 | +71 | 0.3152 | +57 |

TABLE 11

THE CALCULATED POSITION OF THE NEUTRAL POINTS

$$\tau_1 = 0.15$$

| $\theta_0$ | $\varphi$ | Position Angles of the Neutral Points | | |
|---|---|---|---|---|
| | | Babinet | Brewster | Arago |
| 90.0° | 0° | 70.7° | | 70.7° |
| | 10° | 68.3° | | 68.3° |
| | 20° | 61.2° | | 61.2° |
| | 30° | 49.6° | | 49.6° |
| | 40° | 30.1° | | 30.1° |
| 85.4° | 0° | 65.4° | | 73.7° |
| | 10° | 62.8° | | 71.8° |
| | 20° | 55.7° | | 64.5° |
| | 30° | 44.2° | | 53.6° |
| | 40° | 24.8° | | 84.9° |
| 76.1° | 0° | 57.5° | | 83.1° |
| | 10° | 54.9° | | 80.9° |
| | 20° | 47.8° | | 74.7° |
| | 30° | 36.2° | | 64.8° |
| | 40° | 17.5° | | 48.7° |
| 58.7° | 0° | 43.9° | 76.6° | |
| | 10° | 41.5° | 78.6° | |
| | 20° | 34.7° | 83.3° | |
| | 30° | 24.6° | | 89.6° |
| | 40° | 10.4° | | 80.6° |
| | 50° | | | 66.0°;13.6° |
| 50.1° | 0° | 37.8° | 65.8° | |
| | 10° | 35.5° | 67.4° | |
| | 20° | 29.2° | 71.5° | |
| | 30° | 20.1° | 76.6° | |
| | 40° | 8.89° | 82.7° | |
| | 50° | | 89.6° | |
| | 60° | | | 9.07° |
| | | | | 32.1°;77.9° |

| $\theta_0$ | $\varphi$ | Position Angles of the Neutral Points | | |
|---|---|---|---|---|
| | | Babinet | Brewster | Arago |
| 43.9° | 0° | 33.4° | 57.5° | |
| | 10° | 31.2° | 58.9° | |
| | 20° | 25.4° | 62.2° | |
| | 30° | 17.3° | 66.2° | |
| | 40° | 7.6 | 70.5° | |
| | 50° | | 74.8° | 6.4° |
| | 60° | | 78.8° | 22.1° |
| | 70° | | 83.1° | 41.6° |
| | 80° | | 87.7° | 65.8° |
| 36.9° | 0° | 28.4° | 47.9° | |
| | 10° | 26.6° | 48.9° | |
| | 20° | 21.3° | 51.6° | |
| | 30° | 14.5° | 54.7° | |
| | 40° | 7.02° | 57.3° | |
| | 50° | | 59.3° | 7.69° |
| | 60° | | 60.2° | 15.7° |
| | 70° | | 59.5° | 27.9° |
| | 80° | | 55.9° | 37.5° |
| | 90° | | 49.5° | 49.5° |
| 19.95° | 20° | 13.3° | 25.6° | |
| | 30° | 10.1° | 27.5° | |
| | 40° | 6.69° | 28.3° | |
| | 50° | | 28.4° | 3.14° |
| | 60° | | 27.6° | 6.88° |
| | 70° | | 26.4° | 11.0° |
| | 80° | | 23.6° | 15.6° |
| | 90° | | 19.7° | 19.7° |

$$\tau_1 = 0.10$$

$$\tau_1 \pm 0.20$$

| $\theta_0$ | $\varphi$ | Position Angles of the Neutral Points | | |
|---|---|---|---|---|
| | | Babinet | Brewster | Arago |
| 90.0° | 0° | 72.9° | | 72.9° |
| 80.8° | 0° | 63.6° | | 80.8° |
| 60.0° | 0° | 46.5° | 75.8° | |
| 30.7° | 0° | 24.9° | 37.8° | |
| | 90° | | 36.2° | 36.2° |

| $\theta_0$ | $\varphi$ | Position Angles of the Neutral Points | | |
|---|---|---|---|---|
| | | Babinet | Brewster | Arago |
| 90.0° | 0° | 69.1° | | 69.1° |
| 80.8° | 0° | 59.8° | | 76.5° |
| 60.0° | 0° | 43.9° | 80.1° | |
| 30.7° | 0° | 25.4° | 38.8° | |
| | 90° | | 35.8° | 35.8° |

## TABLE 12

### SUPPLEMENTARY TABLES OF THE FUNCTIONS $\psi, \phi, \chi, \zeta, \xi, \eta, \sigma, \theta, X^{(1)}, Y^{(1)}, X^{(2)}, Y^{(2)},$
### $\chi_l$ AND $\chi_n$ OBTAINED ON THE CORRECTED SECOND APPROXIMATION FOR VARIOUS VALUES OF $\zeta$

$$\zeta_1 = 0.01; \quad \zeta = 0.00916$$

| $\mu$ | $\psi$ | $\phi$ | $\chi$ | $\zeta$ | $\xi$ | $\eta$ | $\sigma$ | $\theta$ | $X^{(1)}$ | $Y^{(1)}$ | $X^{(2)}$ | $Y^{(2)}$ | $\chi_l$ | $\chi_n$ |
|---|---|---|---|---|---|---|---|---|---|---|---|---|---|---|
| 0 | 0 | 1.00000 | 1.00000 | 0 | 0 | 0 | 0 | 0 | 1.00000 | 0 | 1.00000 | 0 | 0.48348 | 0.48961 |
| 0.010 | 0.00134 | 1.02222 | 1.01341 | 0.00062 | 0.00127 | 0.38785 | 0.38013 | 0.00062 | 1.01279 | 0.37951 | 1.00784 | 0.37513 | 0.68301 | 0.68289 |
| 0.025 | 0.00225 | 1.02749 | 1.01700 | 0.00081 | 0.00204 | 0.69680 | 0.66670 | 0.00080 | 1.01619 | 0.65590 | 1.00997 | 0.67998 | 0.85576 | 0.85494 |
| 0.050 | 0.00432 | 1.02802 | 1.01852 | 0.00089 | 0.00386 | 0.84653 | 0.85691 | 0.00089 | 1.01768 | 0.83502 | 1.01088 | 0.82944 | 0.91051 | 0.90980 |
| 0.075 | 0.00756 | 1.02569 | 1.01907 | 0.00092 | 0.00685 | 0.90110 | 0.89401 | 0.00091 | 1.01815 | 0.89509 | 1.01121 | 0.88627 | 0.93862 | 0.98756 |
| 0.1 | 0.01204 | 1.02168 | 1.01935 | 0.00093 | 0.01108 | 0.92707 | 0.98401 | 0.00095 | 1.01842 | 0.92808 | 1.01138 | 0.91613 | 0.95348 | 0.95240 |
| 0.2 | 0.04265 | 0.99138 | 1.01979 | 0.00095 | 0.04067 | 0.94459 | 0.97095 | 0.00095 | 1.01884 | 0.96998 | 1.01164 | 0.96282 | 0.97673 | 0.97552 |
| 0.3 | 0.09955 | 0.94002 | 1.01994 | 0.00096 | 0.09059 | 0.91007 | 0.98710 | 0.00096 | 1.01898 | 0.96613 | 1.01173 | 0.97891 | 0.98475 | 0.98560 |
| 0.4 | 0.16485 | 0.86789 | 1.02002 | 0.00097 | 0.16687 | 0.84706 | 0.99528 | 0.00097 | 1.01905 | 0.98411 | 1.01177 | 0.98706 | 0.98877 | 0.98765 |
| 0.5 | 0.25847 | 0.77906 | 1.02006 | 0.00097 | 0.25151 | 0.76015 | 1.00022 | 0.00097 | 1.01909 | 0.99925 | 1.01180 | 0.99198 | 0.99128 | 0.99009 |
| 0.6 | 0.36848 | 0.66136 | 1.02009 | 0.00097 | 0.36251 | 0.65995 | 1.00353 | 0.00097 | 1.01912 | 1.00256 | 1.01182 | 0.99528 | 0.99287 | 0.99290 |
| 0.7 | 0.50085 | 0.52740 | 1.02011 | 0.00097 | 0.49388 | 0.52013 | 1.00590 | 0.00097 | 1.01914 | 1.00498 | 1.01183 | 0.99765 | 0.99405 | 0.99378 |
| 0.8 | 0.65859 | 0.37258 | 1.02013 | 0.00097 | 0.64561 | 0.36808 | 1.00768 | 0.00097 | 1.01916 | 1.00671 | 1.01184 | 0.99941 | 0.99495 | 0.99447 |
| 0.9 | 0.82669 | 0.19711 | 1.02014 | 0.00097 | 0.81771 | 0.19499 | 1.00907 | 0.00097 | 1.01917 | 1.00810 | 1.01185 | 1.00079 | 0.99562 | 0.99447 |
| 1.0 | 1.02015 | 0.00098 | 1.02015 | 0.00097 | 1.01017 | 0.00097 | 1.01018 | 0.00097 | 1.01916 | 1.00921 | 1.01186 | 1.00190 | 0.99617 | 0.99502 |

TABLE 12 (continued)

$\zeta_1 = 0.02; \quad \bar{\delta} = 0.01904$

| μ | ψ | φ | χ | ζ | ξ | η | σ | θ | $X^{(3)}$ | $Y^{(3)}$ | $X^{(2)}$ | $Y^{(2)}$ | $\mathcal{X}$ | $\gamma_\infty$ |
|---|---|---|---|---|---|---|---|---|---|---|---|---|---|---|
| 0 | 0 | 1.00000 | 1.00000 | 0 | 0 | 0 | 0 | 0 | 1.00000 | 0 | 1.00000 | 0 | 0.47088 | 0.46180 |
| 0.010 | 0.00186 | 1.02820 | 1.01718 | 0.00088 | 0.00175 | 0.15757 | 0.14945 | 0.00086 | 1.01680 | 0.14858 | 1.01010 | 0.11489 | 0.56010 | 0.56290 |
| 0.025 | 0.00342 | 1.04117 | 1.02571 | 0.00159 | 0.00506 | 0.48682 | 0.47300 | 0.00138 | 1.02432 | 0.47161 | 1.01526 | 0.46555 | 0.72265 | 0.72359 |
| 0.050 | 0.00590 | 1.04618 | 1.03014 | 0.00166 | 0.00506 | 0.71497 | 0.66928 | 0.00165 | 1.02817 | 0.69756 | 1.01794 | 0.69764 | 0.84455 | 0.84478 |
| 0.075 | 0.00983 | 1.04562 | 1.03187 | 0.00177 | 0.00800 | 0.81118 | 0.79694 | 0.00176 | 1.03010 | 0.79517 | 1.01899 | 0.78449 | 0.88269 | 0.88279 |
| 0.1 | 0.01595 | 1.04246 | 1.03280 | 0.00182 | 0.01215 | 0.68169 | 0.89085 | 0.00182 | 1.03096 | 0.84902 | 1.01955 | 0.88794 | 0.90920 | 0.90926 |
| 0.2 | 0.04506 | 1.03818 | 1.03425 | 0.00191 | 0.04125 | 0.92210 | 0.95874 | 0.00191 | 1.03235 | 0.93661 | 1.02044 | 0.92510 | 0.95287 | 0.95239 |
| 0.3 | 0.09668 | 1.01515 | 1.03476 | 0.00195 | 0.09085 | 0.90207 | 0.97005 | 0.00194 | 1.03280 | 0.96007 | 1.02075 | 0.95613 | 0.96775 | 0.96774 |
| 0.4 | 0.16691 | 0.96119 | 1.03502 | 0.00196 | 0.16108 | 0.84648 | 0.98606 | 0.00196 | 1.03304 | 0.98109 | 1.02090 | 0.97204 | 0.97550 | 0.97561 |
| 0.5 | 0.26176 | 0.88774 | 1.03517 | 0.00197 | 0.25192 | 0.76840 | 0.99582 | 0.00197 | 1.03319 | 0.99385 | 1.02100 | 0.98171 | 0.98069 | 0.98069 |
| 0.6 | 0.37524 | 0.79502 | 1.03528 | 0.00198 | 0.36589 | 0.65598 | 1.00257 | 0.00198 | 1.03329 | 1.00038 | 1.02106 | 0.98821 | 0.98860 | 0.98861 |
| 0.7 | 0.50985 | 0.67711 | 1.03535 | 0.00198 | 0.49550 | 0.52557 | 1.00708 | 0.00198 | 1.03336 | 1.00509 | 1.02111 | 0.99289 | 0.98591 | 0.98592 |
| 0.8 | 0.66409 | 0.54004 | 1.03541 | 0.00199 | 0.64823 | 0.37287 | 1.01065 | 0.00199 | 1.03341 | 1.00868 | 1.02114 | 0.99640 | 0.98765 | 0.98766 |
| 0.9 | 0.89947 | 0.38182 | 1.03545 | 0.00199 | 0.82164 | 0.19827 | 1.01389 | 0.00199 | 1.03345 | 1.01139 | 1.02117 | 0.99915 | 0.98901 | 0.98901 |
| 1.0 | 1.05949 | 0.20217 | 1.03549 | 0.00199 | 1.00561 | 0.00199 | 1.01561 | 0.00199 | 1.03349 | 1.01361 | 1.02119 | 1.00135 | 0.99010 | 0.99010 |

$\zeta_1 = 0.05; \quad \bar{\delta} = 0.02805$

| μ | ψ | φ | χ | ζ | ξ | η | σ | θ | $X^{(3)}$ | $Y^{(3)}$ | $X^{(2)}$ | $Y^{(2)}$ | $\mathcal{X}$ | $\gamma_\infty$ |
|---|---|---|---|---|---|---|---|---|---|---|---|---|---|---|
| 0 | 0 | 1.00000 | 1.00000 | 0 | 0 | 0 | 0 | 0 | 1.00000 | 0 | 1.00000 | 0 | 0.46068 | 0.47495 |
| 0.010 | 0.00209 | 1.05083 | 1.03851 | 0.00099 | 0.00194 | 0.07087 | 0.06654 | 0.00096 | 1.01753 | 0.06256 | 1.01090 | 0.05826 | 0.50922 | 0.51482 |
| 0.025 | 0.00428 | 1.04934 | 1.03102 | 0.00181 | 0.00379 | 0.34846 | 0.32888 | 0.00179 | 1.02920 | 0.32657 | 1.01850 | 0.51774 | 0.64547 | 0.64780 |
| 0.050 | 0.00727 | 1.05928 | 1.03872 | 0.00255 | 0.00610 | 0.64451 | 0.58502 | 0.00231 | 1.03687 | 0.58267 | 1.02321 | 0.57074 | 0.77273 | 0.77335 |
| 0.075 | 0.01096 | 1.06698 | 1.04195 | 0.00255 | 0.00907 | 0.72958 | 0.71044 | 0.00254 | 1.03938 | 0.70787 | 1.02519 | 0.69457 | 0.84431 | 0.84463 |
| 0.1 | 0.01575 | 1.05300 | 1.04578 | 0.00267 | 0.01313 | 0.79960 | 0.78810 | 0.00266 | 1.04103 | 0.78040 | 1.02628 | 0.76666 | 0.86993 | 0.87010 |
| 0.2 | 0.04740 | 1.03125 | 1.04660 | 0.00287 | 0.04179 | 0.89606 | 0.90655 | 0.00286 | 1.04371 | 0.90364 | 1.02804 | 0.88885 | 0.93023 | 0.93027 |
| 0.3 | 0.09966 | 0.97915 | 1.04762 | 0.00294 | 0.09104 | 0.89158 | 0.95195 | 0.00294 | 1.04465 | 0.94896 | 1.02867 | 0.93324 | 0.95287 | 0.95238 |
| 0.4 | 0.17273 | 0.90472 | 1.04815 | 0.00297 | 0.16110 | 0.84855 | 0.97548 | 0.00297 | 1.04514 | 0.97246 | 1.02899 | 0.95653 | 0.96585 | 0.96385 |
| 0.5 | 0.26664 | 0.80848 | 1.04846 | 0.00300 | 0.25199 | 0.76481 | 0.98990 | 0.00299 | 1.04543 | 0.98687 | 1.02918 | 0.97078 | 0.97087 | 0.97087 |
| 0.6 | 0.38140 | 0.69056 | 1.04868 | 0.00301 | 0.35874 | 0.65899 | 0.99968 | 0.00301 | 1.04558 | 0.99659 | 1.02952 | 0.98041 | 0.97561 | 0.97561 |
| 0.7 | 0.51703 | 0.55105 | 1.04888 | 0.00302 | 0.49586 | 0.52389 | 1.00664 | 0.00302 | 1.04578 | 1.00559 | 1.02941 | 0.98784 | 0.97902 | 0.97902 |
| 0.8 | 0.67852 | 0.58992 | 1.04895 | 0.00303 | 0.64984 | 0.37651 | 1.01198 | 0.00303 | 1.04588 | 1.00887 | 1.02948 | 0.99257 | 0.98160 | 0.98160 |
| 0.9 | 0.85089 | 0.20726 | 1.04904 | 0.00304 | 0.82118 | 0.20095 | 1.01607 | 0.00304 | 1.04597 | 1.01300 | 1.02954 | 0.99666 | 0.98561 | 0.98561 |
| 1.0 | 1.04912 | 0.00505 | 1.04911 | 0.00304 | 1.01940 | 0.00505 | 1.01989 | 0.00304 | 1.04603 | 1.01651 | 1.02958 | 0.99994 | 0.98525 | 0.98522 |

TABLE 12 (continued)

$\tau_1 = 0.04; \quad \bar{s} = 0.08678$

| $\mu$ | $\psi$ | $\phi$ | $\chi$ | $\zeta$ | $\xi$ | $\eta$ | $\sigma$ | $\theta$ | $X^{(1)}$ | $Y^{(1)}$ | $X^{(2)}$ | $Y^{(2)}$ | $\tilde{\chi}$ | $\gamma_n$ |
|---|---|---|---|---|---|---|---|---|---|---|---|---|---|---|
| 0 | 0 | 1.00000 | 1.00000 | 0 | 0 | 0 | 0 | 0 | 1.00000 | 0 | 1.00000 | 0 | 0.45166 | 0.46670 |
| 0.010 | 0.00222 | 1.03125 | 1.01906 | 0.00105 | 0.00202 | 0.03772 | 0.05127 | 0.00101 | 1.01802 | 0.03022 | 1.01121 | 0.02689 | 0.48476 | 0.49822 |
| 0.025 | 0.00491 | 1.05450 | 1.03441 | 0.00212 | 0.00431 | 0.24572 | 0.25057 | 0.00207 | 1.03228 | 0.22845 | 1.02057 | 0.21953 | 0.59154 | 0.59479 |
| 0.050 | 0.00845 | 1.06917 | 1.04530 | 0.00291 | 0.00700 | 0.51198 | 0.49057 | 0.00287 | 1.04286 | 0.48763 | 1.02727 | 0.47452 | 0.72133 | 0.72248 |
| 0.075 | 0.01245 | 1.07835 | 1.05920 | 0.00827 | 0.01105 | 0.65629 | 0.63878 | 0.00324 | 1.04688 | 0.65047 | 1.03029 | 0.61586 | 0.79166 | 0.79223 |
| 0.1 | 0.01745 | 1.07274 | 1.05297 | 0.00847 | 0.01409 | 0.74160 | 0.72088 | 0.00345 | 1.04942 | 0.71781 | 1.03197 | 0.70106 | 0.84419 | 0.84451 |
| 0.2 | 0.04967 | 1.04701 | 1.05754 | 0.00881 | 0.04284 | 0.87047 | 0.87492 | 0.00380 | 1.05367 | 0.87105 | 1.03482 | 0.85286 | 0.90911 | 0.90919 |
| 0.3 | 0.10254 | 0.99498 | 1.05920 | 0.00894 | 0.09121 | 0.87974 | 0.93444 | 0.00893 | 1.05520 | 0.92945 | 1.03595 | 0.91055 | 0.93748 | 0.93751 |
| 0.4 | 0.17687 | 0.91985 | 1.06005 | 0.00900 | 0.16102 | 0.88875 | 0.96418 | 0.00899 | 1.05599 | 0.96012 | 1.03688 | 0.94085 | 0.95237 | 0.95258 |
| 0.5 | 0.27125 | 0.82283 | 1.06058 | 0.00404 | 0.25187 | 0.76682 | 0.98812 | 0.00406 | 1.05647 | 0.97902 | 1.03670 | 0.95952 | 0.96154 | 0.96155 |
| 0.6 | 0.38718 | 0.70266 | 1.06098 | 0.00407 | 0.36378 | 0.66078 | 0.99595 | 0.00406 | 1.05680 | 0.99182 | 1.03692 | 0.97218 | 0.96775 | 0.96774 |
| 0.7 | 0.52418 | 0.56097 | 1.06118 | 0.00408 | 0.49676 | 0.58222 | 1.00528 | 0.00408 | 1.05703 | 1.00108 | 1.03707 | 0.98132 | 0.97228 | 0.97222 |
| 0.8 | 0.68226 | 0.39729 | 1.06187 | 0.00410 | 0.65081 | 0.57947 | 1.01224 | 0.00409 | 1.05721 | 1.00807 | 1.03719 | 0.98624 | 0.97562 | 0.97561 |
| 0.9 | 0.86112 | 0.21167 | 1.06152 | 0.00411 | 0.82594 | 0.20828 | 1.01772 | 0.00410 | 1.05784 | 1.01355 | 1.03729 | 0.99865 | 0.97827 | 0.97826 |
| 1.0 | 1.06166 | 0.00412 | 1.06164 | 0.00412 | 1.02215 | 0.00411 | 1.02214 | 0.00411 | 1.05745 | 1.01795 | 1.03736 | 0.99800 | 0.98041 | 0.98059 |

$\tau_1 = 0.05; \quad \bar{s} = 0.04518$

| $\mu$ | $\psi$ | $\phi$ | $\chi$ | $\zeta$ | $\xi$ | $\eta$ | $\sigma$ | $\theta$ | $X^{(1)}$ | $Y^{(1)}$ | $X^{(2)}$ | $Y^{(2)}$ | $\tilde{\chi}$ | $\gamma_n$ |
|---|---|---|---|---|---|---|---|---|---|---|---|---|---|---|
| 0 | 0 | 1.00000 | 1.00000 | 0 | 0 | 0 | 0 | 0 | 1.00000 | 0 | 1.00000 | 0 | 0.44841 | 0.46291 |
| 0.010 | 0.00280 | 1.03170 | 1.01984 | 0.00109 | 0.00207 | 0.02462 | 0.01890 | 0.00108 | 1.01825 | 0.01785 | 1.01135 | 0.01140 | 0.47041 | 0.48155 |
| 0.025 | 0.00658 | 1.05788 | 1.03664 | 0.00285 | 0.00468 | 0.17901 | 0.16487 | 0.00227 | 1.03428 | 0.16201 | 1.02192 | 0.15334 | 0.55310 | 0.55801 |
| 0.050 | 0.00948 | 1.07682 | 1.05046 | 0.00341 | 0.00777 | 0.44510 | 0.44256 | 0.00355 | 1.04700 | 0.40911 | 1.03045 | 0.39535 | 0.67888 | 0.68025 |
| 0.075 | 0.01381 | 1.08387 | 1.05707 | 0.00392 | 0.01096 | 0.59088 | 0.56600 | 0.00388 | 1.05907 | 0.56201 | 1.03454 | 0.51565 | 0.75385 | 0.75480 |
| 0.1 | 0.01904 | 1.08132 | 1.06089 | 0.00422 | 0.01500 | 0.67831 | 0.66879 | 0.00419 | 1.05659 | 0.65948 | 1.03692 | 0.64157 | 0.80154 | 0.80211 |
| 0.2 | 0.05187 | 1.06101 | 1.06740 | 0.00474 | 0.04290 | 0.84483 | 0.84415 | 0.00472 | 1.06256 | 0.85951 | 1.04096 | 0.81870 | 0.88894 | 0.88908 |
| 0.3 | 0.10535 | 1.00929 | 1.06981 | 0.00498 | 0.09136 | 0.86707 | 0.91486 | 0.00492 | 1.06477 | 0.90983 | 1.04245 | 0.88820 | 0.92305 | 0.92310 |
| 0.4 | 0.17989 | 0.93562 | 1.07106 | 0.00508 | 0.16090 | 0.88513 | 0.95246 | 0.00502 | 1.06591 | 0.94782 | 1.04325 | 0.92517 | 0.94115 | 0.94117 |
| 0.5 | 0.27566 | 0.85501 | 1.07182 | 0.00509 | 0.25168 | 0.76283 | 0.97577 | 0.00508 | 1.06661 | 0.97057 | 1.04571 | 0.94809 | 0.95237 | 0.95237 |
| 0.6 | 0.39268 | 0.71380 | 1.07234 | 0.00513 | 0.36861 | 0.66166 | 0.99164 | 0.00512 | 1.06709 | 0.98489 | 1.04405 | 0.96869 | 0.96000 | 0.95999 |
| 0.7 | 0.53096 | 0.57016 | 1.07271 | 0.00516 | 0.49685 | 0.54483 | 1.00031 | 0.00515 | 1.06743 | 0.99785 | 1.04426 | 0.97499 | 0.96552 | 0.96551 |
| 0.8 | 0.69050 | 0.40415 | 1.07299 | 0.00519 | 0.65135 | 0.85191 | 1.01184 | 0.00518 | 1.06769 | 1.00654 | 1.04445 | 0.98556 | 0.96971 | 0.96969 |
| 0.9 | 0.87155 | 0.25584 | 1.07821 | 0.00520 | 0.82718 | 0.20534 | 1.01866 | 0.00519 | 1.06789 | 1.01535 | 1.04457 | 0.99027 | 0.97299 | 0.97297 |
| 1.0 | 1.07843 | 0.00525 | 1.07839 | 0.00522 | 1.02819 | 0.00518 | 1.02216 | 0.00521 | 1.06805 | 1.01885 | 1.04468 | 0.99568 | 0.97568 | 0.97561 |

TABLE 12 (continued)

$\zeta_1 = 0.06; \quad \bar{\xi} = 0.05389$

| $\mu$ | $\psi$ | $\phi$ | $\chi$ | $\zeta$ | $\xi$ | $\eta$ | $\sigma$ | $\theta$ | $X^{(1)}$ | $Y^{(1)}$ | $X^{(2)}$ | $Y^{(2)}$ | $u$ | $\varkappa$ |
|---|---|---|---|---|---|---|---|---|---|---|---|---|---|---|
| 0 | 0 | 1.00000 | 1.00000 | 0 | 0 | 0 | 0 | 0 | 1.00000 | 0 | 1.00000 | 0 | 0.43585 | 0.45748 |
| 0.010 | 0.00287 | 1.03201 | 1.03951 | 0.00113 | 0.00809 | 0.01912 | 0.01897 | 0.00104 | 1.04839 | 0.01288 | 1.01113 | 0.00977 | 0.46017 | 0.47574 |
| 0.025 | 0.00574 | 1.06016 | 1.04816 | 0.00252 | 0.00494 | 0.13337 | 0.11949 | 0.00211 | 1.05361 | 0.11696 | 1.02282 | 0.10868 | 0.52508 | 0.53175 |
| 0.050 | 0.01087 | 1.06284 | 1.05457 | 0.00384 | 0.00842 | 0.57116 | 0.34815 | 0.00375 | 1.05965 | 0.34485 | 1.03298 | 0.33023 | 0.64232 | 0.64503 |
| 0.075 | 0.01506 | 1.09281 | 1.06287 | 0.00455 | 0.01178 | 0.53262 | 0.50614 | 0.00445 | 1.05824 | 0.50452 | 1.03811 | 0.48433 | 0.72029 | 0.72170 |
| 0.1 | 0.02054 | 1.09431 | 1.06780 | 0.00495 | 0.01586 | 0.58812 | 0.61168 | 0.00487 | 1.06275 | 0.60653 | 1.04127 | 0.58744 | 0.77178 | 0.77258 |
| 0.2 | 0.05401 | 1.07860 | 1.07640 | 0.00565 | 0.04246 | 0.81946 | 0.81134 | 0.00562 | 1.07059 | 0.80854 | 1.04859 | 0.78590 | 0.86968 | 0.86989 |
| 0.3 | 0.10806 | 1.02288 | 1.07968 | 0.00598 | 0.09151 | 0.85587 | 0.89685 | 0.00590 | 1.07854 | 0.89026 | 1.04964 | 0.86628 | 0.90906 | 0.90915 |
| 0.4 | 0.18831 | 0.94635 | 1.08135 | 0.00607 | 0.16074 | 0.82679 | 0.94048 | 0.00605 | 1.07509 | 0.93425 | 1.04960 | 0.90955 | 0.93021 | 0.93022 |
| 0.5 | 0.27998 | 0.84677 | 1.08287 | 0.00616 | 0.23111 | 0.76011 | 0.95808 | 0.00611 | 1.07604 | 0.96470 | 1.05000 | 0.98657 | 0.94339 | 0.94358 |
| 0.6 | 0.39796 | 0.72418 | 1.08308 | 0.00622 | 0.56329 | 0.66492 | 1.00052 | 0.00624 | 1.07668 | 0.98046 | 1.05074 | 0.95593 | 0.95892 | 0.95989 |
| 0.7 | 0.53743 | 0.57875 | 1.08859 | 0.00686 | 0.49671 | 0.55591 | 1.01090 | 0.00684 | 1.07714 | 0.99409 | 1.05105 | 0.96844 | 0.96888 | 0.96804 |
| 0.8 | 0.69856 | 0.41061 | 1.08398 | 0.00629 | 0.65157 | 0.38897 | 1.01905 | 0.00627 | 1.07750 | 1.00444 | 1.05130 | 0.97862 | 0.96888 | 0.96804 |
| 0.9 | 0.88074 | 0.21980 | 1.08428 | 0.00632 | 0.82789 | 0.20721 | 1.02562 | 0.00630 | 1.07777 | 1.01256 | 1.05148 | 0.98662 | 0.96776 | 0.96775 |
| 1.0 | 1.08458 | 0.00686 | 1.08452 | 0.00634 | 1.02567 | 0.00628 | 1.02562 | 0.00632 | 1.07799 | 1.01911 | 1.05164 | 0.99306 | 0.97091 | 0.97087 |

$\zeta_1 = 0.07; \quad \bar{\xi} = 0.06158$

| $\mu$ | $\psi$ | $\phi$ | $\chi$ | $\zeta$ | $\xi$ | $\eta$ | $\sigma$ | $\theta$ | $X^{(1)}$ | $Y^{(1)}$ | $X^{(2)}$ | $Y^{(2)}$ | $u$ | $\varkappa$ |
|---|---|---|---|---|---|---|---|---|---|---|---|---|---|---|
| 0 | 0 | 1.00000 | 1.00000 | 0 | 0 | 0 | 0 | 0 | 1.00000 | 0 | 1.00000 | 0 | 0.44881 | 0.45285 |
| 0.010 | 0.00242 | 1.03224 | 1.01964 | 0.00115 | 0.00210 | 0.01654 | 0.01185 | 0.00104 | 1.04850 | 0.01078 | 1.01119 | 0.00789 | 0.45178 | 0.46753 |
| 0.025 | 0.00602 | 1.06174 | 1.02921 | 0.00266 | 0.00513 | 0.10202 | 0.08899 | 0.00251 | 1.03652 | 0.08683 | 1.02844 | 0.07852 | 0.50409 | 0.51256 |
| 0.050 | 0.01116 | 1.08768 | 1.05787 | 0.00422 | 0.00899 | 0.31799 | 0.29498 | 0.00409 | 1.05955 | 0.29069 | 1.03500 | 0.27666 | 0.61186 | 0.61552 |
| 0.075 | 0.01621 | 1.09892 | 1.06781 | 0.00508 | 0.01254 | 0.48080 | 0.45380 | 0.00497 | 1.06260 | 0.44811 | 1.04119 | 0.43040 | 0.69041 | 0.69288 |
| 0.1 | 0.02195 | 1.10297 | 1.07887 | 0.00550 | 0.01667 | 0.59232 | 0.56898 | 0.00551 | 1.06810 | 0.55824 | 1.04497 | 0.53820 | 0.74445 | 0.74564 |
| 0.2 | 0.05610 | 1.08948 | 1.08467 | 0.00655 | 0.04402 | 0.79451 | 0.78554 | 0.00649 | 1.07190 | 0.77877 | 1.05178 | 0.75445 | 0.85127 | 0.85158 |
| 0.3 | 0.11072 | 1.08448 | 1.08885 | 0.00691 | 0.09167 | 0.84030 | 0.87795 | 0.00688 | 1.08168 | 0.87082 | 1.05433 | 0.84476 | 0.89548 | 0.89558 |
| 0.4 | 0.18664 | 0.95811 | 1.09100 | 0.00711 | 0.16056 | 0.81990 | 0.92886 | 0.00707 | 1.08454 | 0.92103 | 1.05569 | 0.89407 | 0.91950 | 0.91955 |
| 0.5 | 0.28406 | 0.85777 | 1.09285 | 0.00725 | 0.25098 | 0.75780 | 0.96000 | 0.00720 | 1.08486 | 0.95254 | 1.05653 | 0.92501 | 0.93957 | 0.93456 |
| 0.6 | 0.40806 | 0.73686 | 1.09926 | 0.00731 | 0.36286 | 0.66162 | 0.98170 | 0.00728 | 1.08570 | 0.97416 | 1.05711 | 0.94625 | 0.94489 | 0.94486 |
| 0.7 | 0.54866 | 0.59686 | 1.09898 | 0.00787 | 0.49669 | 0.53708 | 0.99751 | 0.00784 | 1.08680 | 0.98991 | 1.05752 | 0.96172 | 0.95241 | 0.95287 |
| 0.8 | 0.70589 | 0.41673 | 1.09443 | 0.00741 | 0.65158 | 0.38571 | 1.00954 | 0.00739 | 1.08675 | 1.00189 | 1.05784 | 0.97749 | 0.95812 | 0.95307 |
| 0.9 | 0.88974 | 0.22859 | 1.09485 | 0.00745 | 0.82880 | 0.20890 | 1.01900 | 0.00742 | 1.08711 | 1.01131 | 1.05808 | 0.98275 | 0.96262 | 0.96656 |
| 1.0 | 1.09528 | 0.00751 | 1.09514 | 0.00746 | 1.02671 | 0.00787 | 1.02668 | 0.00745 | 1.08739 | 1.01892 | 1.05828 | 0.99022 | 0.96604 | 0.96618 |

TABLE 12 (continued)

$\zeta_1 = 0.08; \quad \bar{\xi} = 0.06918$

| $\mu$ | $\psi$ | $\phi$ | $\chi$ | $\zeta$ | $\xi$ | $\eta$ | $\sigma$ | $\theta$ | $X^{(1)}$ | $Y^{(1)}$ | $X^{(2)}$ | $Y^{(2)}$ | $\gamma_e$ | $\gamma$ |
|---|---|---|---|---|---|---|---|---|---|---|---|---|---|---|
| 0 | 0 | 1.00000 | 1.00000 | 0 | 0 | 0 | 0 | 0 | 1.00000 | 0 | 1.00000 | 0 | 0.42222 | 0.44746 |
| 0.010 | 0.00248 | 1.03243 | 1.03974 | 0.00118 | 0.00212 | 0.00514 | 0.01081 | 0.00105 | 1.01858 | 0.00967 | 1.01154 | 0.00705 | 0.44456 | 0.46207 |
| 0.025 | 0.00626 | 1.06288 | 1.03997 | 0.00278 | 0.00527 | 0.00840 | 0.06821 | 0.00259 | 1.03716 | 0.05044 | 1.02587 | 0.05010 | 0.48788 | 0.49813 |
| 0.050 | 0.01184 | 1.09149 | 1.06955 | 0.00456 | 0.00947 | 0.27877 | 0.25107 | 0.00438 | 1.05588 | 0.24644 | 1.03668 | 0.23258 | 0.58598 | 0.59067 |
| 0.075 | 0.01726 | 1.10495 | 1.07206 | 0.00598 | 0.01332 | 0.44476 | 0.40668 | 0.00543 | 1.06680 | 0.40096 | 1.04571 | 0.38299 | 0.66374 | 0.66685 |
| 0.1 | 0.02827 | 1.11054 | 1.07928 | 0.00628 | 0.01743 | 0.55919 | 0.58050 | 0.00610 | 1.07278 | 0.51409 | 1.04830 | 0.49342 | 0.72941 | 0.72202 |
| 0.2 | 0.05812 | 1.09590 | 1.09281 | 0.00742 | 0.04458 | 0.77011 | 0.75777 | 0.00785 | 1.08460 | 0.75909 | 1.05349 | 0.72428 | 0.83566 | 0.83408 |
| 0.3 | 0.11332 | 1.04566 | 1.09742 | 0.00790 | 0.09185 | 0.82665 | 0.85989 | 0.00784 | 1.08921 | 0.85170 | 1.05570 | 0.82585 | 0.88262 | 0.88146 |
| 0.4 | 0.18989 | 0.96921 | 1.10014 | 0.00830 | 0.16087 | 0.81259 | 0.91618 | 0.00810 | 1.09166 | 0.90778 | 1.06011 | 0.87674 | 0.90904 | 0.90908 |
| 0.5 | 0.28809 | 0.86813 | 1.10188 | 0.00841 | 0.25050 | 0.75408 | 0.95177 | 0.00826 | 1.09318 | 0.94815 | 1.06046 | 0.91847 | 0.92591 | 0.92590 |
| 0.6 | 0.40801 | 0.74414 | 1.10298 | 0.00849 | 0.36234 | 0.66089 | 0.97628 | 0.00887 | 1.09421 | 0.96755 | 1.06018 | 0.93740 | 0.93751 | 0.93747 |
| 0.7 | 0.54969 | 0.59456 | 1.10381 | 0.00849 | 0.49592 | 0.58780 | 0.99419 | 0.00845 | 1.09496 | 0.98538 | 1.06177 | 0.95488 | 0.94598 | 0.94592 |
| 0.8 | 0.71815 | 0.42257 | 1.10414 | 0.00855 | 0.65129 | 0.48719 | 1.00784 | 0.00851 | 1.09553 | 0.99897 | 1.06110 | 0.96820 | 0.95243 | 0.95286 |
| 0.9 | 0.89840 | 0.22726 | 1.10494 | 0.00859 | 0.82844 | 0.21046 | 1.01859 | 0.00855 | 1.09597 | 1.00967 | 1.06144 | 0.97870 | 0.95751 | 0.95743 |
| 1.0 | 1.10546 | 0.00868 | 1.10535 | 0.00868 | 1.02789 | 0.00848 | 1.02727 | 0.00859 | 1.09635 | 1.01832 | 1.06166 | 0.98719 | 0.96161 | 0.96155 |

$\zeta_1 = 0.09; \quad \bar{\xi} = 0.07698$

| $\mu$ | $\psi$ | $\phi$ | $\chi$ | $\zeta$ | $\xi$ | $\eta$ | $\sigma$ | $\theta$ | $X^{(1)}$ | $Y^{(1)}$ | $X^{(2)}$ | $Y^{(2)}$ | $\gamma_e$ | $\gamma$ |
|---|---|---|---|---|---|---|---|---|---|---|---|---|---|---|
| 0 | 0 | 1.00000 | 1.00000 | 0 | 0 | 0 | 0 | 0 | 1.00000 | 0 | 1.00000 | 0 | 0.41660 | 0.44280 |
| 0.010 | 0.00253 | 1.03260 | 1.04084 | 0.00120 | 0.00212 | 0.00425 | 0.01020 | 0.00105 | 1.01866 | 0.00905 | 1.01158 | 0.00661 | 0.43756 | 0.45704 |
| 0.025 | 0.00615 | 1.06873 | 1.06054 | 0.00287 | 0.00537 | 0.06598 | 0.05398 | 0.00264 | 1.03765 | 0.05113 | 1.02819 | 0.04423 | 0.47496 | 0.48694 |
| 0.050 | 0.01215 | 1.09461 | 1.06275 | 0.00485 | 0.00987 | 0.28699 | 0.21480 | 0.00462 | 1.05776 | 0.20987 | 1.03796 | 0.19651 | 0.56885 | 0.56963 |
| 0.075 | 0.01828 | 1.11010 | 1.07572 | 0.00605 | 0.01384 | 0.59387 | 0.36556 | 0.00585 | 1.06946 | 0.35985 | 1.04606 | 0.31130 | 0.63987 | 0.64314 |
| 0.1 | 0.03452 | 1.11718 | 1.08899 | 0.00682 | 0.01815 | 0.53147 | 0.46008 | 0.00665 | 1.07690 | 0.47879 | 1.05124 | 0.45270 | 0.69644 | 0.69859 |
| 0.2 | 0.06009 | 1.10510 | 1.09941 | 0.00828 | 0.04514 | 0.74652 | 0.78104 | 0.00817 | 1.09076 | 0.72843 | 1.06092 | 0.69540 | 0.81681 | 0.81756 |
| 0.3 | 0.11587 | 1.05611 | 1.10554 | 0.00887 | 0.09200 | 0.81282 | 0.84205 | 0.00879 | 1.09626 | 0.83281 | 1.06477 | 0.80334 | 0.86954 | 0.86978 |
| 0.4 | 0.19807 | 0.97961 | 1.10885 | 0.00918 | 0.16017 | 0.84497 | 0.90898 | 0.00912 | 1.09921 | 0.89441 | 1.06684 | 0.86860 | 0.89882 | 0.89888 |
| 0.5 | 0.29202 | 0.87792 | 1.11088 | 0.00938 | 0.25004 | 0.75087 | 0.94339 | 0.00932 | 1.10104 | 0.93560 | 1.06812 | 0.90196 | 0.91741 | 0.91740 |
| 0.6 | 0.41282 | 0.75187 | 1.11227 | 0.00952 | 0.36175 | 0.65979 | 0.97064 | 0.00946 | 1.10250 | 0.96071 | 1.06900 | 0.92850 | 0.93025 | 0.93020 |
| 0.7 | 0.55553 | 0.60189 | 1.11329 | 0.00962 | 0.49535 | 0.53826 | 0.99060 | 0.00956 | 1.10320 | 0.98057 | 1.06961 | 0.94791 | 0.93964 | 0.93957 |
| 0.8 | 0.72018 | 0.42815 | 1.11406 | 0.00969 | 0.65087 | 0.38846 | 1.00584 | 0.00964 | 1.10389 | 0.99574 | 1.07012 | 0.96280 | 0.94681 | 0.94672 |
| 0.9 | 0.90677 | 0.23078 | 1.11467 | 0.00975 | 0.82833 | 0.21192 | 1.01787 | 0.00970 | 1.10443 | 1.00770 | 1.07050 | 0.97451 | 0.95247 | 0.95237 |
| 1.0 | 1.11532 | 0.00985 | 1.11516 | 0.00980 | 1.02775 | 0.00960 | 1.02759 | 0.00974 | 1.10487 | 1.01788 | 1.07081 | 0.98899 | 0.95704 | 0.95698 |

TABLE 12 (continued)

$v = 0.10;\ \bar{v} = 0.08420$

| $\mu$ | $\psi$ | $\phi$ | $\chi$ | $\zeta$ | $\xi$ | $\eta$ | $\sigma$ | $\theta$ | $X^{(1)}$ | $Y^{(1)}$ | $X^{(2)}$ | $Y^{(2)}$ | $\mathcal{N}$ | $\mathcal{N}$ |
|---|---|---|---|---|---|---|---|---|---|---|---|---|---|---|
| 0 | 0 | 1.00000 | 1.00000 | 0 | 0 | 0 | 0 | 0 | 1.00000 | 0 | 1.00000 | 0 | 0.41014 | 0.44882 |
| 0.010 | 0.00257 | 1.03275 | 1.03995 | 0.00128 | 0.00213 | 0.03357 | 0.00978 | 0.00106 | 1.01872 | 0.00862 | 1.01162 | 0.00683 | 0.43120 | 0.45228 |
| 0.025 | 0.00662 | 1.04458 | 1.04097 | 0.00295 | 0.00544 | 0.05486 | 0.04418 | 0.00268 | 1.03798 | 0.04127 | 1.02442 | 0.03478 | 0.44453 | 0.47795 |
| 0.050 | 0.01299 | 1.09717 | 1.06456 | 0.00511 | 0.01022 | 0.20687 | 0.18482 | 0.00482 | 1.05929 | 0.17968 | 1.03905 | 0.16643 | 0.54480 | 0.55170 |
| 0.075 | 0.01912 | 1.11452 | 1.07890 | 0.00648 | 0.01440 | 0.57756 | 0.52928 | 0.00622 | 1.07217 | 0.32262 | 1.04800 | 0.30468 | 0.61845 | 0.62244 |
| 0.10 | 0.02570 | 1.12305 | 1.08825 | 0.00788 | 0.01882 | 0.47591 | 0.44465 | 0.00716 | 1.08053 | 0.43702 | 1.05385 | 0.43569 | 0.67530 | 0.67785 |
| 0.15 | 0.04155 | 1.12491 | 1.09951 | 0.00848 | 0.09006 | 0.68291 | 0.60408 | 0.00851 | 1.09062 | 0.59520 | 1.06093 | 0.56966 | 0.75233 | 0.75358 |
| 0.2 | 0.06201 | 1.11396 | 1.10602 | 0.00912 | 0.04570 | 0.72218 | 0.70535 | 0.00896 | 1.09645 | 0.69581 | 1.06508 | 0.66776 | 0.80069 | 0.80158 |
| 0.3 | 0.11886 | 1.06591 | 1.11322 | 0.00983 | 0.09217 | 0.79897 | 0.82453 | 0.00972 | 1.10288 | 0.81421 | 1.06956 | 0.78335 | 0.85714 | 0.85738 |
| 0.4 | 0.19619 | 0.98945 | 1.11711 | 0.01022 | 0.15998 | 0.79709 | 0.89182 | 0.01013 | 1.10634 | 0.88111 | 1.07201 | 0.84867 | 0.88883 | 0.88890 |
| 0.5 | 0.29586 | 0.88720 | 1.11954 | 0.01046 | 0.24956 | 0.74640 | 0.98490 | 0.01038 | 1.10851 | 0.92595 | 1.07354 | 0.89051 | 0.90907 | 0.90905 |
| 0.6 | 0.41752 | 0.76021 | 1.12121 | 0.01065 | 0.36110 | 0.65888 | 0.96482 | 0.01055 | 1.10999 | 0.95467 | 1.07459 | 0.93953 | 0.92210 | 0.92305 |
| 0.7 | 0.56122 | 0.60891 | 1.12242 | 0.01075 | 0.49467 | 0.58845 | 0.98679 | 0.01068 | 1.11107 | 0.97552 | 1.07535 | 0.94098 | 0.93389 | 0.93530 |
| 0.8 | 0.72700 | 0.43554 | 1.12334 | 0.01085 | 0.65030 | 0.38952 | 1.00560 | 0.01077 | 1.11189 | 0.99224 | 1.07592 | 0.95728 | 0.94127 | 0.94115 |
| 0.9 | 0.91468 | 0.28422 | 1.12406 | 0.01092 | 0.82803 | 0.21326 | 1.01689 | 0.01085 | 1.11258 | 1.00545 | 1.07658 | 0.97020 | 0.94748 | 0.94785 |
| 1.0 | 1.12487 | 0.01104 | 1.12465 | 0.01098 | 1.02785 | 0.01072 | 1.02765 | 0.01090 | 1.11305 | 1.01615 | 1.07674 | 0.98066 | 0.95251 | 0.95237 |

$v = 0.11;\ \bar{v} = 0.09146$

| $\mu$ | $\psi$ | $\phi$ | $\chi$ | $\zeta$ | $\xi$ | $\eta$ | $\sigma$ | $\theta$ | $X^{(1)}$ | $Y^{(1)}$ | $X^{(2)}$ | $Y^{(2)}$ | $\mathcal{N}$ | $\mathcal{N}$ |
|---|---|---|---|---|---|---|---|---|---|---|---|---|---|---|
| 0 | 0 | 1.00000 | 1.00000 | 0 | 0 | 0 | 0 | 0 | 1.00000 | 0 | 1.00000 | 0 | 0.40455 | 0.45402 |
| 0.010 | 0.00262 | 1.03288 | 1.04132 | 0.00125 | 0.00214 | 0.03304 | 0.00946 | 0.00106 | 1.01879 | 0.00828 | 1.01166 | 0.00612 | 0.42519 | 0.44774 |
| 0.025 | 0.00677 | 1.04489 | 1.06608 | 0.00305 | 0.00550 | 0.04741 | 0.08758 | 0.00270 | 1.03826 | 0.04441 | 1.02460 | 0.02851 | 0.45531 | 0.47049 |
| 0.050 | 0.01348 | 1.09928 | 1.06608 | 0.00534 | 0.01051 | 0.18086 | 0.16002 | 0.00500 | 1.06055 | 0.15460 | 1.05994 | 0.11182 | 0.52829 | 0.55683 |
| 0.075 | 0.01995 | 1.11854 | 1.08167 | 0.00687 | 0.01497 | 0.32532 | 0.29728 | 0.00656 | 1.07450 | 0.29021 | 1.04968 | 0.27239 | 0.59916 | 0.60391 |
| 0.10 | 0.02650 | 1.12825 | 1.09208 | 0.00790 | 0.01944 | 0.44828 | 0.41166 | 0.00762 | 1.08375 | 0.40847 | 1.05618 | 0.58204 | 0.65585 | 0.65890 |
| 0.15 | 0.04309 | 1.13198 | 1.10475 | 0.00918 | 0.03072 | 0.60498 | 0.57464 | 0.00896 | 1.09508 | 0.56595 | 1.06418 | 0.53897 | 0.73470 | 0.73622 |
| 0.2 | 0.06587 | 1.13215 | 1.11220 | 0.00994 | 0.04625 | 0.70071 | 0.68064 | 0.00976 | 1.10171 | 0.67021 | 1.06887 | 0.64132 | 0.78523 | 0.78608 |
| 0.3 | 0.12081 | 1.07511 | 1.12051 | 0.01079 | 0.09295 | 0.78515 | 0.80734 | 0.01065 | 1.10909 | 0.79599 | 1.07410 | 0.76887 | 0.84508 | 0.84539 |
| 0.4 | 0.19924 | 0.99873 | 1.12508 | 0.01125 | 0.15978 | 0.78901 | 0.87971 | 0.01113 | 1.11510 | 0.86786 | 1.07694 | 0.83895 | 0.87905 | 0.87914 |
| 0.5 | 0.29962 | 0.89605 | 1.12786 | 0.01155 | 0.24907 | 0.74215 | 0.92685 | 0.01144 | 1.11562 | 0.91418 | 1.07873 | 0.87911 | 0.90087 | 0.90065 |
| 0.6 | 0.42210 | 0.76818 | 1.12981 | 0.01175 | 0.36042 | 0.65668 | 0.95386 | 0.01165 | 1.11784 | 0.94647 | 1.07995 | 0.91066 | 0.91605 | 0.91598 |
| 0.7 | 0.56676 | 0.61565 | 1.13123 | 0.01190 | 0.49892 | 0.53839 | 0.98280 | 0.01180 | 1.11860 | 0.97026 | 1.08085 | 0.93358 | 0.92722 | 0.92710 |
| 0.8 | 0.78356 | 0.48772 | 1.13251 | 0.01201 | 0.49762 | 0.59098 | 1.00116 | 0.01192 | 1.11955 | 0.98851 | 1.08153 | 0.95169 | 0.93578 | 0.93554 |
| 0.9 | 0.92276 | 0.23755 | 1.13516 | 0.01210 | 0.82756 | 0.21448 | 1.01569 | 0.01201 | 1.12050 | 1.00295 | 1.08206 | 0.96578 | 0.94254 | 0.94258 |
| 1.0 | 1.13413 | 0.01224 | 1.13585 | 0.01217 | 1.02774 | 0.01180 | 1.02747 | 0.01208 | 1.12091 | 1.01465 | 1.08249 | 0.97721 | 0.94802 | 0.94785 |

| $\mu$ | $\psi$ | $\phi$ | $\chi$ | $\zeta$ | $\xi$ | $\eta$ | $\sigma$ | $\theta$ | $X^{(1)}$ | $Y^{(1)}$ | $X^{(2)}$ | $Y^{(2)}$ | $\gamma_l$ | $\hat{\lambda}$ |
|---|---|---|---|---|---|---|---|---|---|---|---|---|---|---|
| 0 | 0 | 1.00000 | 1.00000 | 0 | 0 | 0 | 0 | 0 | 1.00000 | 0 | 1.00000 | 0 | 0.59925 | 0.42987 |
| 0.010 | 0.00266 | 1.03301 | 1.02009 | 0.00127 | 0.00224 | 0.01260 | 0.00918 | 0.00106 | 1.01884 | 0.00800 | 1.01169 | 0.00595 | 0.61950 | 0.44838 |
| 0.025 | 0.00691 | 1.06531 | 1.04161 | 0.00309 | 0.00554 | 0.04198 | 0.03257 | 0.00272 | 1.05847 | 0.02955 | 1.02475 | 0.02882 | 0.44741 | 0.46408 |
| 0.050 | 0.01391 | 1.10105 | 1.06735 | 0.00555 | 0.01076 | 0.15958 | 0.18950 | 0.00514 | 1.06159 | 0.13587 | 1.04068 | 0.12153 | 0.51388 | 0.52805 |
| 0.075 | 0.02071 | 1.12164 | 1.08409 | 0.00723 | 0.01586 | 0.29671 | 0.26904 | 0.00685 | 1.07652 | 0.26159 | 1.05115 | 0.24104 | 0.50475 | 0.58729 |
| 0.10 | 0.02785 | 1.13287 | 1.09545 | 0.00840 | 0.02002 | 0.41334 | 0.38158 | 0.00805 | 1.08661 | 0.37286 | 1.05827 | 0.35116 | 0.69786 | 0.64119 |
| 0.15 | 0.04457 | 1.13845 | 1.10960 | 0.00986 | 0.03135 | 0.57885 | 0.54692 | 0.00959 | 1.09916 | 0.58659 | 1.06716 | 0.51012 | 0.71603 | 0.71985 |
| 0.2 | 0.06569 | 1.12975 | 1.11799 | 0.01074 | 0.04679 | 0.67894 | 0.66692 | 0.01051 | 1.10659 | 0.64561 | 1.07245 | 0.61602 | 0.77045 | 0.77116 |
| 0.3 | 0.12821 | 1.08677 | 1.12744 | 0.01174 | 0.09234 | 0.77142 | 0.79059 | 0.01156 | 1.11495 | 0.77810 | 1.07840 | 0.74488 | 0.83339 | 0.83575 |
| 0.4 | 0.20025 | 1.00757 | 1.13261 | 0.01228 | 0.15959 | 0.78079 | 0.86169 | 0.01213 | 1.11952 | 0.85469 | 1.08166 | 0.81946 | 0.86990 | 0.86961 |
| 0.5 | 0.30531 | 0.90449 | 1.13587 | 0.01265 | 0.24855 | 0.78169 | 0.92175 | 0.01249 | 1.12239 | 0.90437 | 1.08372 | 0.86785 | 0.89284 | 0.89281 |
| 0.6 | 0.42659 | 0.77582 | 1.13811 | 0.01287 | 0.35968 | 0.65177 | 0.95278 | 0.01274 | 1.12437 | 0.93915 | 1.08513 | 0.90174 | 0.90913 | 0.90905 |
| 0.7 | 0.57218 | 0.62214 | 1.13975 | 0.01305 | 0.49309 | 0.58815 | 0.97866 | 0.01292 | 1.12581 | 0.96484 | 1.08616 | 0.92678 | 0.92114 | 0.92100 |
| 0.8 | 0.74012 | 0.44575 | 1.14100 | 0.01318 | 0.64882 | 0.39113 | 0.99954 | 0.01306 | 1.12691 | 0.98458 | 1.08694 | 0.94603 | 0.93037 | 0.93019 |
| 0.9 | 0.93044 | 0.24082 | 1.14198 | 0.01329 | 0.82692 | 0.21566 | 1.01129 | 0.01317 | 1.12777 | 1.00022 | 1.08756 | 0.96127 | 0.93767 | 0.93747 |
| 1.0 | 1.14515 | 0.01548 | 1.14278 | 0.01358 | 1.02740 | 0.01294 | 1.02708 | 0.01325 | 1.12847 | 1.01295 | 1.08806 | 0.97366 | 0.94860 | 0.94838 |

| $\mu$ | $\psi$ | $\phi$ | $\chi$ | $\zeta$ | $\xi$ | $\eta$ | $\sigma$ | $\theta$ | $X^{(1)}$ | $Y^{(1)}$ | $X^{(2)}$ | $Y^{(2)}$ | $\gamma_l$ | $\hat{\lambda}$ |
|---|---|---|---|---|---|---|---|---|---|---|---|---|---|---|
| 0 | 0 | 1.00000 | 1.00000 | 0 | 0 | 0 | 0 | 0 | 1.00000 | 0 | 1.00000 | 0 | 0.39414 | 0.42586 |
| 0.010 | 0.00270 | 1.03312 | 1.02016 | 0.00129 | 0.00235 | 0.01221 | 0.00894 | 0.00106 | 1.01890 | 0.00775 | 1.01172 | 0.00580 | 0.41107 | 0.43918 |
| 0.025 | 0.00708 | 1.06567 | 1.04185 | 0.00315 | 0.00557 | 0.05819 | 0.02924 | 0.00274 | 1.05866 | 0.02618 | 1.02487 | 0.02074 | 0.44044 | 0.45846 |
| 0.050 | 0.01431 | 1.10249 | 1.06842 | 0.00574 | 0.01097 | 0.11180 | 0.12248 | 0.00526 | 1.06245 | 0.11668 | 1.04130 | 0.10480 | 0.50119 | 0.53149 |
| 0.075 | 0.02153 | 1.12412 | 1.08680 | 0.00759 | 0.01588 | 0.26459 | 0.24092 | 0.00715 | 1.07827 | 0.23299 | 1.05242 | 0.21741 | 0.56858 | 0.57102 |
| 0.10 | 0.02885 | 1.13698 | 1.09852 | 0.00886 | 0.02056 | 0.38587 | 0.35415 | 0.00845 | 1.08915 | 0.34493 | 1.06013 | 0.32365 | 0.62125 | 0.62596 |
| 0.15 | 0.04599 | 1.14458 | 1.11408 | 0.01052 | 0.03197 | 0.55511 | 0.52079 | 0.01018 | 1.10290 | 0.49973 | 1.06992 | 0.43299 | 0.70224 | 0.70438 |
| 0.2 | 0.06746 | 1.13682 | 1.12815 | 0.01152 | 0.04782 | 0.65787 | 0.68416 | 0.01124 | 1.11111 | 0.62199 | 1.07580 | 0.59182 | 0.75623 | 0.75745 |
| 0.3 | 0.12556 | 1.09196 | 1.13404 | 0.01267 | 0.09275 | 0.75778 | 0.77400 | 0.01245 | 1.12048 | 0.76056 | 1.08249 | 0.72689 | 0.82201 | 0.82245 |
| 0.4 | 0.20520 | 1.01598 | 1.13989 | 0.01331 | 0.15989 | 0.77245 | 0.85576 | 0.01312 | 1.12562 | 0.84162 | 1.08618 | 0.80519 | 0.86015 | 0.86028 |
| 0.5 | 0.30698 | 0.91258 | 1.14459 | 0.01372 | 0.24605 | 0.78502 | 0.90912 | 0.01354 | 1.12887 | 0.89454 | 1.08851 | 0.85666 | 0.88495 | 0.88490 |
| 0.6 | 0.43100 | 0.78317 | 1.14611 | 0.01400 | 0.35992 | 0.65264 | 0.94661 | 0.01385 | 1.13111 | 0.93172 | 1.09012 | 0.89285 | 0.90230 | 0.90218 |
| 0.7 | 0.57748 | 0.62881 | 1.14801 | 0.01421 | 0.49220 | 0.58770 | 0.97437 | 0.01405 | 1.13275 | 0.95927 | 1.09130 | 0.91966 | 0.91514 | 0.91497 |
| 0.8 | 0.74644 | 0.44865 | 1.14944 | 0.01436 | 0.64795 | 0.39172 | 0.99574 | 0.01421 | 1.13399 | 0.98048 | 1.09219 | 0.94031 | 0.92502 | 0.92480 |
| 0.9 | 0.93792 | 0.24405 | 1.15057 | 0.01449 | 0.82615 | 0.21673 | 1.01271 | 0.01433 | 1.13498 | 0.99781 | 1.09290 | 0.95669 | 0.93284 | 0.93261 |
| 1.0 | 1.15392 | 0.01474 | 1.15148 | 0.01459 | 1.02689 | 0.01404 | 1.02649 | 0.01445 | 1.13577 | 1.01100 | 1.09347 | 0.97002 | 0.93920 | 0.93895 |

TABLE 12 (continued)

$\zeta = 0.14;\ \mathfrak{f} = 0.1125$

| $\mu$ | $\psi$ | $\phi$ | $\chi$ | $\zeta$ | $\xi$ | $\eta$ | $\sigma$ | $\theta$ | $X^{(1)}$ | $Y^{(1)}$ | $X^{(2)}$ | $Y^{(2)}$ | $\nu$ | $\varkappa$ |
|---|---|---|---|---|---|---|---|---|---|---|---|---|---|---|
| 0 | 0 | 1.00000 | 1.00000 | 0 | 0 | 0 | 0 | 0 | 1.00000 | 0 | 1.00000 | 0 | 0.38926 | 0.42198 |
| 0.010 | 0.00274 | 1.03323 | 1.02023 | 0.00131 | 0.00215 | 0.00186 | 0.00878 | 0.00107 | 1.01894 | 0.00752 | 1.01176 | 0.00566 | 0.40888 | 0.43512 |
| 0.025 | 0.00715 | 1.06597 | 1.04207 | 0.00342 | 0.00559 | 0.03532 | 0.02681 | 0.00275 | 1.04881 | 0.02371 | 1.02497 | 0.01855 | 0.44407 | 0.45356 |
| 0.050 | 0.01467 | 1.10878 | 1.06994 | 0.00592 | 0.01115 | 0.12694 | 0.10857 | 0.00536 | 1.06818 | 0.10341 | 1.04182 | 0.09098 | 0.48995 | 0.50186 |
| 0.075 | 0.02208 | 1.12702 | 1.08811 | 0.00789 | 0.01614 | 0.24873 | 0.22212 | 0.00735 | 1.07980 | 0.21400 | 1.05354 | 0.19713 | 0.55165 | 0.55882 |
| 0.10 | 0.02976 | 1.14066 | 1.10430 | 0.00980 | 0.02106 | 0.35069 | 0.32915 | 0.00882 | 1.09143 | 0.31946 | 1.06181 | 0.29687 | 0.60587 | 0.61067 |
| 0.15 | 0.04786 | 1.14982 | 1.11682 | 0.01115 | 0.03256 | 0.52916 | 0.49615 | 0.01074 | 1.10635 | 0.48140 | 1.07846 | 0.45750 | 0.68729 | 0.68976 |
| 0.2 | 0.06917 | 1.14810 | 1.12854 | 0.01229 | 0.04785 | 0.63749 | 0.61282 | 0.01194 | 1.11537 | 0.59950 | 1.07894 | 0.56869 | 0.74261 | 0.74403 |
| 0.3 | 0.12787 | 1.09969 | 1.14034 | 0.01360 | 0.09293 | 0.74427 | 0.75787 | 0.01332 | 1.12571 | 0.74359 | 1.08688 | 0.70888 | 0.81096 | 0.81148 |
| 0.4 | 0.20809 | 1.02400 | 1.14688 | 0.01433 | 0.13921 | 0.76402 | 0.84396 | 0.01410 | 1.13144 | 0.82867 | 1.09051 | 0.79117 | 0.85100 | 0.85115 |
| 0.5 | 0.31049 | 0.92085 | 1.15104 | 0.01480 | 0.24151 | 0.72818 | 0.90049 | 0.01459 | 1.13597 | 0.88469 | 1.09313 | 0.84558 | 0.87717 | 0.87714 |
| 0.6 | 0.43532 | 0.79027 | 1.15394 | 0.01512 | 0.35816 | 0.65924 | 0.94035 | 0.01494 | 1.13761 | 0.92417 | 1.09497 | 0.88397 | 0.89557 | 0.89544 |
| 0.7 | 0.58267 | 0.68445 | 1.15608 | 0.01536 | 0.49127 | 0.53708 | 0.96698 | 0.01517 | 1.13942 | 0.95357 | 1.09627 | 0.91252 | 0.90922 | 0.90901 |
| 0.8 | 0.75268 | 0.45335 | 1.15765 | 0.01555 | 0.64697 | 0.39216 | 0.99282 | 0.01536 | 1.14082 | 0.97622 | 1.09729 | 0.93455 | 0.91978 | 0.91948 |
| 0.9 | 0.94522 | 0.28715 | 1.15892 | 0.01570 | 0.83528 | 0.21771 | 1.01097 | 0.01553 | 1.14195 | 0.99425 | 1.09808 | 0.95205 | 0.92807 | 0.92779 |
| 1.0 | 1.16048 | 0.01599 | 1.15995 | 0.01582 | 1.02628 | 0.01531 | 1.02574 | 0.01568 | 1.14282 | 1.00888 | 1.09873 | 0.96680 | 0.93485 | 0.93455 |

$\zeta = 0.15;\ \mathfrak{f} = 0.1190$

| $\mu$ | $\psi$ | $\phi$ | $\chi$ | $\zeta$ | $\xi$ | $\eta$ | $\sigma$ | $\theta$ | $X^{(1)}$ | $Y^{(1)}$ | $X^{(2)}$ | $Y^{(2)}$ | $\nu$ | $\varkappa$ |
|---|---|---|---|---|---|---|---|---|---|---|---|---|---|---|
| 0 | 0 | 1.00000 | 1.00000 | 0 | 0 | 0 | 0 | 0 | 1.00000 | 0 | 1.00000 | 0 | 0.38458 | 0.41822 |
| 0.010 | 0.00278 | 1.03332 | 1.03029 | 0.00133 | 0.00216 | 0.00156 | 0.00853 | 0.00107 | 1.01899 | 0.00782 | 1.01179 | 0.00554 | 0.40391 | 0.43120 |
| 0.025 | 0.00726 | 1.06624 | 1.04226 | 0.00326 | 0.00561 | 0.00314 | 0.02502 | 0.00276 | 1.03895 | 0.02190 | 1.02506 | 0.01698 | 0.42819 | 0.44867 |
| 0.050 | 0.01500 | 1.11077 | 1.07011 | 0.00607 | 0.01130 | 0.01448 | 0.09664 | 0.00545 | 1.06579 | 0.09054 | 1.04227 | 0.07955 | 0.47990 | 0.49240 |
| 0.075 | 0.02269 | 1.12921 | 1.08979 | 0.00818 | 0.01647 | 0.22868 | 0.20269 | 0.00757 | 1.08113 | 0.19428 | 1.05452 | 0.17781 | 0.53860 | 0.54659 |
| 0.10 | 0.03004 | 1.14895 | 1.10582 | 0.00972 | 0.02158 | 0.33760 | 0.30485 | 0.00915 | 1.09346 | 0.29692 | 1.06832 | 0.27598 | 0.59160 | 0.59701 |
| 0.15 | 0.04868 | 1.15482 | 1.12211 | 0.01176 | 0.03512 | 0.50645 | 0.47295 | 0.01128 | 1.10949 | 0.46053 | 1.07461 | 0.43553 | 0.67811 | 0.67594 |
| 0.2 | 0.07095 | 1.14955 | 1.13895 | 0.01508 | 0.04866 | 0.61781 | 0.59188 | 0.01262 | 1.11982 | 0.57755 | 1.08189 | 0.54657 | 0.72955 | 0.73218 |
| 0.3 | 0.13014 | 1.10702 | 1.14685 | 0.01453 | 0.09318 | 0.78093 | 0.74211 | 0.01418 | 1.13066 | 0.78659 | 1.09009 | 0.69086 | 0.80022 | 0.80082 |
| 0.4 | 0.21095 | 1.03164 | 1.15562 | 0.01581 | 0.15903 | 0.75554 | 0.83229 | 0.01506 | 1.13699 | 0.61584 | 1.09467 | 0.77787 | 0.84205 | 0.84223 |
| 0.5 | 0.31899 | 0.92775 | 1.15825 | 0.01588 | 0.24698 | 0.72822 | 0.89187 | 0.01562 | 1.14102 | 0.87484 | 1.09759 | 0.83460 | 0.86955 | 0.86951 |
| 0.6 | 0.43956 | 0.79704 | 1.16417 | 0.01685 | 0.35784 | 0.64784 | 0.93407 | 0.01601 | 1.14381 | 0.91664 | 1.09961 | 0.87517 | 0.88895 | 0.88880 |
| 0.7 | 0.58777 | 0.64029 | 1.16588 | 0.01653 | 0.49090 | 0.58688 | 0.96549 | 0.01629 | 1.14585 | 0.94776 | 1.10110 | 0.90538 | 0.90357 | 0.90311 |
| 0.8 | 0.75970 | 0.45793 | 1.16558 | 0.01674 | 0.64598 | 0.39951 | 0.98978 | 0.01651 | 1.14742 | 0.97188 | 1.10223 | 0.92875 | 0.91450 | 0.91422 |
| 0.9 | 0.95289 | 0.25920 | 1.16706 | 0.01691 | 0.82140 | 0.21866 | 1.00911 | 0.01668 | 1.14865 | 0.99098 | 1.10312 | 0.94735 | 0.92835 | 0.92303 |
| 1.0 | 1.16587 | 0.01724 | 1.16822 | 0.01705 | 1.02944 | 0.01621 | 1.02685 | 0.01681 | 1.14965 | 1.00659 | 1.10385 | 0.96251 | 0.93055 | 0.98020 |

Table 12 (continued)

$\zeta_1 = 0.163; \quad \zeta = 0.1255$

| $\mu$ | $\psi$ | $\phi$ | $\chi$ | $\varsigma$ | $\xi$ | $\eta$ | $\sigma$ | $\theta$ | $X^{(1)}$ | $Y^{(1)}$ | $X^{(2)}$ | $Y^{(2)}$ | $\mathscr{n}$ | $\mathscr{x}$ |
|---|---|---|---|---|---|---|---|---|---|---|---|---|---|---|
| 0 | 0 | 1.00000 | 1.00000 | 0 | 0 | 0 | 0 | 0 | 1.00000 | 0 | 1.00000 | 0 | 0.38009 | 0.41157 |
| 0.010 | 0.00282 | 1.03841 | 1.02066 | 0.00135 | 0.00216 | 0.01128 | 0.00835 | 0.00713 | 1.01908 | 0.00713 | 1.01181 | 0.00543 | 0.39914 | 0.42739 |
| 0.025 | 0.00787 | 1.06648 | 1.04244 | 0.00351 | 0.00563 | 0.03145 | 0.02868 | 0.00277 | 1.03907 | 0.08053 | 1.02511 | 0.01581 | 0.42270 | 0.44429 |
| 0.050 | 0.01530 | 1.10567 | 1.07085 | 0.00622 | 0.01143 | 0.10402 | 0.08688 | 0.00552 | 1.04631 | 0.08065 | 1.04264 | 0.07009 | 0.47087 | 0.48441 |
| 0.075 | 0.02327 | 1.13114 | 1.09127 | 0.00845 | 0.01676 | 0.22085 | 0.18553 | 0.00775 | 1.08280 | 0.17684 | 1.05539 | 0.16079 | 0.52667 | 0.55548 |
| 0.10 | 0.03148 | 1.14690 | 1.10611 | 0.01012 | 0.02195 | 0.31642 | 0.28555 | 0.00946 | 1.09529 | 0.27501 | 1.06468 | 0.25447 | 0.57884 | 0.59438 |
| 0.15 | 0.04996 | 1.15843 | 1.13571 | 0.01285 | 0.03566 | 0.48492 | 0.45305 | 0.00179 | 1.11289 | 0.43798 | 1.07699 | 0.41101 | 0.65966 | 0.66284 |
| 0.2 | 0.07248 | 1.15527 | 1.13790 | 0.01576 | 0.04886 | 0.59880 | 0.57180 | 0.00387 | 1.12801 | 0.55564 | 1.08467 | 0.52548 | 0.71700 | 0.71885 |
| 0.3 | 0.13286 | 1.11897 | 1.15211 | 0.01541 | 0.09834 | 0.72175 | 0.72670 | 0.00592 | 1.13536 | 0.72017 | 1.09862 | 0.67681 | 0.78978 | 0.79046 |
| 0.4 | 0.21575 | 1.08896 | 1.16011 | 0.00685 | 0.15585 | 0.74701 | 0.82075 | 0.00602 | 1.14229 | 0.80316 | 1.09866 | 0.76582 | 0.83330 | 0.83350 |
| 0.5 | 0.31743 | 0.99490 | 1.16528 | 0.00696 | 0.24644 | 0.71812 | 0.88827 | 0.00665 | 1.14678 | 0.86592 | 1.10189 | 0.82875 | 0.86606 | 0.86601 |
| 0.6 | 0.44372 | 0.80561 | 1.16880 | 0.00738 | 0.35652 | 0.64520 | 0.92778 | 0.00709 | 1.14980 | 0.90901 | 1.10413 | 0.86689 | 0.88243 | 0.88285 |
| 0.7 | 0.59276 | 0.64596 | 1.17141 | 0.00769 | 0.48927 | 0.55342 | 0.96092 | 0.00742 | 1.15206 | 0.94186 | 1.10578 | 0.89824 | 0.89764 | 0.89733 |
| 0.8 | 0.76468 | 0.46289 | 1.17342 | 0.00794 | 0.64482 | 0.39272 | 0.98662 | 0.00766 | 1.15379 | 0.96781 | 1.10704 | 0.92292 | 0.90985 | 0.90901 |
| 0.9 | 0.95989 | 0.25519 | 1.17592 | 0.00813 | 0.82321 | 0.23959 | 1.00711 | 0.00785 | 1.15516 | 0.98981 | 1.10803 | 0.94260 | 0.91869 | 0.91651 |
| 1.0 | 1.17704 | 0.01851 | 1.17603 | 0.00829 | 1.02848 | 0.07726 | 1.02381 | 0.00801 | 1.15626 | 1.00816 | 1.10884 | 0.95866 | 0.92650 | 0.92599 |

$\zeta_1 = 0.171; \quad \zeta = 0.1319$

| $\mu$ | $\psi$ | $\phi$ | $\chi$ | $\varsigma$ | $\xi$ | $\eta$ | $\sigma$ | $\theta$ | $X^{(1)}$ | $Y^{(1)}$ | $X^{(2)}$ | $Y^{(2)}$ | $\mathscr{n}$ | $\mathscr{x}$ |
|---|---|---|---|---|---|---|---|---|---|---|---|---|---|---|
| 0 | 0 | 1.00000 | 1.00000 | 0 | 0 | 0 | 0 | 0 | 1.00000 | 0 | 1.00000 | 0 | 0.37575 | 0.41102 |
| 0.010 | 0.00286 | 1.03350 | 1.02042 | 0.00136 | 0.00216 | 0.01103 | 0.00819 | 0.00107 | 1.01907 | 0.00696 | 1.01184 | 0.00532 | 0.39454 | 0.42869 |
| 0.025 | 0.00747 | 1.06670 | 1.04260 | 0.00356 | 0.00564 | 0.03012 | 0.02865 | 0.00277 | 1.03918 | 0.01947 | 1.02522 | 0.01499 | 0.41752 | 0.44015 |
| 0.050 | 0.01559 | 1.10644 | 1.07145 | 0.00656 | 0.01154 | 0.09521 | 0.07875 | 0.00558 | 1.04476 | 0.07240 | 1.04297 | 0.06226 | 0.46267 | 0.47728 |
| 0.075 | 0.02380 | 1.13283 | 1.09260 | 0.00871 | 0.01702 | 0.19501 | 0.17085 | 0.00792 | 1.08335 | 0.16111 | 1.05615 | 0.14530 | 0.51571 | 0.52534 |
| 0.10 | 0.03227 | 1.14956 | 1.10819 | 0.01049 | 0.02285 | 0.29700 | 0.26659 | 0.00974 | 1.09695 | 0.25565 | 1.06593 | 0.25545 | 0.54599 | 0.57266 |
| 0.15 | 0.05119 | 1.16867 | 1.12907 | 0.01292 | 0.04118 | 0.46452 | 0.43043 | 0.00227 | 1.11508 | 0.41674 | 1.07901 | 0.38984 | 0.64688 | 0.65044 |
| 0.2 | 0.07606 | 1.16065 | 1.14129 | 0.01447 | 0.04935 | 0.59247 | 0.55306 | 0.00390 | 1.12846 | 0.53661 | 1.08727 | 0.50521 | 0.70494 | 0.70702 |
| 0.3 | 0.13855 | 1.13056 | 1.15763 | 0.00680 | 0.09855 | 0.70476 | 0.71166 | 0.00585 | 1.13982 | 0.69411 | 1.09700 | 0.65721 | 0.77961 | 0.78089 |
| 0.4 | 0.21652 | 1.04596 | 1.16688 | 0.00785 | 0.15969 | 0.78468 | 0.80995 | 0.00696 | 1.14787 | 0.79062 | 1.10251 | 0.75950 | 0.82672 | 0.82895 |
| 0.5 | 0.32083 | 0.94177 | 1.17200 | 0.00803 | 0.24592 | 0.71298 | 0.87471 | 0.00761 | 1.15221 | 0.85532 | 1.10604 | 0.81301 | 0.85469 | 0.85465 |
| 0.6 | 0.44785 | 0.80996 | 1.17592 | 0.00851 | 0.35569 | 0.64246 | 0.92236 | 0.00817 | 1.15558 | 0.90134 | 1.10851 | 0.85766 | 0.87600 | 0.87580 |
| 0.7 | 0.59769 | 0.65145 | 1.17881 | 0.00886 | 0.48824 | 0.53440 | 0.95628 | 0.00853 | 1.15806 | 0.93588 | 1.11032 | 0.89110 | 0.89192 | 0.89160 |
| 0.8 | 0.77049 | 0.46575 | 1.18103 | 0.00914 | 0.64368 | 0.59286 | 0.98837 | 0.00881 | 1.15996 | 0.96269 | 1.11117 | 0.91707 | 0.90424 | 0.90386 |
| 0.9 | 0.96629 | 0.25613 | 1.18278 | 0.00936 | 0.82207 | 0.22082 | 1.00500 | 0.00908 | 1.16146 | 0.98410 | 1.11281 | 0.93781 | 0.91407 | 0.91364 |
| 1.0 | 1.18511 | 0.01979 | 1.18421 | 0.00954 | 1.02846 | 0.08084 | 1.02266 | 0.00920 | 1.16267 | 1.00158 | 1.11370 | 0.95475 | 0.92208 | 0.92261 |

TABLE 12 (continued)

$\zeta_1 = 0.18; \ \mathfrak{Z} = 0.1582$

| μ | ψ | φ | χ | ζ | ξ | η | σ | θ | X" | Y" | X^α | Y^α | ν | χ |
|---|---|---|---|---|---|---|---|---|---|---|---|---|---|---|
| 0 | 0 | 1.00000 | 1.00000 | 0 | 0 | 0 | 0 | 0 | 1.00000 | 0 | 1.00000 | 0 | 0.87157 | 0.40757 |
| 0.010 | 0.00289 | 1.05858 | 1.02047 | 0.00138 | 0.00217 | 0.01080 | 0.00804 | 0.00107 | 1.01911 | 0.00680 | 1.01187 | 0.00522 | 0.59012 | 0.42010 |
| 0.025 | 0.00757 | 1.06690 | 1.04276 | 0.00341 | 0.00565 | 0.02904 | 0.02184 | 0.00278 | 1.03928 | 0.01864 | 1.02529 | 0.01455 | 0.41260 | 0.43619 |
| 0.050 | 0.01586 | 1.10711 | 1.07197 | 0.00648 | 0.01165 | 0.08779 | 0.07195 | 0.00568 | 1.06515 | 0.06550 | 1.04325 | 0.05575 | 0.45519 | 0.47072 |
| 0.075 | 0.02451 | 1.13435 | 1.09580 | 0.00895 | 0.01726 | 0.18090 | 0.15693 | 0.00807 | 1.08428 | 0.14776 | 1.05683 | 0.13258 | 0.50954 | 0.51608 |
| 0.10 | 0.03502 | 1.15195 | 1.11010 | 0.01085 | 0.02271 | 0.27920 | 0.24929 | 0.01000 | 1.09811 | 0.23798 | 1.06702 | 0.21815 | 0.55448 | 0.56178 |
| 0.15 | 0.05287 | 1.16760 | 1.13220 | 0.01347 | 0.03467 | 0.44530 | 0.41100 | 0.01273 | 1.11755 | 0.39671 | 1.08089 | 0.36994 | 0.64474 | 0.63869 |
| 0.2 | 0.07561 | 1.16565 | 1.14626 | 0.01516 | 0.04982 | 0.56279 | 0.53862 | 0.01451 | 1.12969 | 0.51739 | 1.08972 | 0.48589 | 0.69336 | 0.69558 |
| 0.3 | 0.13669 | 1.12688 | 1.16293 | 0.01718 | 0.09376 | 0.69197 | 0.66698 | 0.01666 | 1.14405 | 0.67843 | 1.10022 | 0.61107 | 0.76978 | 0.77060 |
| 0.4 | 0.21923 | 1.05267 | 1.17243 | 0.01885 | 0.15852 | 0.72995 | 0.79811 | 0.01789 | 1.15222 | 0.77823 | 1.10620 | 0.73743 | 0.81682 | 0.81659 |
| 0.5 | 0.32416 | 0.94889 | 1.17857 | 0.01910 | 0.24539 | 0.70765 | 0.86619 | 0.01869 | 1.15748 | 0.84546 | 1.11006 | 0.80239 | 0.84746 | 0.84741 |
| 0.6 | 0.45187 | 0.81610 | 1.18295 | 0.01964 | 0.35485 | 0.68958 | 0.91496 | 0.01925 | 1.16115 | 0.89365 | 1.11275 | 0.84898 | 0.86967 | 0.86944 |
| 0.7 | 0.60252 | 0.65679 | 1.18502 | 0.02003 | 0.48717 | 0.58825 | 0.95157 | 0.01965 | 1.16385 | 0.92985 | 1.11474 | 0.88398 | 0.88630 | 0.88595 |
| 0.8 | 0.77622 | 0.47099 | 1.18845 | 0.02034 | 0.62248 | 0.89289 | 0.98008 | 0.01996 | 1.16592 | 0.95797 | 1.11626 | 0.91121 | 0.89920 | 0.89877 |
| 0.9 | 0.97802 | 0.25902 | 1.19088 | 0.02058 | 0.82084 | 0.22106 | 1.00279 | 0.02021 | 1.16756 | 0.98047 | 1.11747 | 0.93299 | 0.90950 | 0.90901 |
| 1.0 | 1.19297 | 0.02108 | 1.19196 | 0.02078 | 1.02232 | 0.01759 | 1.02159 | 0.02040 | 1.16889 | 0.99887 | 1.11844 | 0.95080 | 0.91791 | 0.91738 |

$\zeta_1 = 0.19; \ \mathfrak{Z} = 0.1445$

| μ | ψ | φ | χ | ζ | ξ | η | σ | θ | X" | Y" | X^α | Y^α | ν | χ |
|---|---|---|---|---|---|---|---|---|---|---|---|---|---|---|
| 0 | 0 | 1.00000 | 1.00000 | 0 | 0 | 0 | 0 | 0 | 1.00000 | 0 | 1.00000 | 0 | 0.86753 | 0.40421 |
| 0.010 | 0.00295 | 1.05365 | 1.02053 | 0.00110 | 0.00217 | 0.01059 | 0.00790 | 0.00107 | 1.01915 | 0.00666 | 1.01189 | 0.00513 | 0.58585 | 0.41660 |
| 0.025 | 0.00766 | 1.06708 | 1.04290 | 0.00346 | 0.00566 | 0.02815 | 0.02118 | 0.00278 | 1.03938 | 0.01796 | 1.02536 | 0.01384 | 0.40791 | 0.43240 |
| 0.050 | 0.01611 | 1.10769 | 1.07245 | 0.00660 | 0.01170 | 0.08151 | 0.06625 | 0.00567 | 1.06549 | 0.05972 | 1.04350 | 0.05034 | 0.44851 | 0.46478 |
| 0.075 | 0.02479 | 1.13565 | 1.09487 | 0.00918 | 0.01747 | 0.16835 | 0.14506 | 0.00820 | 1.08504 | 0.13568 | 1.05743 | 0.12095 | 0.49654 | 0.50757 |
| 0.10 | 0.03568 | 1.15217 | 1.11179 | 0.01117 | 0.02296 | 0.26671 | 0.25580 | 0.01021 | 1.09974 | 0.22001 | 1.06805 | 0.20052 | 0.54511 | 0.55235 |
| 0.15 | 0.05552 | 1.17122 | 1.13513 | 0.01400 | 0.03514 | 0.42690 | 0.39270 | 0.01316 | 1.11984 | 0.37783 | 1.08268 | 0.35124 | 0.62819 | 0.62753 |
| 0.2 | 0.07711 | 1.17036 | 1.15010 | 0.01583 | 0.05029 | 0.54574 | 0.51595 | 0.01509 | 1.13272 | 0.49897 | 1.09204 | 0.46744 | 0.68221 | 0.68478 |
| 0.3 | 0.18879 | 1.13280 | 1.16601 | 0.01805 | 0.09897 | 0.67998 | 0.68266 | 0.01745 | 1.14808 | 0.66511 | 1.10350 | 0.62568 | 0.76011 | 0.76108 |
| 0.4 | 0.22191 | 1.05911 | 1.17829 | 0.01983 | 0.15986 | 0.72142 | 0.78703 | 0.01881 | 1.15688 | 0.76601 | 1.10976 | 0.72459 | 0.80810 | 0.80840 |
| 0.5 | 0.32745 | 0.95478 | 1.18495 | 0.02017 | 0.24485 | 0.70229 | 0.85771 | 0.01969 | 1.16256 | 0.85575 | 1.11395 | 0.79190 | 0.84035 | 0.84029 |
| 0.6 | 0.45581 | 0.82205 | 1.18961 | 0.02076 | 0.35899 | 0.68660 | 0.90856 | 0.02051 | 1.16653 | 0.88593 | 1.11687 | 0.84057 | 0.86843 | 0.86517 |
| 0.7 | 0.60727 | 0.66198 | 1.19306 | 0.02120 | 0.48607 | 0.53200 | 0.94682 | 0.02076 | 1.16946 | 0.92872 | 1.11905 | 0.87687 | 0.88075 | 0.88036 |
| 0.8 | 0.78185 | 0.47514 | 1.19571 | 0.02155 | 0.64122 | 0.89288 | 0.97662 | 0.02111 | 1.17171 | 0.95316 | 1.12069 | 0.90533 | 0.89422 | 0.89373 |
| 0.9 | 0.97964 | 0.26188 | 1.19783 | 0.02182 | 0.83955 | 0.22174 | 1.00048 | 0.02138 | 1.17350 | 0.97674 | 1.12201 | 0.92815 | 0.90498 | 0.90443 |
| 1.0 | 1.20059 | 0.02240 | 1.19955 | 0.02204 | 1.02106 | 0.02041 | 1.02001 | 0.02160 | 1.17494 | 0.99605 | 1.12308 | 0.94679 | 0.91378 | 0.91318 |

TABLE 12 (continued)

$\zeta = 0.20; \quad \beta = 0.1504$

| $\mu$ | $\psi$ | $\phi$ | $\chi$ | $\iota$ | $\varepsilon$ | $\eta$ | $\sigma$ | $\theta$ | $X''$ | $Y''$ | $X^{(3)}$ | $Y^{(3)}$ | $\gamma_i$ | $\gamma_a$ |
|---|---|---|---|---|---|---|---|---|---|---|---|---|---|---|
| 0 | 0 | 1.00000 | 1.00000 | 0 | 0 | 0 | 0 | 0 | 1.00000 | 0 | 1.00000 | 0 | 0.36862 | 0.40098 |
| 0.025 | 0.00775 | 1.06725 | 1.04304 | 0.00350 | 0.00566 | 0.02789 | 0.02065 | 0.00278 | 1.05947 | 0.01789 | 1.02542 | 0.01842 | 0.40341 | 0.42874 |
| 0.050 | 0.01684 | 1.11821 | 1.07288 | 0.00672 | 0.01176 | 0.07619 | 0.06117 | 0.00570 | 1.06578 | 0.05485 | 1.04972 | 0.04538 | 0.44194 | 0.45930 |
| 0.075 | 0.02524 | 1.16685 | 1.09584 | 0.00989 | 0.01765 | 0.15717 | 0.12454 | 0.00831 | 1.08575 | 0.12496 | 1.05797 | 0.11065 | 0.48772 | 0.49978 |
| 0.10 | 0.03442 | 1.15505 | 1.11844 | 0.01151 | 0.02386 | 0.24788 | 0.21909 | 0.01045 | 1.11094 | 0.20712 | 1.06695 | 0.18806 | 0.53566 | 0.54222 |
| 0.15 | 0.05468 | 1.17458 | 1.13788 | 0.01453 | 0.03559 | 0.40957 | 0.37345 | 0.01357 | 1.12196 | 0.36002 | 1.08425 | 0.33565 | 0.61220 | 0.61698 |
| 0.2 | 0.07657 | 1.17478 | 1.15875 | 0.01649 | 0.05078 | 0.52934 | 0.49905 | 0.01566 | 1.13556 | 0.48129 | 1.09422 | 0.44978 | 0.67150 | 0.67432 |
| 0.3 | 0.11086 | 1.15849 | 1.17289 | 0.01890 | 0.09418 | 0.66700 | 0.66669 | 0.01822 | 1.15392 | 0.61616 | 1.10625 | 0.61011 | 0.75075 | 0.75185 |
| 0.4 | 0.22455 | 1.06529 | 1.18895 | 0.02051 | 0.15821 | 0.71292 | 0.77611 | 0.01971 | 1.16134 | 0.75395 | 1.11120 | 0.71198 | 0.80005 | 0.80038 |
| 0.5 | 0.35070 | 0.96095 | 1.19114 | 0.02189 | 0.24432 | 0.69687 | 0.84980 | 0.02069 | 1.16745 | 0.82609 | 1.11171 | 0.78153 | 0.83336 | 0.83329 |
| 0.6 | 0.45976 | 0.82781 | 1.19619 | 0.02238 | 0.55213 | 0.68353 | 0.90215 | 0.02137 | 1.17173 | 0.87821 | 1.12087 | 0.83181 | 0.85729 | 0.85699 |
| 0.7 | 0.61196 | 0.66702 | 1.19995 | 0.02238 | 0.48095 | 0.58065 | 0.94202 | 0.02187 | 1.17489 | 0.91756 | 1.12312 | 0.86979 | 0.87528 | 0.87484 |
| 0.8 | 0.78741 | 0.47919 | 1.20281 | 0.02276 | 0.68992 | 0.59269 | 0.97835 | 0.02225 | 1.17735 | 0.94828 | 1.12501 | 0.89945 | 0.88929 | 0.88875 |
| 0.9 | 0.98617 | 0.26468 | 1.20511 | 0.02306 | 0.81816 | 0.22237 | 0.99810 | 0.02255 | 1.17925 | 0.97292 | 1.12644 | 0.92325 | 0.90051 | 0.89989 |
| 1.0 | 1.20880 | 0.02371 | 1.20698 | 0.02351 | 1.01972 | 0.02143 | 1.01854 | 0.02279 | 1.18082 | 0.99812 | 1.12760 | 0.94275 | 0.90969 | 0.90902 |

$\zeta = 0.25; \quad \beta = 0.1790$

| $\mu$ | $\psi$ | $\phi$ | $\chi$ | $\iota$ | $\varepsilon$ | $\eta$ | $\sigma$ | $\theta$ | $X''$ | $Y''$ | $X^{(3)}$ | $Y^{(3)}$ | $\gamma_i$ | $\gamma_a$ |
|---|---|---|---|---|---|---|---|---|---|---|---|---|---|---|
| 0 | 0 | 1.00000 | 1.00000 | 0 | 0 | 0 | 0 | 0 | 1.00000 | 0 | 1.00000 | 0 | 0.34581 | 0.38566 |
| 0.025 | 0.00818 | 1.06795 | 1.04865 | 0.00571 | 0.00568 | 0.02478 | 0.01877 | 0.00279 | 1.05983 | 0.01544 | 1.02570 | 0.01209 | 0.38322 | 0.41197 |
| 0.050 | 0.01788 | 1.11006 | 1.07456 | 0.00721 | 0.01195 | 0.05918 | 0.04660 | 0.00580 | 1.06685 | 0.03967 | 1.04454 | 0.03213 | 0.43565 | 0.43587 |
| 0.075 | 0.02717 | 1.14107 | 1.09958 | 0.01031 | 0.01830 | 0.11691 | 0.09780 | 0.00872 | 1.08882 | 0.08694 | 1.05996 | 0.07465 | 0.45256 | 0.46815 |
| 0.10 | 0.03789 | 1.16857 | 1.11974 | 0.01291 | 0.02454 | 0.18956 | 0.16875 | 0.01128 | 1.10548 | 0.15043 | 1.07246 | 0.13349 | 0.49175 | 0.50334 |
| 0.15 | 0.05962 | 1.18902 | 1.14990 | 0.01682 | 0.03750 | 0.35586 | 0.30306 | 0.01529 | 1.13046 | 0.28513 | 1.09086 | 0.26032 | 0.56443 | 0.57119 |
| 0.2 | 0.08534 | 1.19818 | 1.16945 | 0.01955 | 0.05276 | 0.45500 | 0.42451 | 0.01814 | 1.14740 | 0.40335 | 1.10843 | 0.37259 | 0.62857 | 0.62778 |
| 0.3 | 0.15067 | 1.16821 | 1.19471 | 0.02800 | 0.09521 | 0.60847 | 0.60896 | 0.02181 | 1.16856 | 0.57867 | 1.11922 | 0.53990 | 0.70749 | 0.70913 |
| 0.4 | 0.25717 | 1.09280 | 1.20978 | 0.02509 | 0.15750 | 0.67112 | 0.72895 | 0.02405 | 1.18114 | 0.69618 | 1.12863 | 0.65241 | 0.76219 | 0.76270 |
| 0.5 | 0.34626 | 0.98882 | 1.21975 | 0.02649 | 0.24172 | 0.66921 | 0.80820 | 0.02551 | 1.18943 | 0.77882 | 1.13485 | 0.73164 | 0.80013 | 0.80004 |
| 0.6 | 0.47857 | 0.85219 | 1.22684 | 0.02748 | 0.34879 | 0.61704 | 0.87020 | 0.02655 | 1.19531 | 0.83968 | 1.13926 | 0.79006 | 0.82785 | 0.82739 |
| 0.7 | 0.63442 | 0.69037 | 1.23224 | 0.02824 | 0.47924 | 0.52266 | 0.91761 | 0.02732 | 1.19968 | 0.88625 | 1.14255 | 0.83178 | 0.84898 | 0.84824 |
| 0.8 | 0.81396 | 0.49818 | 1.23625 | 0.02885 | 0.63504 | 0.39090 | 0.95497 | 0.02792 | 1.20307 | 0.92297 | 1.14510 | 0.87005 | 0.86550 | 0.86464 |
| 0.9 | 1.01781 | 0.27808 | 1.23954 | 0.02930 | 0.81064 | 0.22476 | 0.98518 | 0.02859 | 1.20577 | 0.95264 | 1.14713 | 0.89857 | 0.87884 | 0.87787 |
| 1.0 | 1.24453 | 0.03058 | 1.24224 | 0.02970 | 1.01208 | 0.02622 | 1.00999 | 0.02876 | 1.20797 | 0.97710 | 1.14878 | 0.92209 | 0.88982 | 0.88876 |

TABLE 12 (continued)

$\zeta_1 = 0.30; \quad \beta = 0.2053$

| μ | ψ | φ | χ | ζ | ξ | η | σ | θ | X''' | Y'' | X''' | Y''' | 𝒳 | 𝓍 |
|---|---|---|---|---|---|---|---|---|---|---|---|---|---|---|
| 0 | 0 | 1.00000 | 1.00000 | 0 | 0 | 0 | 0 | 0 | 1.00000 | 0 | 1.00000 | 0 | 0.33084 | 0.37194 |
| 0.025 | 0.00856 | 1.06847 | 1.04416 | 0.00389 | 0.00567 | 0.02814 | 0.01755 | 0.00278 | 1.04011 | 0.01117 | 1.02593 | 0.01124 | 0.36590 | 0.39709 |
| 0.050 | 0.01825 | 1.11122 | 1.07578 | 0.00768 | 0.01200 | 0.05087 | 0.03965 | 0.00583 | 1.06753 | 0.03355 | 1.04509 | 0.02607 | 0.39511 | 0.41930 |
| 0.075 | 0.02873 | 1.14560 | 1.10211 | 0.01105 | 0.01862 | 0.09874 | 0.07844 | 0.00892 | 1.08988 | 0.06557 | 1.06121 | 0.05496 | 0.42646 | 0.44498 |
| 0.10 | 0.03981 | 1.16795 | 1.12412 | 0.01405 | 0.02527 | 0.15139 | 0.12883 | 0.01179 | 1.10834 | 0.11102 | 1.07476 | 0.09906 | 0.46004 | 0.47434 |
| 0.15 | 0.06886 | 1.19784 | 1.15785 | 0.01880 | 0.03891 | 0.27998 | 0.24920 | 0.01696 | 1.13658 | 0.22926 | 1.09559 | 0.20636 | 0.52625 | 0.54498 |
| 0.2 | 0.09130 | 1.20661 | 1.18181 | 0.02225 | 0.05440 | 0.39552 | 0.36447 | 0.02015 | 1.15618 | 0.34035 | 1.11044 | 0.31100 | 0.58855 | 0.58908 |
| 0.3 | 0.15964 | 1.18291 | 1.21497 | 0.02682 | 0.09611 | 0.55562 | 0.54715 | 0.02497 | 1.18180 | 0.51744 | 1.12977 | 0.47897 | 0.66945 | 0.67171 |
| 0.4 | 0.24892 | 1.11564 | 1.23211 | 0.02966 | 0.15686 | 0.68105 | 0.67592 | 0.02799 | 1.19786 | 0.64276 | 1.14165 | 0.59832 | 0.72793 | 0.72866 |
| 0.5 | 0.35086 | 1.01258 | 1.24501 | 0.03159 | 0.28921 | 0.59982 | 0.76905 | 0.03003 | 1.20796 | 0.73357 | 1.14961 | 0.68498 | 0.76953 | 0.76940 |
| 0.6 | 0.49625 | 0.87714 | 1.25429 | 0.03500 | 0.34441 | 0.51812 | 0.88687 | 0.03150 | 1.21549 | 0.80176 | 1.15534 | 0.75212 | 0.80041 | 0.79976 |
| 0.7 | 0.66550 | 0.71108 | 1.26129 | 0.03407 | 0.47810 | 0.58764 | 0.89298 | 0.03259 | 1.22114 | 0.85461 | 1.15964 | 0.80666 | 0.82417 | 0.82819 |
| 0.8 | 0.85885 | 0.52537 | 1.26675 | 0.04191 | 0.62564 | 0.72606 | 0.98594 | 0.03384 | 1.22553 | 0.89669 | 1.16298 | 0.84095 | 0.84500 | 0.84177 |
| 0.9 | 1.04644 | 0.29063 | 1.27115 | 0.04560 | 0.80223 | 0.89053 | 0.97091 | 0.03411 | 1.22905 | 0.93095 | 1.16566 | 0.87872 | 0.85326 | 0.85586 |
| 1.0 | 1.27895 | 0.05716 | 1.27477 | 0.04617 | 1.00501 | 0.08053 | 0.99989 | 0.03465 | 1.23195 | 0.95937 | 1.16786 | 0.90092 | 0.87088 | 0.86695 |

$\zeta_1 = 0.35; \quad \beta = 0.2297$

| μ | ψ | φ | χ | ζ | ξ | η | σ | θ | X''' | Y'' | X''' | Y''' | 𝒳 | 𝓍 |
|---|---|---|---|---|---|---|---|---|---|---|---|---|---|---|
| 0 | 0 | 1.00000 | 1.00000 | 0 | 0 | 0 | 0 | 0 | 1.00000 | 0 | 1.00000 | 0 | 0.31669 | 0.35947 |
| 0.025 | 0.00891 | 1.06887 | 1.04461 | 0.00406 | 0.00565 | 0.02198 | 0.01661 | 0.00278 | 1.04033 | 0.01319 | 1.02612 | 0.01057 | 0.35970 | 0.38868 |
| 0.050 | 0.01901 | 1.11204 | 1.07676 | 0.00800 | 0.01199 | 0.04681 | 0.06592 | 0.00585 | 1.06801 | 0.02867 | 1.04552 | 0.02800 | 0.37798 | 0.40633 |
| 0.075 | 0.03006 | 1.14527 | 1.10899 | 0.01169 | 0.01875 | 0.06001 | 0.06486 | 0.00901 | 1.09088 | 0.05314 | 1.06208 | 0.04586 | 0.40556 | 0.42651 |
| 0.10 | 0.04184 | 1.17097 | 1.12784 | 0.01502 | 0.02566 | 0.12618 | 0.10534 | 0.01209 | 1.11022 | 0.09032 | 1.07684 | 0.07710 | 0.48594 | 0.45366 |
| 0.15 | 0.06751 | 1.20389 | 1.16489 | 0.02050 | 0.03990 | 0.28752 | 0.20902 | 0.01748 | 1.14056 | 0.18744 | 1.09905 | 0.16521 | 0.49509 | 0.50567 |
| 0.20 | 0.09657 | 1.21701 | 1.19172 | 0.02466 | 0.05569 | 0.34611 | 0.31608 | 0.02174 | 1.16277 | 0.28939 | 1.11584 | 0.26181 | 0.54976 | 0.55664 |
| 0.25 | 0.12982 | 1.21442 | 1.21289 | 0.02785 | 0.07487 | 0.48789 | 0.44314 | 0.02507 | 1.17947 | 0.38248 | 1.12855 | 0.34948 | 0.59651 | 0.60401 |
| 0.3 | 0.16786 | 1.19879 | 1.22845 | 0.03095 | 0.09694 | 0.59681 | 0.49744 | 0.02771 | 1.19241 | 0.46666 | 1.13842 | 0.42618 | 0.64581 | 0.63871 |
| 0.4 | 0.25988 | 1.15482 | 1.25166 | 0.03401 | 0.15624 | 0.59820 | 0.68194 | 0.03158 | 1.21108 | 0.59362 | 1.15258 | 0.54931 | 0.69648 | 0.69777 |
| 0.5 | 0.37458 | 1.08507 | 1.26756 | 0.03654 | 0.23678 | 0.61866 | 0.78203 | 0.03423 | 1.22373 | 0.69062 | 1.16242 | 0.61118 | 0.74127 | 0.70111 |
| 0.6 | 0.51293 | 0.89785 | 1.27913 | 0.03840 | 0.34008 | 0.58092 | 0.80845 | 0.03619 | 1.23292 | 0.76485 | 1.16949 | 0.71210 | 0.77478 | 0.77894 |
| 0.7 | 0.67542 | 0.72965 | 1.28793 | 0.03985 | 0.46698 | 0.59247 | 0.86888 | 0.03765 | 1.23987 | 0.82312 | 1.17484 | 0.76762 | 0.80086 | 0.79955 |
| 0.8 | 0.86286 | 0.55113 | 1.29485 | 0.04096 | 0.61795 | 0.88521 | 0.91650 | 0.03878 | 1.24531 | 0.86997 | 1.17903 | 0.81228 | 0.82168 | 0.82005 |
| 0.9 | 1.07892 | 0.30245 | 1.30044 | 0.04189 | 0.79827 | 0.22669 | 0.95590 | 0.03968 | 1.24967 | 0.90839 | 1.18239 | 0.84893 | 0.83867 | 0.83680 |
| 1.0 | 1.51022 | 0.04410 | 1.30596 | 0.04267 | 0.99814 | 0.08428 | 0.98878 | 0.04041 | 1.25326 | 0.94044 | 1.18516 | 0.87951 | 0.85279 | 0.85074 |

TABLE 12 (continued)

$\zeta = 0.40; \ \bar{\delta} = 0.2525$

| $\mu$ | $\psi$ | $\phi$ | $\chi$ | $\zeta$ | $\xi$ | $\eta$ | $\sigma$ | $\theta$ | $X''$ | $Y''$ | $X^{(2)}$ | $Y^{(2)}$ | $u$ | $\varkappa$ |
|---|---|---|---|---|---|---|---|---|---|---|---|---|---|---|
| 0 | 0 | 1.00000 | 1.00000 | 0 | 0 | 0 | 0 | 0 | 1.00000 | 0 | 1.00000 | 0 | 0.50149 | 0.54808 |
| 0.050 | 0.01971 | 1.11265 | 1.07761 | 0.00085 | 0.01198 | 0.04451 | 0.08357 | 0.00561 | 1.06888 | 0.02625 | 1.04537 | 0.02216 | 0.56695 | 0.59101 |
| 0.075 | 0.02122 | 1.14680 | 1.10559 | 0.01224 | 0.01875 | 0.07157 | 0.05701 | 0.00908 | 1.09158 | 0.04557 | 1.06272 | 0.03785 | 0.58822 | 0.61106 |
| 0.10 | 0.04359 | 1.17288 | 1.12988 | 0.01585 | 0.02584 | 0.10981 | 0.09016 | 0.01225 | 1.11150 | 0.07464 | 1.07747 | 0.06987 | 0.61458 | 0.63311 |
| 0.15 | 0.07069 | 1.20853 | 1.16956 | 0.02199 | 0.04055 | 0.20582 | 0.17889 | 0.01813 | 1.14857 | 0.15502 | 1.10162 | 0.13695 | 0.64922 | 0.66145 |
| 0.20 | 0.10135 | 1.23470 | 1.19981 | 0.02681 | 0.05564 | 0.30562 | 0.27708 | 0.02296 | 1.16675 | 0.24818 | 1.12005 | 0.22215 | 0.52092 | 0.53907 |
| 0.25 | 0.13556 | 1.28492 | 1.22822 | 0.03060 | 0.07531 | 0.39486 | 0.37080 | 0.02690 | 1.18685 | 0.33667 | 1.13431 | 0.30525 | 0.54659 | 0.57801 |
| 0.3 | 0.17531 | 1.32167 | 1.24172 | 0.03868 | 0.09757 | 0.46628 | 0.45899 | 0.03007 | 1.20099 | 0.41652 | 1.14557 | 0.38001 | 0.60591 | 0.60946 |
| 0.4 | 0.27012 | 1.33302 | 1.26991 | 0.04813 | 0.13568 | 0.55778 | 0.59177 | 0.03482 | 1.22240 | 0.54860 | 1.16611 | 0.50496 | 0.66848 | 0.66967 |
| 0.5 | 0.38753 | 1.09207 | 1.28985 | 0.04867 | 0.28442 | 0.59670 | 0.69719 | 0.03815 | 1.23728 | 0.65908 | 1.17760 | 0.60101 | 0.71510 | 0.71492 |
| 0.6 | 0.52874 | 0.93537 | 1.30178 | 0.04867 | 0.33588 | 0.56228 | 0.77915 | 0.04060 | 1.24807 | 0.72917 | 1.18201 | 0.67601 | 0.75080 | 0.74975 |
| 0.7 | 0.69484 | 0.74643 | 1.31245 | 0.04530 | 0.46088 | 0.49109 | 0.84421 | 0.04246 | 1.25633 | 0.79209 | 1.18848 | 0.78575 | 0.77886 | 0.77728 |
| 0.8 | 0.88471 | 0.54568 | 1.32093 | 0.04696 | 0.61035 | 0.57786 | 0.89692 | 0.04371 | 1.26285 | 0.84016 | 1.19848 | 0.78427 | 0.80145 | 0.79940 |
| 0.9 | 1.10006 | 0.31866 | 1.32778 | 0.04616 | 0.78408 | 0.25536 | 0.94040 | 0.04507 | 1.26807 | 0.88584 | 1.19757 | 0.82458 | 0.81999 | 0.81766 |
| 1.0 | 1.34052 | 0.09105 | 1.33348 | 0.04917 | 0.98275 | 0.08783 | 0.97681 | 0.04601 | 1.27259 | 0.92075 | 1.20092 | 0.85704 | 0.83548 | 0.83288 |

$\zeta = 0.45; \ \bar{\delta} = 0.2784$

| $\mu$ | $\psi$ | $\phi$ | $\chi$ | $\zeta$ | $\xi$ | $\eta$ | $\sigma$ | $\theta$ | $X''$ | $Y''$ | $X^{(2)}$ | $Y^{(2)}$ | $u$ | $\varkappa$ |
|---|---|---|---|---|---|---|---|---|---|---|---|---|---|---|
| 0 | 0 | 1.00000 | 1.00000 | 0 | 0 | 0 | 0 | 0 | 1.00000 | 0 | 1.00000 | 0 | 0.29349 | 0.35748 |
| 0.050 | 0.02084 | 1.11312 | 1.07886 | 0.00844 | 0.01185 | 0.04162 | 0.08149 | 0.00577 | 1.06867 | 0.02852 | 1.04617 | 0.01988 | 0.34977 | 0.37888 |
| 0.075 | 0.08286 | 1.14708 | 1.10676 | 0.01274 | 0.01867 | 0.06617 | 0.05227 | 0.00901 | 1.09209 | 0.04069 | 1.06328 | 0.03826 | 0.37827 | 0.39748 |
| 0.10 | 0.04524 | 1.17428 | 1.13186 | 0.01659 | 0.02595 | 0.09787 | 0.07999 | 0.01231 | 1.11239 | 0.06904 | 1.07833 | 0.05494 | 0.39711 | 0.41756 |
| 0.15 | 0.07849 | 1.21186 | 1.17874 | 0.02830 | 0.04092 | 0.18098 | 0.15616 | 0.01856 | 1.14575 | 0.13829 | 1.10258 | 0.11497 | 0.44786 | 0.46011 |
| 0.20 | 0.10544 | 1.25052 | 1.20652 | 0.02874 | 0.05728 | 0.27255 | 0.24544 | 0.02895 | 1.17155 | 0.21477 | 1.12356 | 0.19089 | 0.49606 | 0.50586 |
| 0.25 | 0.14154 | 1.28538 | 1.23242 | 0.03312 | 0.07598 | 0.35752 | 0.33406 | 0.02840 | 1.19179 | 0.29782 | 1.13898 | 0.26014 | 0.54028 | 0.54658 |
| 0.3 | 0.18285 | 1.32218 | 1.25330 | 0.03667 | 0.09901 | 0.42896 | 0.41604 | 0.03208 | 1.20795 | 0.37524 | 1.15151 | 0.34058 | 0.57992 | 0.58387 |
| 0.4 | 0.27970 | 1.34460 | 1.28425 | 0.04204 | 0.15494 | 0.53487 | 0.55516 | 0.03570 | 1.23196 | 0.50746 | 1.17020 | 0.46485 | 0.64253 | 0.64401 |
| 0.5 | 0.39976 | 1.06642 | 1.30632 | 0.04590 | 0.23309 | 0.56086 | 0.66453 | 0.04178 | 1.24884 | 0.61196 | 1.18338 | 0.56843 | 0.69038 | 0.69065 |
| 0.6 | 0.54875 | 0.98127 | 1.32256 | 0.04677 | 0.38164 | 0.54866 | 0.75106 | 0.04472 | 1.26130 | 0.69497 | 1.19818 | 0.64183 | 0.72880 | 0.72705 |
| 0.7 | 0.71287 | 0.78168 | 1.33518 | 0.05106 | 0.45480 | 0.47917 | 0.82059 | 0.04700 | 1.27066 | 0.76175 | 1.20068 | 0.70510 | 0.75907 | 0.75610 |
| 0.8 | 0.90604 | 0.55907 | 1.34533 | 0.05288 | 0.60229 | 0.67175 | 0.87142 | 0.04880 | 1.27842 | 0.81650 | 1.20656 | 0.75698 | 0.78222 | 0.77978 |
| 0.9 | 1.12592 | 0.32833 | 1.35344 | 0.05488 | 0.77659 | 0.22453 | 0.91461 | 0.05081 | 1.28455 | 0.86206 | 1.21157 | 0.80017 | 0.80216 | 0.79929 |
| 1.0 | 1.36948 | 0.09802 | 1.36028 | 0.05566 | 0.97139 | 0.08975 | 0.94485 | 0.05111 | 1.28962 | 0.90059 | 1.21534 | 0.83664 | 0.81889 | 0.81572 |

TABLE 12 (continued)

$\zeta = 0.50; \ \bar{\beta} = 0.2921$

| μ | ψ | φ | χ | ζ | ξ̇ | η | σ | θ | X" | Y"' | X^∞ | Y^∞ | η_c | η_s |
|---|---|---|---|---|---|---|---|---|---|---|---|---|---|---|
| 0 | 0 | 1.00000 | 1.00000 | 0 | 0 | 0 | 0 | 0 | 1.00000 | 0 | 1.00000 | 0 | 0.28349 | 0.52767 |
| 0.050 | 0.03095 | 1.11350 | 1.07905 | 0.00895 | 0.01176 | 0.04024 | 0.08056 | 0.00578 | 1.06890 | 0.02318 | 1.04343 | 0.01889 | 0.33779 | 0.56770 |
| 0.075 | 0.04521 | 1.14766 | 1.10787 | 0.01330 | 0.01854 | 0.06956 | 0.04900 | 0.00896 | 1.09247 | 0.03784 | 1.06865 | 0.03094 | 0.36005 | 0.58530 |
| 0.10 | 0.04653 | 1.17519 | 1.13857 | 0.01725 | 0.02576 | 0.08998 | 0.07276 | 0.01280 | 1.11504 | 0.05568 | 1.07900 | 0.04704 | 0.38255 | 0.60959 |
| 0.15 | 0.07599 | 1.21427 | 1.17721 | 0.02448 | 0.04106 | 0.16217 | 0.13888 | 0.01885 | 1.14786 | 0.11426 | 1.10509 | 0.09835 | 0.42856 | 0.64842 |
| 0.20 | 0.10921 | 1.24498 | 1.21217 | 0.03049 | 0.05766 | 0.24557 | 0.21982 | 0.02464 | 1.17446 | 0.18769 | 1.12598 | 0.16550 | 0.47440 | 0.68473 |
| 0.25 | 0.14668 | 1.25983 | 1.24081 | 0.03542 | 0.07641 | 0.32607 | 0.30384 | 0.02960 | 1.19610 | 0.26488 | 1.14280 | 0.25694 | 0.51700 | 0.52810 |
| 0.30 | 0.18874 | 1.28079 | 1.26625 | 0.03942 | 0.09825 | 0.39607 | 0.38889 | 0.03577 | 1.21362 | 0.35915 | 1.15647 | 0.30610 | 0.55524 | 0.55999 |
| 0.35 | 0.25598 | 1.30294 | 1.28215 | 0.04288 | 0.12897 | 0.45348 | 0.45590 | 0.05728 | 1.22802 | 0.40773 | 1.16775 | 0.37025 | 0.58906 | 0.57295 |
| 0.4 | 0.28866 | 1.17658 | 1.29800 | 0.04574 | 0.15419 | 0.49445 | 0.52418 | 0.04026 | 1.24005 | 0.46994 | 1.17717 | 0.42860 | 0.61888 | 0.60049 |
| 0.5 | 0.41132 | 1.06010 | 1.32297 | 0.05060 | 0.22980 | 0.53607 | 0.63897 | 0.04499 | 1.25886 | 0.57625 | 1.17198 | 0.52854 | 0.66835 | 0.66808 |
| 0.6 | 0.55804 | 0.94555 | 1.34178 | 0.05576 | 0.32758 | 0.52528 | 0.72420 | 0.04854 | 1.27290 | 0.66211 | 1.20804 | 0.60952 | 0.70717 | 0.70571 |
| 0.7 | 0.72960 | 0.77554 | 1.35584 | 0.05649 | 0.44880 | 0.46696 | 0.79759 | 0.05128 | 1.28374 | 0.73224 | 1.21160 | 0.67569 | 0.73839 | 0.74509 |
| 0.8 | 0.92647 | 0.57162 | 1.36604 | 0.05970 | 0.59447 | 0.56906 | 0.85810 | 0.05345 | 1.29237 | 0.79020 | 1.21842 | 0.73246 | 0.76691 | 0.76698 |
| 0.9 | 1.14896 | 0.38453 | 1.37765 | 0.06094 | 0.76510 | 0.22258 | 0.90868 | 0.05519 | 1.29958 | 0.83877 | 1.22296 | 0.77640 | 0.78510 | 0.78171 |
| 1.0 | 1.39728 | 0.06592 | 1.38569 | 0.06609 | 0.96106 | 0.04455 | 0.95150 | 0.05561 | 1.30520 | 0.87998 | 1.23856 | 0.81540 | 0.80297 | 0.79921 |

$\zeta = 0.60; \ \bar{\beta} = 0.3291$

| μ | ψ | φ | χ | ζ | ξ̇ | η | σ | θ | X" | Y"' | X^∞ | Y^∞ | η_c | η_s |
|---|---|---|---|---|---|---|---|---|---|---|---|---|---|---|
| 0 | 0 | 1.00000 | 1.00000 | 0 | 0 | 0 | 0 | 0 | 1.00000 | 0 | 1.00000 | 0 | 0.26592 | 0.50997 |
| 0.050 | 0.02200 | 1.11407 | 1.08022 | 0.00944 | 0.01158 | 0.03839 | 0.02848 | 0.00566 | 1.06924 | 0.03505 | 1.04686 | 0.01783 | 0.31685 | 0.54765 |
| 0.075 | 0.03451 | 1.14847 | 1.10977 | 0.01402 | 0.01819 | 0.05825 | 0.04470 | 0.00882 | 1.09301 | 0.04305 | 1.06451 | 0.02721 | 0.33784 | 0.56685 |
| 0.10 | 0.04697 | 1.17641 | 1.13668 | 0.01842 | 0.02535 | 0.08047 | 0.06680 | 0.01218 | 1.11890 | 0.04755 | 1.08000 | 0.03920 | 0.35726 | 0.58006 |
| 0.15 | 0.08028 | 1.21730 | 1.18269 | 0.02551 | 0.04085 | 0.13696 | 0.11584 | 0.01902 | 1.14944 | 0.08978 | 1.10723 | 0.07623 | 0.39767 | 0.61482 |
| 0.20 | 0.11572 | 1.24098 | 1.22117 | 0.03852 | 0.05776 | 0.20562 | 0.18191 | 0.02349 | 1.17844 | 0.14764 | 1.12980 | 0.12843 | 0.43845 | 0.65343 |
| 0.25 | 0.15555 | 1.24929 | 1.25315 | 0.04949 | 0.07659 | 0.27648 | 0.25314 | 0.03127 | 1.20228 | 0.21517 | 1.14851 | 0.18846 | 0.47767 | 0.64612 |
| 0.30 | 0.20009 | 1.24871 | 1.27987 | 0.04454 | 0.09814 | 0.34171 | 0.32855 | 0.03661 | 1.22205 | 0.27994 | 1.16412 | 0.25014 | 0.51404 | 0.51982 |
| 0.35 | 0.24978 | 1.22543 | 1.30289 | 0.04685 | 0.12830 | 0.39866 | 0.39866 | 0.04067 | 1.23868 | 0.34434 | 1.17725 | 0.30995 | 0.54709 | 0.55561 |
| 0.4 | 0.30494 | 1.19539 | 1.32157 | 0.05255 | 0.15240 | 0.44082 | 0.46695 | 0.04445 | 1.25267 | 0.40472 | 1.18839 | 0.36622 | 0.57686 | 0.57896 |
| 0.5 | 0.43263 | 1.10286 | 1.35238 | 0.05856 | 0.22521 | 0.49022 | 0.57879 | 0.05059 | 1.27597 | 0.51168 | 1.20623 | 0.46623 | 0.62754 | 0.62730 |
| 0.6 | 0.58464 | 0.97029 | 1.37598 | 0.06322 | 0.31948 | 0.48977 | 0.67429 | 0.05580 | 1.29208 | 0.60121 | 1.21980 | 0.55020 | 0.66850 | 0.66668 |
| 0.7 | 0.76188 | 0.80025 | 1.39465 | 0.06695 | 0.45705 | 0.44218 | 0.75378 | 0.05900 | 1.30541 | 0.67609 | 1.23045 | 0.62057 | 0.70200 | 0.69906 |
| 0.8 | 0.96494 | 0.59452 | 1.40976 | 0.07002 | 0.74614 | 0.55084 | 0.82045 | 0.06195 | 1.31611 | 0.73915 | 1.23901 | 0.67990 | 0.72977 | 0.72559 |
| 0.9 | 1.19420 | 0.35561 | 1.42280 | 0.07258 | 0.74614 | 0.21670 | 0.87690 | 0.06485 | 1.32468 | 0.79275 | 1.24605 | 0.73058 | 0.75309 | 0.74868 |
| 1.0 | 1.44992 | 0.07684 | 1.43287 | 0.07478 | 0.98896 | 0.04522 | 0.93518 | 0.06661 | 1.33220 | 0.83875 | 1.25190 | 0.77872 | 0.77293 | 0.76802 |

| μ | ψ | φ | χ | ζ | ξ | η | σ | θ | X⁽ⁱ⁾ | Y⁽ⁱ⁾ | X⁽ᵃ⁾ | Y⁽ᵃ⁾ | ʋ | ϰ |
|---|---|---|---|---|---|---|---|---|---|---|---|---|---|---|
| 0 | 0 | 1.00000 | 1.00000 | 0 | 0 | 0 | 0 | 0 | 1.00000 | 0 | 1.00000 | 0 | 0.25089 | 0.29435 |
| 0.050 | 0.02295 | 1.11447 | 1.08124 | 0.00990 | 0.01126 | 0.08728 | 0.02687 | 0.00552 | 1.06946 | 0.01994 | 1.04719 | 0.01618 | 0.29902 | 0.32997 |
| 0.075 | 0.03640 | 1.14900 | 1.11118 | 0.01473 | 0.01776 | 0.05585 | 0.04182 | 0.00864 | 1.09835 | 0.03026 | 1.06482 | 0.02490 | 0.31828 | 0.34520 |
| 0.10 | 0.05108 | 1.17718 | 1.13869 | 0.01942 | 0.02478 | 0.07548 | 0.05847 | 0.01196 | 1.11440 | 0.04227 | 1.08071 | 0.03492 | 0.33559 | 0.36012 |
| 0.15 | 0.08859 | 1.21897 | 1.18694 | 0.02825 | 0.04018 | 0.12200 | 0.10095 | 0.01892 | 1.15061 | 0.07489 | 1.10864 | 0.06502 | 0.37296 | 0.39070 |
| 0.20 | 0.12119 | 1.24454 | 1.22809 | 0.03609 | 0.05717 | 0.17795 | 0.15682 | 0.02579 | 1.18081 | 0.13080 | 1.13283 | 0.10892 | 0.40967 | 0.42278 |
| 0.25 | 0.16507 | 1.25537 | 1.26217 | 0.04296 | 0.07601 | 0.24056 | 0.20212 | 0.03218 | 1.20622 | 0.17579 | 1.15241 | 0.15801 | 0.44569 | 0.45316 |
| 0.30 | 0.20985 | 1.25255 | 1.29315 | 0.04895 | 0.09783 | 0.29995 | 0.28686 | 0.03978 | 1.22772 | 0.28166 | 1.16956 | 0.20779 | 0.47798 | 0.48452 |
| 0.35 | 0.26180 | 1.25698 | 1.51884 | 0.05412 | 0.12135 | 0.35855 | 0.55287 | 0.04804 | 1.24641 | 0.29574 | 1.18427 | 0.26201 | 0.51175 | 0.51407 |
| 0.4 | 0.51928 | 1.20943 | 1.81107 | 0.05864 | 0.15020 | 0.39596 | 0.41608 | 0.04755 | 1.26179 | 0.35089 | 1.19688 | 0.31527 | 0.54097 | 0.54845 |
| 0.5 | 0.45177 | 1.12085 | 1.57741 | 0.06612 | 0.22053 | 0.46947 | 0.58082 | 0.05905 | 1.28782 | 0.45566 | 1.21785 | 0.41277 | 0.57187 | 0.59162 |
| 0.6 | 0.60886 | 0.99062 | 1.40578 | 0.07204 | 0.31166 | 0.45650 | 0.62925 | 0.06094 | 1.30705 | 0.54689 | 1.23835 | 0.49790 | 0.60401 | 0.63186 |
| 0.7 | 0.79158 | 0.82124 | 1.44851 | 0.07485 | 0.44562 | 0.44758 | 0.71301 | 0.06562 | 1.32270 | 0.62412 | 1.24538 | 0.57026 | 0.66908 | 0.66555 |
| 0.8 | 1.00060 | 0.61458 | 1.44716 | 0.08068 | 0.56694 | 0.58441 | 0.78445 | 0.06941 | 1.33539 | 0.69076 | 1.25615 | 0.63278 | 0.69955 | 0.69898 |
| 0.9 | 1.25687 | 0.37116 | 1.46275 | 0.08421 | 0.72754 | 0.20083 | 0.84567 | 0.07250 | 1.34587 | 0.74818 | 1.26461 | 0.66665 | 0.72257 | 0.71419 |
| 1.0 | 1.49922 | 0.09288 | 1.47600 | 0.08712 | 0.91708 | 0.04245 | 0.89857 | 0.07595 | 1.35468 | 0.79799 | 1.27178 | 0.78945 | 0.74508 | 0.74901 |

| μ | ψ | φ | χ | ζ | ξ | η | σ | θ | X⁽ⁱ⁾ | Y⁽ⁱ⁾ | X⁽ᵃ⁾ | Y⁽ᵃ⁾ | ʋ | ϰ |
|---|---|---|---|---|---|---|---|---|---|---|---|---|---|---|
| 0 | 0 | 1.00000 | 1.00000 | 0 | 0 | 0 | 0 | 0 | 1.00000 | 0 | 1.00000 | 0 | 0.23788 | 0.28058 |
| 0.050 | 0.02879 | 1.11476 | 1.08214 | 0.01031 | 0.01097 | 0.08648 | 0.02555 | 0.00549 | 1.06960 | 0.01846 | 1.04745 | 0.01515 | 0.28855 | 0.31124 |
| 0.075 | 0.08774 | 1.14989 | 1.11279 | 0.01537 | 0.01728 | 0.05487 | 0.03961 | 0.00844 | 1.09356 | 0.02826 | 1.06521 | 0.02335 | 0.30172 | 0.32867 |
| 0.10 | 0.05294 | 1.17761 | 1.14068 | 0.02032 | 0.02411 | 0.07256 | 0.05488 | 0.01170 | 1.11470 | 0.03884 | 1.08125 | 0.03395 | 0.31494 | 0.34567 |
| 0.15 | 0.08698 | 1.21999 | 1.19040 | 0.02970 | 0.03914 | 0.11491 | 0.09226 | 0.01860 | 1.15129 | 0.06666 | 1.10968 | 0.05517 | 0.35285 | 0.37091 |
| 0.20 | 0.12538 | 1.24668 | 1.28366 | 0.03880 | 0.05611 | 0.16187 | 0.13866 | 0.02572 | 1.18225 | 0.10254 | 1.13406 | 0.08788 | 0.38591 | 0.39972 |
| 0.25 | 0.16956 | 1.25915 | 1.27126 | 0.04597 | 0.07485 | 0.24175 | 0.19487 | 0.03258 | 1.20874 | 0.14860 | 1.15519 | 0.12880 | 0.41911 | 0.42927 |
| 0.30 | 0.21829 | 1.25867 | 1.50896 | 0.05277 | 0.09592 | 0.26908 | 0.25466 | 0.03884 | 1.25154 | 0.19999 | 1.17848 | 0.17558 | 0.45121 | 0.45841 |
| 0.35 | 0.27285 | 1.24538 | 1.83269 | 0.05918 | 0.11996 | 0.51709 | 0.31610 | 0.04458 | 1.25125 | 0.25341 | 1.18966 | 0.22458 | 0.48160 | 0.48648 |
| 0.4 | 0.33197 | 1.22001 | 1.85748 | 0.06409 | 0.14757 | 0.55881 | 0.57645 | 0.04974 | 1.26840 | 0.30658 | 1.20521 | 0.27861 | 0.50996 | 0.51274 |
| 0.5 | 0.46900 | 1.13524 | 1.89900 | 0.07508 | 0.21572 | 0.42586 | 0.58877 | 0.05349 | 1.29665 | 0.40782 | 1.22509 | 0.36695 | 0.56088 | 0.56015 |
| 0.6 | 0.66098 | 1.00753 | 1.45876 | 0.08025 | 0.30897 | 0.59574 | 0.67582 | 0.06551 | 1.31661 | 0.49743 | 1.24409 | 0.45079 | 0.60905 | 0.60062 |
| 0.7 | 0.81899 | 0.83939 | 1.45976 | 0.08615 | 0.44156 | 0.51789 | 0.73029 | 0.07119 | 1.33610 | 0.57646 | 1.25757 | 0.52452 | 0.64915 | 0.64511 |
| 0.8 | 1.08877 | 0.68227 | 1.44096 | 0.09112 | 0.54966 | 0.19840 | 0.77637 | 0.07582 | 1.35117 | 0.64539 | 1.27045 | 0.58966 | 0.66967 | 0.66460 |
| 0.9 | 1.27589 | 0.48788 | 1.49968 | 0.09587 | 0.70947 | 0.19810 | 0.81587 | 0.07964 | 1.36580 | 0.70561 | 1.28031 | 0.64532 | 0.69622 | 0.68998 |
| 1.0 | 1.51565 | 0.10553 | 1.51572 | 0.09905 | 0.89571 | 0.08948 | 0.87211 | 0.08282 | 1.37854 | 0.75941 | 1.28867 | 0.69477 | 0.71901 | 0.71199 |

727

TABLE 12 (continued)

ζ = 0.90; ℑ = 0.4162

| μ | ψ | φ | χ | ζ | ξ | η | σ | θ | X⁽¹⁾ | Y⁽¹⁾ | X⁽²⁾ | Y⁽²⁾ | 𝒵 | 𝒴 |
|---|---|---|---|---|---|---|---|---|---|---|---|---|---|---|
| 0 | 0 | 1.0000 | 1.00000 | 0 | 0 | 0 | 0 | 0 | 1.00000 | 0 | 1.00000 | 0 | 0.22665 | 0.26781 |
| 0.050 | 0.02455 | 1.11498 | 1.08294 | 0.01067 | 0.01067 | 0.05599 | 0.02845 | 0.00526 | 1.06969 | 0.01752 | 1.04766 | 0.01414 | 0.25998 | 0.30009 |
| 0.075 | 0.03694 | 1.14968 | 1.11405 | 0.01595 | 0.01678 | 0.05847 | 0.08781 | 0.00828 | 1.09369 | 0.02672 | 1.06552 | 0.02195 | 0.28722 | 0.31585 |
| 0.10 | 0.05462 | 1.17795 | 1.14244 | 0.02111 | 0.02839 | 0.07084 | 0.05805 | 0.01141 | 1.11489 | 0.03689 | 1.08167 | 0.02996 | 0.30055 | 0.32710 |
| 0.15 | 0.08979 | 1.22054 | 1.19643 | 0.03103 | 0.03812 | 0.10754 | 0.08050 | 0.01825 | 1.15166 | 0.05924 | 1.11084 | 0.04926 | 0.33485 | 0.35826 |
| 0.20 | 0.13098 | 1.24802 | 1.23882 | 0.04026 | 0.05478 | 0.14957 | 0.12604 | 0.02549 | 1.18010 | 0.08990 | 1.13528 | 0.07597 | 0.36578 | 0.37995 |
| 0.25 | 0.17528 | 1.28135 | 1.27799 | 0.04861 | 0.07035 | 0.19819 | 0.17517 | 0.03249 | 1.21084 | 0.12870 | 1.15713 | 0.11056 | 0.39965 | 0.40714 |
| 0.30 | 0.22570 | 1.26697 | 1.31308 | 0.05816 | 0.09408 | 0.24486 | 0.22959 | 0.03849 | 1.24510 | 0.17888 | 1.17682 | 0.15095 | 0.42666 | 0.44485 |
| 0.35 | 0.28162 | 1.33570 | 1.34402 | 0.06892 | 0.11761 | 0.28886 | 0.28643 | 0.04545 | 1.25497 | 0.22126 | 1.19519 | 0.19462 | 0.45556 | 0.46066 |
| 0.4 | 0.34326 | 1.22816 | 1.37150 | 0.06899 | 0.14456 | 0.32825 | 0.34950 | 0.05115 | 1.27322 | 0.27012 | 1.22007 | 0.25990 | 0.48290 | 0.48597 |
| 0.5 | 0.44858 | 1.14690 | 1.44784 | 0.07934 | 0.21076 | 0.38624 | 0.45302 | 0.06404 | 1.30877 | 0.36577 | 1.23298 | 0.32771 | 0.53289 | 0.52214 |
| 0.6 | 0.65128 | 1.02189 | 1.45523 | 0.08782 | 0.29588 | 0.89796 | 0.55344 | 0.06913 | 1.32810 | 0.45396 | 1.25287 | 0.40949 | 0.57510 | 0.57245 |
| 0.7 | 0.84456 | 0.85530 | 1.48599 | 0.09489 | 0.40585 | 0.57102 | 0.64059 | 0.07577 | 1.34785 | 0.55305 | 1.26904 | 0.48806 | 0.61100 | 0.60785 |
| 0.8 | 1.06177 | 0.64841 | 1.51174 | 0.10087 | 0.53529 | 0.60117 | 0.71821 | 0.08125 | 1.36418 | 0.60824 | 1.28242 | 0.58890 | 0.64481 | 0.63753 |
| 0.9 | 1.31305 | 0.40241 | 1.53568 | 0.10608 | 0.69206 | 0.78830 | 0.76620 | 0.05021 | 1.37787 | 0.66587 | 1.29361 | 0.60650 | 0.67078 | 0.66878 |
| 1.0 | 1.58957 | 0.11832 | 1.55352 | 0.11055 | 0.87506 | 0.08464 | 0.84612 | 0.08962 | 1.38950 | 0.72812 | 1.30018 | 0.65797 | 0.69464 | 0.68678 |

ζ = 1.00; ℑ = 0.4442

| μ | ψ | φ | χ | ζ | ξ | η | σ | θ | X⁽¹⁾ | Y⁽¹⁾ | X⁽²⁾ | Y⁽²⁾ | 𝒵 | 𝒴 |
|---|---|---|---|---|---|---|---|---|---|---|---|---|---|---|
| 0 | 0 | 1.0000 | 1.00000 | 0 | 0 | 0 | 0 | 0 | 1.00000 | 0 | 1.00000 | 0 | 0.21607 | 0.25940 |
| 0.075 | 0.04003 | 1.14971 | 1.11519 | 0.01647 | 0.01686 | 0.05984 | 0.08686 | 0.00801 | 1.09877 | 0.02548 | 1.06576 | 0.02082 | 0.27421 | 0.30089 |
| 0.10 | 0.05634 | 1.17825 | 1.14402 | 0.02184 | 0.02265 | 0.06975 | 0.04979 | 0.01110 | 1.11499 | 0.03452 | 1.08199 | 0.02851 | 0.28986 | 0.31804 |
| 0.15 | 0.09228 | 1.22096 | 1.19607 | 0.03221 | 0.03692 | 0.10612 | 0.08006 | 0.01781 | 1.15190 | 0.05489 | 1.11087 | 0.04540 | 0.30784 | 0.35774 |
| 0.20 | 0.13856 | 1.24889 | 1.21285 | 0.04195 | 0.05014 | 0.11469 | 0.11679 | 0.02494 | 1.18160 | 0.08096 | 1.13616 | 0.06789 | 0.34485 | 0.34785 |
| 0.25 | 0.18024 | 1.26878 | 1.28074 | 0.05095 | 0.07134 | 0.18890 | 0.16059 | 0.03216 | 1.21135 | 0.11411 | 1.15554 | 0.09709 | 0.37708 | 0.38785 |
| 0.30 | 0.23226 | 1.26605 | 1.32078 | 0.05917 | 0.09177 | 0.22558 | 0.20986 | 0.03916 | 1.23582 | 0.15290 | 1.17859 | 0.13199 | 0.40537 | 0.41819 |
| 0.35 | 0.28989 | 1.25621 | 1.55393 | 0.06664 | 0.11489 | 0.26656 | 0.26285 | 0.04578 | 1.25748 | 0.19566 | 1.19605 | 0.17074 | 0.44281 | 0.44812 |
| 0.4 | 0.35336 | 1.23847 | 1.38565 | 0.07840 | 0.14122 | 0.30625 | 0.31604 | 0.05194 | 1.27672 | 0.24082 | 1.22177 | 0.22134 | 0.45906 | 0.46228 |
| 0.5 | 0.49958 | 1.15946 | 1.44343 | 0.08512 | 0.20568 | 0.35590 | 0.42260 | 0.06882 | 1.30928 | 0.38016 | 1.23842 | 0.29112 | 0.50784 | 0.50708 |
| 0.6 | 0.66977 | 1.04410 | 1.47608 | 0.09485 | 0.28675 | 0.37278 | 0.51987 | 0.07190 | 1.32948 | 0.44551 | 1.25999 | 0.37801 | 0.54974 | 0.54691 |
| 0.7 | 0.86784 | 0.86912 | 1.51065 | 0.10307 | 0.39818 | 0.44960 | 0.60867 | 0.07845 | 1.35708 | 0.49875 | 1.27771 | 0.44559 | 0.58670 | 0.58785 |
| 0.8 | 1.09869 | 0.66868 | 1.53991 | 0.11009 | 0.51371 | 0.28458 | 0.68790 | 0.08576 | 1.37498 | 0.54815 | 1.29759 | 0.51122 | 0.61889 | 0.61251 |
| 0.9 | 1.34797 | 0.41640 | 1.56920 | 0.11619 | 0.58171 | 0.17787 | 0.75828 | 0.09104 | 1.39013 | 0.62762 | 1.30498 | 0.57018 | 0.64703 | 0.64993 |
| 1.0 | 1.68118 | 0.13048 | 1.58678 | 0.12255 | 0.85509 | 0.02828 | 0.82078 | 0.09549 | 1.40808 | 0.68429 | 1.31566 | 0.62804 | 0.67174 | 0.66806 |

www.ingramcontent.com/pod-product-compliance
Lightning Source LLC
Chambersburg PA
CBHW081335190326
41458CB00018B/6010